P9-EEA-108

The
Ubiquity
of
CHAOS

American Association for the Advancement of Science

Library of Congress Cataloging-in-Publication Data

Krasner, Saul
The Ubiquity of Chaos / Saul Krasner
p. cm.
Includes bibliographic references.
ISBN 0-87168-350-4
1. Chaotic behavior in systems. 2. Nonlinear theories.

I. Krasner, Saul.

Q172.5.C45U25 1990
003—dc20 90-31235
CIP

Publication No. 89-15S

1333 H Street, N.W., Washington, D.C. 20005

Contents

Foreword

Chaos – an interesting word, or rather concept. What meaning does it conjure in the mind of the reader? Does chaos mean the same thing to the economist examining the past 50 years of data on the quarterly gross national product as it does to the researcher studying turbulent flow? In the first case – and in many others – the data do not originate from controlled experiments, while in the case of turbulent flow, they do. Should one speak not of chaos but of relative degrees of order? Are the analytical techniques applied indiscriminately to the wide range of available data sets because the study of chaos is currently fashionable, or does one gain some new insight or fundamental understanding of the problem? The very wide range of perspectives represented here will help clarify the issues and posit answers to such questions, but, as with most research, some questions will remain and some new ones will be generated.

This volume is, for the most part, a compilation of material presented during six three-hour sessions at the 1989 AAAS Annual Meeting and largely unpublished elsewhere. For a variety of reasons, not all the papers presented appear in this volume. Some participants could not prepare manuscripts, while some of the meeting material had already been published in the open literature. Therefore, a bibliography of material that would have been included had this volume been simply a proceedings of the meeting is given in an appendix on page 241.

The chapters are arranged in much the same order as when presented at the AAAS sessions devoted to "Chaos and Dynamical Systems," "Biological Systems – Physiology and Medicine," "Studies of Turbulence," "Quantized Systems," "Global Affairs – Economics, and the Arms Race," and "Celestial Systems" rather than being specifically grouped together. One paper presented at the 1990 AAAS Annual Meeting is also included. (It will be left as a challenge to the reader to determine which chapter it is.) *The Ubiquity of Chaos* is realized when the various chapters are considered as a single unified, integrated set.

An interdisciplinary atmosphere existed at the AAAS sessions where people generally interested in chaos realized how similar techniques have been applied in very diverse specific fields. Most attendees were stimulated and enlightened. This volume tries to preserve at least part of that atmosphere. The copious references following each chapter should be welcome for a number of reasons. Students as well as researchers will find this volume ideal as a companion to complement a textbook on nonlinear phenomena.

While writing this, I was reminded by a paper in the *American Journal of Physics (1)* of the tendency for subjects of current scientific research to receive full chapter coverage in introductory texts. For example, I have seen this for superconductivity, plasma physics, and semiconductors but not yet for chaos. Fortunately, chaos has been popularized by Gleick's book *(2)* and is taught in undergraduate courses on nonlinear phenomena. Coast Guard Academy cadets are most interested in the guest lectures I arranged on the subject and consider fractals "awesome," to use one of their favorite words. I expect newer editions of introductory texts to devote at least several pages to this subject. Dare I expect this volume to facilitate the work towards that objective?

No book can be published without the capable assistance of many people. This volume could not exist without the untiring and major assistance of Susan Reynolds and the staff at the AAAS Publications Office, especially Susan M. O'Connell, Susan Cherry, Kathy Doucette, and Elisabeth H. Carroll. I am particularly grateful for the help of many individuals when arranging the sessions at the AAAS meeting. Harry Swinney, Joseph Ford, Paul Rapp, and Rick Jensen always seemed to be at the other end of the phone line with timely advice and suggestions. This work also owes its existence to the individual contributors and to the anonymous reviewers for their labor of love. My wife Rachel has been most patient with me, especially when I was home but not altogether there! Many thanks to all my colleagues at the Coast Guard Academy who will welcome me back to the fold with an outstretched arm full of work.

Saul Krasner
U.S. Coast Guard Academy

References

1. De Souza-Machado, S., R. W. Rollins, D. T. Jacobs, J. L. Hartman, *Am. J. Phys.* **58**, 321 (1990).
2. Gleick, J., *Chaos: Making a New Science* (Viking Penguin Inc., New York, 1987).

1

Chaotic Explosions in Simple Dynamical Systems

Robert L. Devaney

Abstract

In this chapter we will investigate the behavior of a certain very special but nevertheless extremely important and beautiful class of dynamical systems known as entire transcendental functions. We will pay special attention to the set of chaotic orbits of this system, the Julia set. We will show how this set may change its shape dramatically as a parameter varies and we will illustrate this by showing the computer graphics experiments that led to this discovery. We begin by describing what a dynamical system is, why such systems are important, and how very simple dynamical systems may behave chaotically. Later, we specialize the discussion to the case of entire functions.

Dynamical Systems

Recent advances in the branch of mathematics known as nonlinear dynamical systems promise to revolutionize the way scientists view many different kinds of evolutionary processes. These processes occur in all branches of science, ranging from the fluctuations of temperature, pressure, wind speed, and so on in meteorology to the ups and downs of the stock market in economics. Any physical, chemical, or biological process that evolves in time is an example of a dynamical system. Some systems are simple, like the motion of a pendulum which gradually damps down to a stable resting position. Others are much more complicated, like the motion of the galaxies in the universe or a fluid tumbling over an obstacle.

It is the basic goal of scientists who work with dynamical systems to develop methods that accurately predict the future behavior of the system. Sometimes this may be accomplished by direct observation or experimentation: years of waking up each morning have convinced us that the sun will rise again tomorrow. But other evolutionary processes like the weather or the stock market are much more difficult to predict. So the scientist seeks other methods to make predictions.

One of the most common methods for making predictions is first to set up a mathematical model of the system at hand and then to solve the resulting equations. The scientist uses accepted laws such as Newton's Laws or Hooke's Law to set up this mathematical model, which is often a differential equation or a difference equation. The solutions of these equations then yield the desired predictions.

Sometimes the solution of these equations is easy. Most often this occurs when the equations are linear, i.e., involve all of the constituent variables in the simplest possible manner: no squares or cubes, logarithms or sines, etc. When the equations are nonlinear, the situation changes drastically. There are very few mathematical techniques available to solve such equations explicitly. Thus, the scientist must use other methods to "solve" the equations. Very often this necessitates the use of the computer and various approximation techniques to gain at least partial insight into the solution. But this approach, as we shall see, is not altogether satisfactory. One of the main advances in recent years in the study of nonlinear dynamics is the recognition that such computer solutions may be totally meaningless. Many systems behave so erratically or unpredictably that the slightest error or approximation used in their formulation or solution leads to completely erroneous predictions. This is the phenomenon known as "chaos."

Chaos

Chaos is the mathematical term for the behavior of a system that is inherently unpredictable. Unpredictable phenomena are readily apparent in all areas of life. Fluctuations in the stock market and the weather make long-term predictions in these areas nearly impossible. One might argue that the many factors that influence economic or meteorological systems are the reason for this unpredictability. But chaos can occur in systems that have few degrees of freedom, too. For example, a damped, forced pendulum is well known to possess motions that are quite chaotic.

The critical ingredient in this example is what mathematicians call sensitive dependence on initial conditions. If one makes even the slightest change in the initial configuration of the system, the resulting behavior may be dramatically different. The system is extremely sensitive to initial measurements.

This should be contrasted with the unforced pendulum example, where, no matter how fast we initially start the pendulum swinging, it eventually winds down to a simple resting position. This dynamical system is not at all chaotic.

When viewed on a computer graphics screen, the solutions of a chaotic system tend to oscillate or wander about randomly, with no apparent order to the motion. If we consider another solution that begins near our original solution, then this solution also seems to behave randomly. However, after an initial period of similarity, the two solutions eventually look completely different from each other.

One of the principal recent discoveries in dynamics is that this type of chaos can be present in completely deterministic systems. That is, even if we can write down the equations that govern the motion explicitly, without recourse to probabilistic terms, the solutions themselves may behave in a seemingly random manner. Moreover, even the simplest mathematical models may possess this complicated behavior. Let us illustrate this using the logistic equation from mathematical biology. The logistic equation is a model for the growth of an idealized population consisting of only one species. This system was put forth as a simple model for population growth by the ecologist Robert May *(1)* in the mid-seventies.

To be precise, let us suppose that there is a single species whose population grows and dwindles over time in a controlled environment. Suppose we measure the population of the species at the end of each generation. Rather than produce the actual count of individuals present in the colony, suppose we measure instead the percentage of some limiting number or maximum population. That is, let us write P_n for the fraction of population after generation n, where $0 \leq P_n \leq 1$. One simple rule that an ecologist may use to model the growth of this population is the logistic equation

$$P_{n+1} = kP_n(1 - P_n)$$

where k is some constant that depends on ecological conditions such as the amount of food present. Using this formula, the population in the succeeding generation may be deduced from a knowledge of only the population in the preceding generation and the constant k.

Note how trivial this formula is. It is a simple quadratic formula in the variable P_n. Given P_n and k, we can compute P_{n+1} exactly. In Table 1 we have listed the populations predicted by this model for various values of k. Note several things. When k is small, the fate of the population seems quite predictable. Indeed, for $k = 0.5$, the population dies out, whereas for $k = 1.2$, 2, and 3, it tends to stabilize or reach a definite limiting value. Above 3, different values of k yield startlingly different results. For $k = 3.1$, the limiting values tend to oscillate between two distinct values. For $k = 3.4$, the limiting values oscillate between four values. And finally, for $k = 4$, there is no apparent pattern to be discerned. One initial value, $P_0 = 0.5$, leads to the disappearance of the species after only two generations, whereas $P_0 = 0.4$ leads to a population count that seems to be completely random.

This is the unpredictable nature of this process. Certain k values such as $k = 1.2$ lead to results which are quite predictable—a fixed or periodically repeating limiting value. But other k values lead to results that are, for all intents and purposes, random.

The reader may object that the quite limited table of values for P_n when $k = 4$ can in no way be interpreted as a proof that the values do behave randomly. Nevertheless, this is a fact which may be proved quite easily with techniques from dynamical systems theory.

This example illustrates one of the major consequences of the discovery of chaos in deterministic systems: the output of the computer can no longer be trusted. Indeed, at virtually each stage of the computation of P_n, the computer makes approximations or round-off errors. As illustrated in Table 1, these small changes may lead to vastly different eventual values for P_n.

The existence of chaos in deterministic systems has a number of important consequences in both mathematics and the physical sciences. First of all, it means that no matter how accurately we measure the physical quantities that determine the system, we may never be able to accurately predict the resulting motion. Second, it indicates that the search for individual, specific solutions to the system may be useless. After all, any small change will produce a vastly different solution, perhaps

Table 1. Values of P_n for various k-values.

$$P_{n+1} = kP_n(1 - P_n)$$

$k = 0.5$	$k = 1.2$	$k = 2$	$k = 3$	$k = 3.1$	$k = 3.4$	$k = 4$	$k = 4$
0.5	0.5	0.5	0.5	0.5	0.5	0.4	0.5
0.125	0.3	0.5	0.675	0.775	0.85	0.96	1
0.055	0.252	0.5	0.592	0.540	0.434	0.154	0
0.026	0.226	0.5	0.652	0.770	0.835	0.520	0
0.013	0.210	0.5	0.613	0.549	0.469	0.998	0
0.006	0.199	0.5	0.641	0.768	0.847	0.006	0
0.003	0.191	0.5	0.622	0.553	0.441	0.025	0
0.002	0.186	0.5	0.635	0.766	0.838	0.099	0
0.001	0.181	0.5	0.626	0.555	0.461	0.358	0
0.000	0.178	0.5	0.632	0.766	0.845	0.919	0
0.000	0.176	0.5	0.628	0.556	0.446	0.298	0
0.000	0.174	0.5	0.631	0.765	0.840	0.837	0
0.000	0.172	0.5	0.629	0.557	0.457	0.547	0
0.000	0.171	0.5	0.630	0.765	0.844	0.991	0
0.000	0.170	0.5	0.629	0.557	0.448	0.035	0
0.000	0.170	0.5	0.630	0.765	0.841	0.135	0
0.000	0.169	0.5	0.629	0.557	0.455	0.466	0
0.000	0.168	0.5	0.630	0.765	0.843	0.996	0
0.000	0.168	0.5	0.629	0.557	0.450	0.018	0
0.000	0.168	0.5	0.630	0.765	0.851	0.070	0
0.000	0.168	0.5	0.630	0.557	0.455	0.261	0
0.000	0.168	0.5	0.630	0.765	0.843	0.773	0

necessitating completely new methods or analysis in order to generate the solution. This may be impractical or even impossible.

What does a mathematician do in the face of chaos? Obviously, by the very nature of a chaotic system, the search for a specific solution of the equations is not especially fruitful. Hence the mathematician takes a more global viewpoint. Instead of seeking specific solutions, the mathematician seeks to describe the totality of all possible solutions. Although the particular behavior of a solution may be unpredictable, the totality of all of these solutions may be identifiable. In a certain sense, this may even be more important than finding specific solutions.

As an example, the meteorologist may not be able to predict whether it will be rainy or sunny on a given day of July in Rome, but he knows that it will not be snowing on that day. Thus, there are limits to the unpredictability of a system, and finding these limits is an important task. Toward that end, there have been a number of remarkable advances in recent years.

Strange Attractors

A profound development in recent years has been the discovery of strange attractors. In nature, all processes seem to tend toward some sort of stable state or equilibrium. This state is, to a mathematician, an attractor—in terms of a dynamical system, an attractor is an object toward which all nearby solutions tend as time moves on.

For years, scientists have believed that the attractors toward which physical systems tend should be quite simple. They should either be rest points or equilibrium points like the rest position of a pendulum, or else they should be limit cycles or repeating configurations such as often occur in population biology or economic systems. Experiments or simulations that did not tend to equilibrium or steady state were often viewed with suspicion. Either the experiment itself was flawed (too much noise) or else the process had been terminated too soon.

Now, however, we know that this is by no means the case. There are other much more complicated attractors—called strange attractors—that often arise. In the early 1960s, Stephen Smale proposed several geometric models of new attractors that were much more complicated than rest points or limit cycles. The dynamics on these attractors was chaotic, not steady state or periodic. Later, R. F. Williams showed that these attractors were stable in the sense that small changes in the basic system could not destroy them.

So strange attractors exist as mathematical objects, but do they exist in nature? While this has not yet been definitely answered, every indication is that the answer is yes. In the early sixties, the meteorologist E. N. Lorenz from MIT suggested a simplified model of meteorological turbulence. His model was a differential equation in three dimensions. When viewed on a computer graphics terminal, each solution curve of the system tends toward the same object—the Lorenz attractor. Once near the attractor, the solution oscillates about one of two lobes of the attractor. The important fact is that the number of oscillations about each lobe seems to be random and depends very much on which solution curve is followed. This, of course, is chaos.

Consequently, the Lorenz system never settles down to equilibrium or periodic behavior, but rather continually cycles chaotically. Since Lorenz' discovery, similar strange attractors have been found in a variety of scientific disciplines. For example, strange attractors may provide the key to understanding turbulence in fluid and aerodynamic problems.

The Chaotic Set

These considerations lead us to try to understand at the outset the set of all chaotic orbits of a dynamical system, rather than the behavior of individual orbits. In general, it is a difficult problem to try to find the set of unpredictable orbits. But there are some classes of systems for which there are techniques available. One such class is the class of complex analytic dynamical systems. These systems have a rich and beautiful history in mathematics. They were first studied in the 1920s by the French mathematicians Fatou *(2)* and Julia *(3)* and their co-workers. But the full beauty of these dynamical systems was not realized until the late 1970s and early 1980s when computer graphics showed the magnificence and complexity of the chaotic set for these systems, the so-called Julia set.

For the remainder of this chapter, we will concentrate on the structure and properties of this class of analytic functions. We will show how to find and to compute the Julia sets for these functions. We will also describe in detail how these chaotic sets may sometimes change in configuration in a quite unexpected manner: sometimes these sets may undergo a burst or explosion as parameters are varied.

We will consider only iterated mappings of the complex plane C such as $E_\lambda(z) = \lambda\exp(z)$ or $S_\lambda(z) = \lambda\sin(z)$. Here λ should be interpreted as a complex parameter $\lambda = \lambda_{re} + i\lambda_{im}$ where λ_{re} and λ_{im} are the real and imaginary parts, respectively, of λ and $i = \sqrt{-1}$. Similarly, the state variable z is complex: $z = x + iy$, with real numbers as the two components of z.

Consider the problem of iterating a complex function F. That is, for a given initial complex number z_0, we compute successively

$$z_1 = F(z_0)$$
$$z_2 = F(z_1) = F(F(z_0)) = F^2(z_0)$$
$$z_3 = F(z_2) = F^3(z_0)$$
$$\vdots$$
$$z_n = F^n(z_0).$$

Note that F^n means that the n-fold iteration of F, not F raised to the nth power. The set of points $\{z_0, z_1, z_2, \ldots\}$ is called the orbit of the initial point z_0. As above, the basic goal of dynamical systems theory is to understand the ultimate fate of all orbits of a given system. This is precisely the problem we encountered in the logistic equation. That is, the question is what happens to $F^n(z_0)$ as n tends to ∞?

There are many possible fates in a given system. For example, an orbit may behave relatively tamely by simply tending to a fixed point, as illustrated by the simple mapping $F(z) = z^2$. If z_0 is a complex number with absolute value less than one, then a simple computation shows that successive squarings yield an orbit that tends to 0 in the limit. That is, all complex numbers inside the circle of radius one tend to 0 under iteration of $F(z) = z^2$. This is stable behavior: all sufficiently nearby initial choices of z_0 lead to the same fate for the orbit.

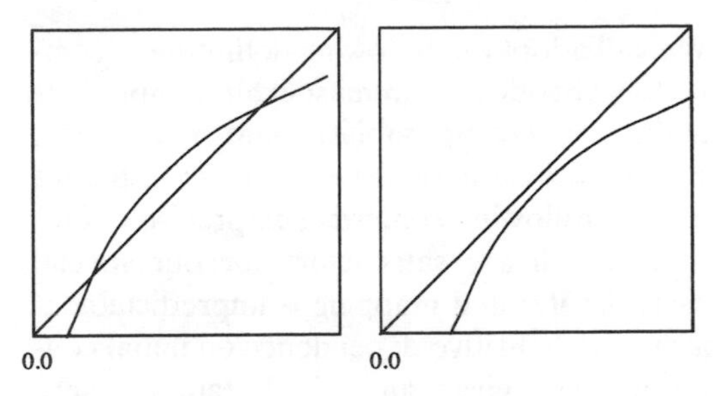

Fig. 1. The graphs of $\lambda \exp z$ for $\lambda > 1/e$ and $\lambda < 1/e$.

As another example, consider the exponential function $E_\lambda(z) = \lambda\exp z$ with $\lambda > 0$. If we use only real values of z, then Fig. 1 shows that the graph of E_λ assumes two different forms, depending upon whether $\lambda > 1/e$ or $\lambda < 1/e$, where $e \approx 2.7128\ldots$ satisfies $\ln e = 1$. In Fig. 1B, E_λ has two fixed points, q_λ and p_λ, i.e., $E_\lambda(q_\lambda) = q_\lambda$ and $E_\lambda(p_\lambda) = p_\lambda$. One may check that all points in the interval $-\infty < x < p_\lambda$ lead to orbits that tend to q_λ as n tends to ∞. On the other hand, in the interval $p_\lambda < x < \infty$, all points have orbits that tend to ∞ under iteration. For values of $\lambda > 1/e$, all points have orbits that tend to ∞. This may be easily checked with a calculator by simply iterating $\lambda \exp(x)$ for various choices of initial x. The rigorous proof is also easy. Thus we see that the dynamical system E_λ has two vastly different behaviors on the real line, depending upon whether $\lambda > 1/e$ or $\lambda < 1/e$. We will see that this is an example of a burst into chaos when viewed as a dynamical system in the complex plane.

Before considering this, we digress to discuss the behavior of a completely chaotic dynamical system. There are a number of different definitions of chaos in the literature.

We will adopt the following definition: a completely chaotic system must exhibit unpredictability, indecomposability, and recurrence. For precise definitions we refer to *(4),* but we list the following somewhat imprecise definitions which are satisfactory for our special case. An iterated mapping is unpredictable if it exhibits sensitive dependence on initial conditions, i.e., given any initial state z_0, there must be a nearby state w_0 whose orbit diverges from that of z_0. That is, the distance between z_n and w_n must eventually be large. We think, therefore, that any numerical computation of the orbit of z_0 must be suspect: a small initial error, perhaps because of roundoff error, may yield a completely different orbit, thus rendering numerical study inaccurate.

The dynamical system is indecomposable if there is an orbit that eventually enters any preassigned region, no matter how small, in the plane. Thus this orbit comes arbitrarily close to any point whatsoever in C, and we cannot separate the given system into two essentially separate subsystems that behave independently.

Finally, a dynamical system exhibits recurrence if, given an initial condition z_0, there is another initial condition w_0 arbitrarily close to z_0 that is periodic. That is, there is an iteration n for which

$$F^n(w_0) = w_0.$$

Consequently, $w_{n+1} = w_1$, $w_{n+2} = w_2$, etc., and the orbit of w_0 is a cycle or periodic orbit. Periodic orbits are usually regarded as among the most important motions in a dynamical system, so our assumption is that they abound.

To summarize, a dynamical system in C is completely chaotic if it exhibits all three of the above properties. This definition is intended to mirror the properties of physical systems that exhibit turbulence.

Spurred on by the computer graphics of Mandelbrot *(5),* D. Sullivan *(6)* has given effectively computable criteria for a dynamical system to be completely chaotic. For example, in the case of systems such as $\lambda \exp(z)$ or $\lambda \sin(z)$, all we need to do is follow the critical orbits of the system. For the exponential, this is the orbit of 0 (the omitted value), and for sine, these are the orbits of the critical points $\pm\pi/2$ (where the maxima and minima occur on the real line). Technically, the critical orbits are defined to be the orbits of the critical and asymptotic values. Critical values are simply the images of the critical points. Note that sine has infinitely many critical points but only two critical values and no asymptotic values, so checking the above criteria is straightforward.

One consequence of Sullivan's recent No Wandering Domains Theorem *(6)* is that, if all critical orbits of these maps tend to infinity, then the dynamical system is completely chaotic on the whole plane. In the example $E_\lambda(z) = \lambda\exp(z)$ where $\lambda > 1/e$, it is easy to check that 0 indeed tends to infinity, so E_λ is completely chaotic in the whole plane when $\lambda > 1/e$.

Now let us contrast this with the case $0 < \lambda < 1/e$. Consider the vertical line $x = 1$ in the complex plane. Via Euler's formula

$$e^{x+iy} = e^x e^{iy} = e^x (\cos y + i \sin y)$$

this vertical line is mapped to the circle Γ of radius $\lambda \cdot e^1 < 1$. Moreover, each point in the plane to the left of this vertical line is mapped inside the circle Γ. Thus, the whole left half plane (real part of $z \leq 1$) is contracted inside itself and, in fact, inside Γ. Now apply E_λ again; Γ is contracted further inside itself. Continuing in this fashion, we see that E_λ cannot be chaotic in this half plane; for example, there cannot be any cycles or periodic points outside Γ. Indeed, one may check that the only cycle to the left of $x = 1$ is the fixed point q_λ discussed previously.

Thus we see that as λ increases through $\lambda = 1/e$, there is a dramatic change in the set of points on which the dynamical system is chaotic. When $\lambda < 1/e$, there are no such points to the left of $x = 1$, whereas when $\lambda > 1/e$, E_λ is completely chaotic on the entire plane. This is the burst into chaos.

The Julia Set

The set of points in the complex plane for

which a dynamical system like $\lambda \exp(z)$ or $\lambda \sin(z)$ is completely chaotic is called the Julia set. It is known that this set is often a fractal *(5)* and so may assume spectacular geometric shapes. There are a number of different techniques for plotting the Julia sets numerically. A procedure that works for polynomials is described in *(7)* and *(8)*. For the transcendental maps that we are considering, there is a special and rather simple algorithm due to John Hubbard that allows for easy plotting of the Julia set. It is known that the Julia set of a map like $\lambda \exp(z)$ or $\lambda \sin(z)$ is the closure of the set of points whose orbit tends to ∞ *(9)*. That is, any point whose orbit tends to ∞ and any limit point of such points lies in the Julia set. Note an apparent contradiction: periodic cycles must occur arbitrarily close to any point in the Julia set by our notion of recurrence, but so too must points whose orbits tend to ∞. Indeed, bounded and unbounded orbits accumulate at all points of the Julia set, giving further indication of the unpredictability of these systems on the Julia set.

Using Hubbard's algorithm, we may thus plot the outline of the Julia set by iterating a grid of points in the plane a preselected number of times N. If the orbit of the point remains bounded for all N iterations, we assume that the point does not lie in the Julia set and color it white. If, however, the orbit escapes to infinity (i.e., becomes too large for the computer), we assume that the point lies in (or, more appropriately, near) the Julia set and color it black.

We have plotted the results for $\lambda \exp(z)$ in Fig. 2. Note the small chaotic region for a particular $\lambda < 1/e$. Indeed, in this picture, almost the entire plane is white. White points never lie in the Julia set; indeed, all of these points are attracted to the fixed point, which we have denoted by q_λ. No matter how large N is chosen, a similar picture results. The two different pictures for $\lambda > 1/e$ are computed with different values of N and different values of λ. We have set N equal to 50 in Fig. 2B; choosing N larger will result in the disappearance of the black region as more points have a chance to escape. N was chosen to be 200 in Fig. 2C.

The results above, together with many similar bursts, were suggested initially by mathematical experimentation. The idea of experimentation is becoming increasingly important in mathematics as the computer becomes the mathematician's laboratory. Experimentation has led to a number of significant new ideas, particularly in dynamical systems. As further examples of this, the above algorithm may be used with minor adjustments to find bursts in other families of complex entire functions. For example, there is a burst that occurs in the family $i\lambda \cos(z)$ when $\lambda \approx 0.67$. There is another burst that occurs at $\lambda = 0$ in the family $(1 + \lambda i) = \sin z$. See *(9)* for color computer graphics images that illustrate these bursts.

Each of these bursts may be proven rigorously to occur *(10)*. For the cosine family, the mechanism that produces the burst is entirely analogous to that which occurs in the exponential family: the elementary saddle-node bifurcation occurs at the critical parameter value and allows the critical orbits to slip away to ∞ (Fig. 3).

For the sine family, however, the mechanism is entirely different. The family $\mu \sin z$ experiences an elementary bifurcation as μ increases through the value 1. This bifurcation is reminiscent of the period-doubling bifurcation as described in *(4)* or *(11)*, although it is technically somewhat different. It is well known that such a bifurcation does not lead to a burst into chaos; rather, the states both before and after the bifurcation are quite stable. Nevertheless, if a different route in parameter space is chosen through the value $\mu = 1$, then a burst is possible. Indeed, one may prove that there are parameter values arbitrarily close to 1 for which the corresponding Julia set is the whole plane *(12)*. In fact, these results suggest that any elementary bifurcation in complex dynamics (for entire transcendental functions) is accompanied by a direction in parameter space that leads to a similar burst.

Fig. 2A. The Julia set of $0.3e^z$.

Fig. 2B. Julia set of $0.4e^z$.

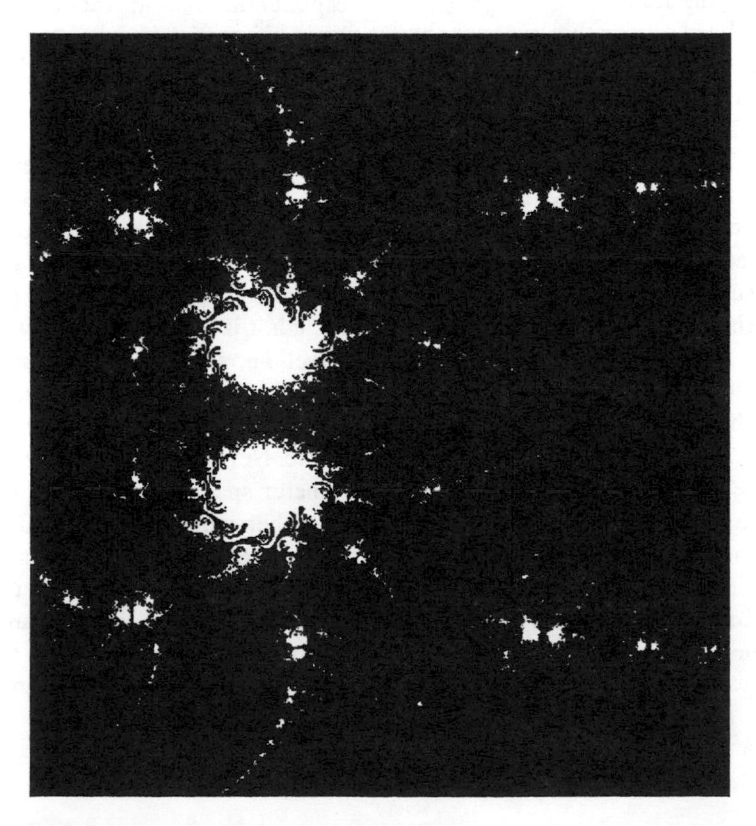

Fig. 2C. Figure 2B recomputed with N = 200.

Fig. 3A. The Julia set of $0.66i \cos z$.

Fig. 3B. The Julia set of $0.68i \cos z$.

References

1. May, R. B., *Nature (London)* **261,** 459 (1976).
2. Fatou, P., *Acta Math.* **47,** 337 (1926).
3. Julia, G., *J. Math. Pures Appl.* **4,** 47 (1918).
4. Devaney, R. L., *An Introduction to Chaotic Dynamical Systems* (Benjamin-Cummings, Menlo Park, CA, 1985).
5. Mandelbrot, B., *The Fractal Geometry of Nature* (Freeman, San Francisco, 1982).
6. Sullivan, D., *Acta Math.* **155,** 243 (1985).
7. Blanchard, P., *Bull. Amer. Math Soc.* **11,** 85 (1984).
8. Douady, A., and J. Hubbard, *C. R. Acad. Sci. Paris* **294,** 123 (1982).
9. Devaney, R. L., *Science* **235,** 342 (1987).
10. Devaney, R. L., *Phys. Lett.* **104,** 385 (1984).
11. Feigenbaum, M., *J. Stat. Phys.* **21,** 669 (1978).
12. Devaney, R. L., in *Chaotic Dynamics and Fractals,* M. Barnsley and S. Demko, Eds. (Academic Press, Orlando, FL, 1986), p. 141.

2

Dynamical Characterization of Brain Electrical Activity

Paul E. Rapp, Theodore R. Bashore
Irwin D. Zimmerman, Jacques M. Martinerie
Alfonso M. Albano, Alistair I. Mees

Abstract

Dynamical systems theory is a branch of mathematics that incorporates elements from the theory of differential equations, nonlinear control theory, and topology. Quantitative measures derived from this theory – for example, fractal dimensions, Lyapunov exponents, and entropy – can be used to characterize time-dependent behavior. These measures assay different aspects of behavior, and they have different strengths and weaknesses. A common element of all of them, however, is an attempt to use mathematics to reconstruct the system generating the observed signal. This contrasts with the classical procedures of signal analysis that focus exclusively on the signal itself. Dynamical analysis has been applied to physical systems such as electronic circuits and chemical reactions. As evidenced by contributions to this volume, the domain of applications has expanded to include economic, social, and biological systems. This chapter focuses on the dynamical analysis of brain electrical activity recorded from healthy, adult subjects. Two general cases are considered: (i) ongoing electroencephalographic (EEG) activity, and (ii) event-related potentials, the electrical responses elicited by strictly defined stimuli.

EEG signals were recorded from subjects at rest and during periods of cognitive activity. The fractal dimension of the signals was found to increase when the subject was mentally active. Event-related potentials (ERPs) characterize the cerebral electrical activity associated with information processing during the first few hundred milliseconds after presentation of a stimulus. In a typical ERP experiment, the subject is instructed to ignore some stimuli (called nontarget stimuli) and to maintain a running count of the presentations of target stimuli. The study summarized here indicates that the dimension of activity drops in response to target stimuli.

While the results presented here encourage further systematic studies, the provisional nature of the available evidence should be recognized. Also, the magnitude of the technical difficulties encountered in the dynamical analysis of biological signals should not be underestimated.

Introduction

Modern hardware has resulted in an extraordinary increase in our capacity to acquire digitized records of bioelectric signals. However, developments in our capacity to extract clinically meaningful understanding from these data have not kept pace with our ever-growing ability to amass data. Most classical interpretive techniques that have been applied to EEG signals, for example, have been based on numerical estimates of spectra. These procedures are certain to remain important, but new methods are available that may complement spectral analysis. These recent developments have two components: (i) topographic-display technologies and (ii) analytical techniques derived from dynamical systems theory. We describe here the application of methods from dynamical systems theory to the analysis of brain electrical activity recorded at the scalp.

Topographic mapping of itself neither introduces the measurement of new variables nor provides new analytic methods. These technologies display quantitative information in a visual form that facilitates its clinical interpretation. Typically, the inputs to the mapping algorithm are voltage measurements. The great value of a topographic display is that information is presented geometrically rather than numerically. The interpretive process can then utilize the most powerful visual image-processing computer presently available, the human visual cortex, to draw inferences.

In the course of our long evolutionary history, we have evolved an impressive talent for processing visually presented geometrical relationships. Numbers are a recent discovery and therefore do not lend themselves to such ready integration. Simply put, we are more clever with pictures than with numbers. Topographic maps exploit this. However, there is a fundamental limiting property of topographic voltage maps. Typically, the input to the mapping algorithm consists of one number from each electrode. Therefore, a small number of input values, usually from 16 to 64, generates the several thousand numbers used to construct the map. The algorithm itself imposes a regularization on the inputs and can produce a spurious order to the resulting map. Indeed, a certain measure of malicious fun can be obtained by inviting one's neurological colleagues to interpret topographic maps generated from random numbers. This is why significance can be attached only to those topographic features that are invariant in time and space.

A computer-based interpretive procedure such as topographic mapping is only as sophisticated as the analytic procedures on which it has been constructed, however. This inescapable cliché has, in part, motivated the search for signal-processing methods based on the mathematics of dynamical systems theory. Quantitative measures of dynamical behavior such as dimensions and Lyapunov exponents can be mapped. Thus, the two approaches are complementary. Dynamical analysis, from which dimensions and Lyapunov exponents are derived, is introduced in the next section.

From Numbers to Geometry

The methods of dynamical analysis can be applied to either a digitized waveform such as an EEG signal or to a record of inter-event time intervals such as those between neuron action potentials or heartbeats. The analysis proceeds more or less identically for each case. For simplicity of presentation, attention is restricted here to the analysis of waveforms digitized at uniform time intervals. A measurement results in a sequential data set of voltages $\{v_1, v_2, \ldots\}$. The first step of the analysis is to embed the signal in a higher dimensional space. Of several mathematically equivalent procedures, the simplest is to construct N-dimensional points X_j from the original data, according to the protocol

$$X_j = (v_j, v_{j+1}, \ldots v_{j+N-1}).$$

The choice of N is not arbitrary [see *(1)* for technical details].

Embedding results in a cloud of points in N-dimensional space. Dynamical analysis, as we use the term, refers to the family of mathematical techniques that will be used to construct quantitative measures characterizing the geometry of this set of points. At first glance, embedding may seem to be an insignificant technicality. This is not the case, however. We began the analysis with a list of numbers corresponding to the measured voltages $\{v_1, v_2, \ldots\}$. We now have a corresponding geometric object, the set of N-dimensional points $\{X_j\}$, that corresponds to this set of numbers. By transforming the problem of analysis to the geometric domain, we can bring to the process a very formidable armamentarium of topological techniques that has been constructed during the last century. But more is involved than just ease of interpretation and the availability of analytical techniques. It can be demonstrated that if certain idealized technical requirements are satisfied, an intimate relationship exists between the set $\{X_j\}$ and the system that generated the signal. [For technical details,

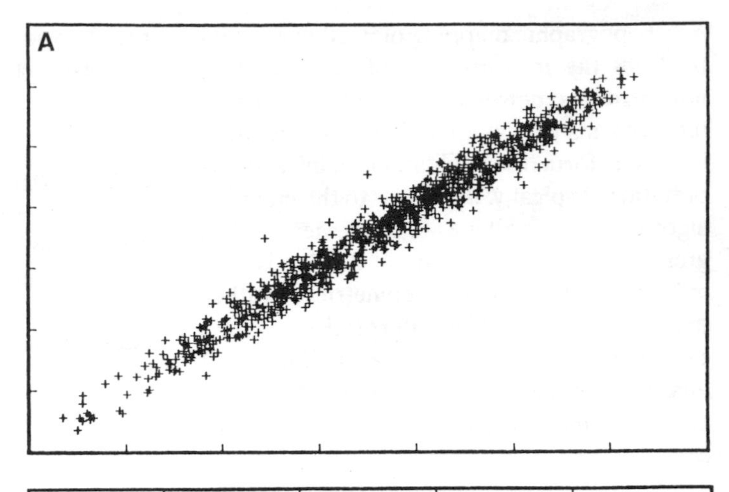

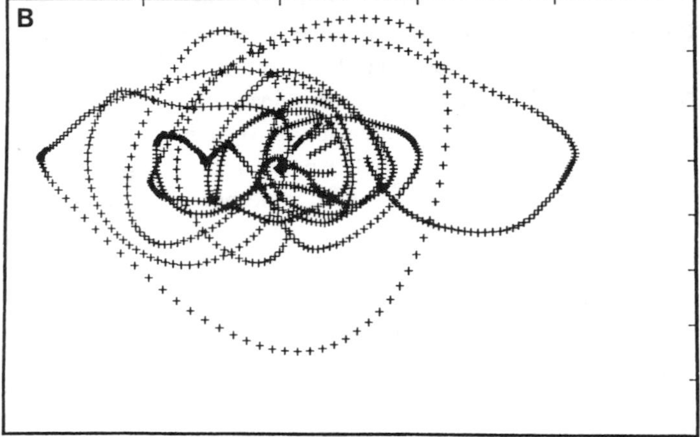

Fig. 1. Two-dimensional projections of embedded electroencephalographic data $\{v_j\}$. The signal was obtained from a healthy adult at site O_z (digitizing frequency 500 Hz, 12-bit digitizer). During the recording, the subject's eyes were closed and she was resting. (**A**) The two-dimensional set obtained from the points (v_j, v_{j+1}). (**B**) The two-dimensional projection obtained from the same data set. In this case, the data were initially embedded in a seven-dimensional space and rotated with the orthogonal transformation obtained from the singular value decomposition of the embedding matrix. [Reproduced with permission from *(1)*, copyright 1989 by Human Sciences Press.]

see *(2, 3)*; a nontechnical description of the underlying ideas is given in *(1)*.] This is the fundamental distinction between dynamical analysis and classical procedures such as spectral analysis. Where spectral analysis quantitatively characterizes properties of a signal, dynamical analysis, in the ideal case, probes the structure of its source.

The potential power of geometrical analysis is shown in Fig. 1. The first diagram, Fig. 1A, shows a two-dimensional set constructed from an EEG time series by plotting the points (v_j, v_{j+1}). Figure 1A is hardly encouraging. While some structure is evident, the points do not fill the plane in a random scatter pattern; fine geometric structure is not apparent. However, this geometric structure does exist. The discouragement obtained from Fig. 1A is an artifact resulting from restricting our view to an untransformed two-dimensional projection. Figure 1B is constructed from the same voltage data. The data were used to construct a seven-dimensional set. To produce the diagram, the original set was rotated in seven-space and projected onto a two-dimensional space. The rotation is the orthogonal transformation obtained by the singular value decomposition of the original set [see *(4)* for details]. Following rotation, a delicate geometrical structure becomes immediately apparent.

Dynamical analysis proceeds by quantitatively describing the structure of sets $\{X_j\}$ that can be accessed when data are embedded in a sufficiently high-dimensional space. The application of different quantitative measures to the analysis of $\{X_j\}$ sets constructed from biological signals is being actively explored by several research groups [*(5–8)*; a more complete bibliography is given in *(1)*]. This chapter focuses on one of these measures, the dimension of set $\{X_j\}$.

The traditional idea of dimension is familiar to all of us. A line is a one-dimensional object. A surface is two dimensional, and a solid is three dimensional. At the turn of the century, the need to generalize the definition of dimension was recognized *(9, 10)*. An informal sense of this generalized dimension can be obtained by reexamining classical one-, two-, and three-dimensional objects. Consider a one-dimensional set of points filling a line segment of length L. Suppose the distance between points were doubled. The set would now occupy a length 2L or, to express it rather clumsily, 2^1L^1. Consider next a two-dimensional object of length L on each side. Its area is L^2. If the distance-doubling experiment were to be conducted with this object, the length of each side would double, and its area would become $4L^2$, that is, 2^2L^2. If the experiment were conducted with a cube of length L on each side, the volume would become 2^3L^3.

The exponentials determined in each experiment, 1, 2, and 3, are central to the expanded definition of dimension. A defining property of a two-dimensional object is that area scales with an exponent of 2. If length was tripled to 3L on each side, area would be 3^2L^2. The area is different, but the exponent remains equal to 2. Therefore, by this abstract definition of dimension, a square is a two-dimensional object. A line is one-dimensional, and a cube is three-dimensional. For the simple objects considered so far – a straight line, a square, and a cube – this may seem an unnecessary complication. However, armed with this generalized definition of dimension, some surprising discoveries were made. It is possible to show that some objects do not scale with integer exponents – that is, it is possible for some sets to have noninteger dimensions.

Mandelbrot *(10)* coined the term "fractals" to describe these very irregular geometries. Reviews of the definitions of dimension and procedures for estimating dimensions from time series have appeared elsewhere *(11–14)* and in chapters in this volume. In dynamics, fractal sets are important because of their intimate relation to chaotic dynamics. For our present purposes, they are important because the dimension provides a sensitive measure of the geometrical complexity of embedding sets $\{X_j\}$ constructed from EEG signals. These sets have noninteger dimensions – that is, they are fractals.

The next operational question to be considered is, How can the dimension of sets $\{X_j\}$ be estimated? A variety of procedures have been tested *(14)*. The results reported in the next section used the Grassberger-Procaccia algorithm *(15, 16)* to estimate the correlation dimensions of EEG signals [for technical details of the algorithm's application to biological signals, see *(1)* and *(7)*]. It is important to note that these reports and others [for example, Babloyantz *(5)* and Grassberger *(17)*] describe the technical difficulties encountered when trying to calculate dimension estimates with noisy signals. In many instances, it is impossible to determine even an approximate value of the dimension. It is possible, however, to impose reliability criteria on these calculations that, at least some of the time, protect against spurious estimates. The results described in the next section encourage a guarded optimism.

Psychological Sensitivity of EEG-Dimension Estimates

The results of Fig. 1 suggest that appropriately embedded EEG signals contain a complex geometric structure. Additionally, the availability of the Grassberger-Procaccia algorithm and other procedures for estimating dimensions from time series indicate that, at least in principle, it should be possible to compute a numerical measure of the structure in Fig. 1B. An additional question must be considered. Are the numerical values of EEG dimension reliably sensitive to changes in the physiological or psychological state of the subject? If they are not, for example, if they never change, then dimension estimates will be of little relevance to either physiological research or clinical practice. A simple, qualitative examination of the geometry of EEG records obtained from subjects in different states suggests that the dimension should be sensitive to changes in cognitive activity.

An example is shown in Fig. 2. The set displayed in Fig. 2A was constructed from data obtained from a subject at rest (eyes closed). A complex structure is observed. The set in Fig. 2B was constructed from data obtained from the same subject during the same experiment. The only difference is that, in this case, the subject was silently counting backwards from 763 in steps of seven. As viewed in Figs. 2A and 2B, these three-dimensional sets are different, but arguably not dramatically so. A marked disparity is observed, however, between Figs. 2C and 2D. These are the sets shown in Figs. 2A and 2B, respectively. They have been rotated in three-dimensional space in order to emphasize the differences in the third axis. The set constructed from data obtained when the subject was counting has more significant behavior in the third component. As a purely qualitative intuition, one would expect this geometrical complexity to be reflected in the dimension. Specifically, one would expect the dimension of the set constructed from data obtained during counting to be greater than the set constructed from data obtained at rest.

The results summarized in Fig. 3 justify this qualitative expectation. This diagram shows median dimension estimates obtained from a series of experiments involving five subjects and the dynamical analysis of a total of 110 EEG data sets. [The details of the experimental design are given in *(1)*.] Each subject was tested in a sequence of five conditions: (i) rest, (ii) performing mental subtraction in steps of seven, (iii) rest, (iv) performing mental addition in steps of two, and (v) rest. The subject's eyes were closed throughout the experiments. As expected, the median values suggest that the dimension increases during periods of cognitive activity. The diagram also suggests that the difference between active and resting dimensions decreases as the experiment continues. However, attention should be directed to the uncertainty in these estimates indicated by the error bars. This degree of uncertainty reflects the previously mentioned technical problems encountered in dimension calculations using noisy biological data and suggests that speculation concerning changes observed over the course of the experiment are premature.

Dynamical Analysis of Event-Related Potentials

Event-related brain potentials (ERPs) characterize the cerebral electrical activity associated with information processing during the first several hundred milliseconds after presentation of a stimulus *(18, 19)*. The easiest way to introduce these potentials is to describe a specific archetypal ERP experiment, the auditory oddball, known most prominently for its elicitation of the P300 component. In this experiment, the subject is seated comfortably and hears one of two auditory signals, a high-pitched tone (e.g., 1500 Hz) and a low-pitched tone (e.g., 1000 Hz), and must keep a running mental count of the occurrences of one of the tones (the 1000 Hz target tone, for example) and refrain from counting the other tone (the nontarget tone). Presentation of the counted or target tone elicits a large positive-going wave, the P300, that is small or absent in the ERP invoked by the nontarget tone.

Examples of ERPs recorded under these experimental conditions are shown in Fig. 4. In the experiment summarized in this figure, 20 presentations of the target tone were randomly distributed among 80 presentations of the nontarget tone. The brain's electrical response to each stimulus presentation was recorded at site P_z [in accordance with the International Electrode Placement System *(20)*]. The average response to each stimulus is displayed in the figure. The differences in the ERPs elicited by the two stimuli are clearly evident. In more complicated information-processing tasks, digital filtering and procedures designed to remove artifacts caused by eye blinking *(21)* are employed to enhance the difference between target and nontarget responses.

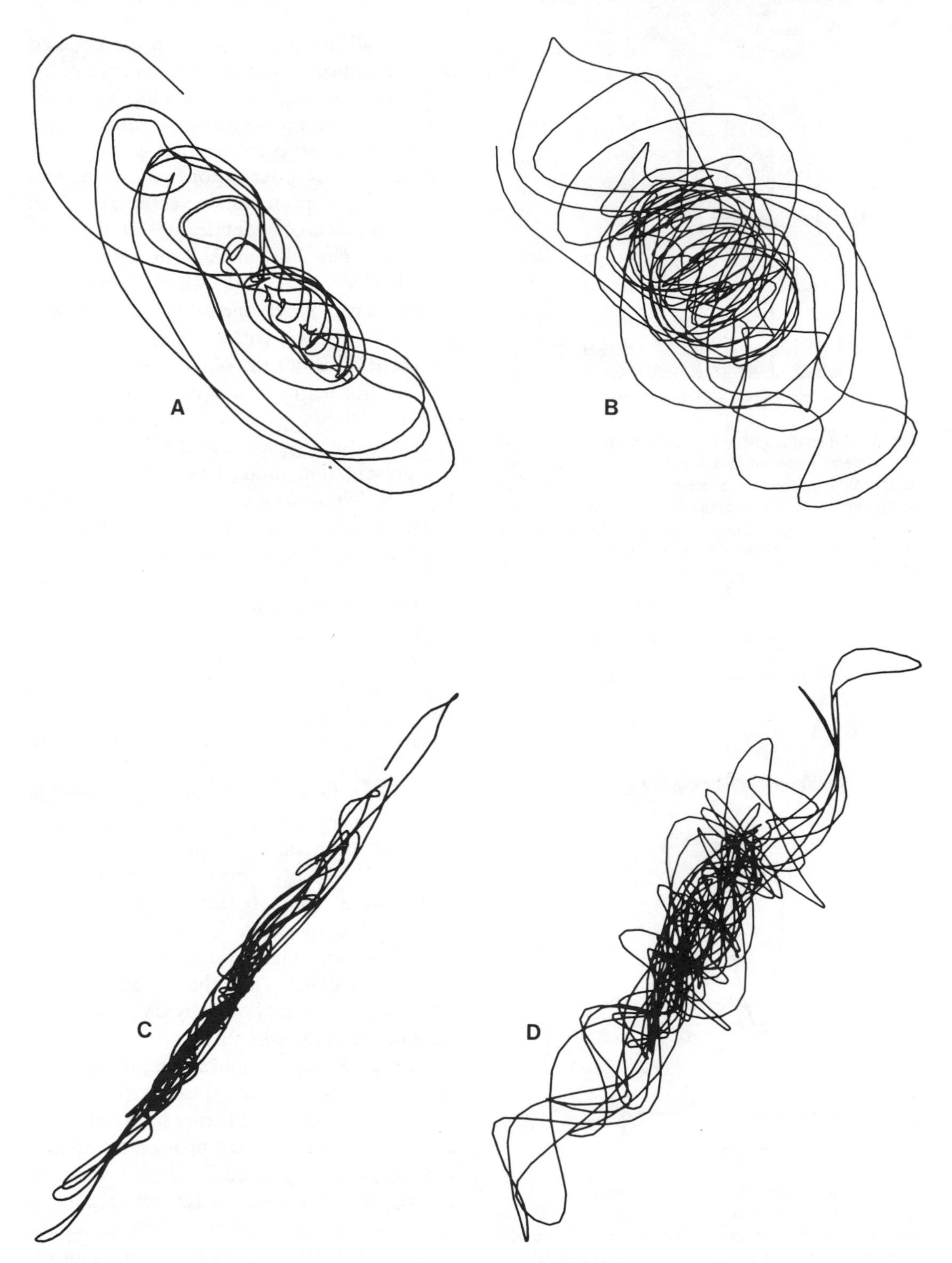

Fig. 2. Three-dimensional sets constructed from electroencephalographic data recorded from a normal adult subject. The recording was obtained from an electrode at site O_Z (500 Hz sampling frequency, 12-bit digitizer). (**A**) A set constructed from a signal obtained from the subject at rest (eyes closed). (**B**) A set constructed from a signal obtained from the same subject performing serial subtractions in steps of seven (eyes closed). (**C**) The set in Fig. 2A after a rotation. (**D**) The set in Fig. 2B after a rotation.

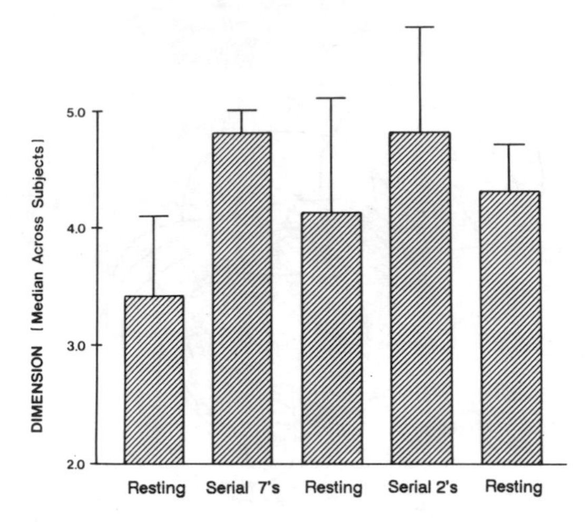

Fig. 3. Median dimension estimates of EEG signals from five subjects recorded in a sequence of five conditions: rest, mental arithmetic in steps of seven, rest, mental arithmetic in steps to two, and rest. The displayed uncertainty is the average absolute deviation from the median. [Reproduced with permission from *(1)*, copyright 1989 by Human Sciences Press.]

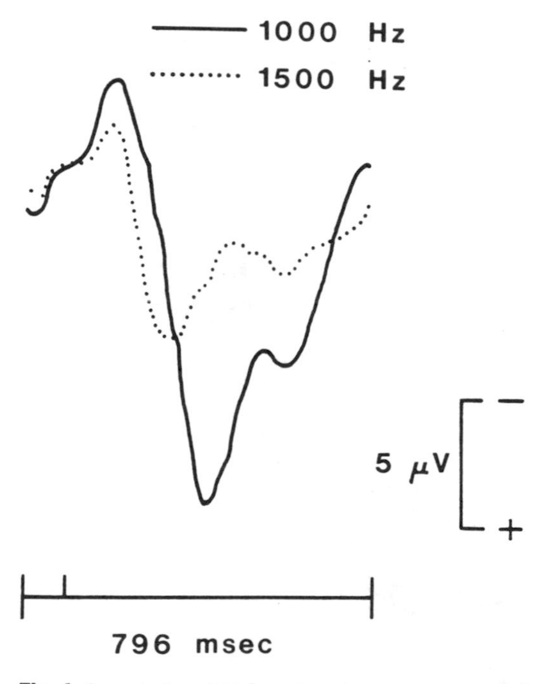

Fig. 4. Average target and nontarget responses recorded at site P_Z in an auditory oddball experiment. Twenty presentations of the target stimulus, a 1000 Hz tone, were randomly distributed among 80 presentations of the nontarget stimulus, a 1500 Hz tone. [Reproduced with permission from *(56)*, copyright 1981 by the Society for Psychophysiological Research.]

The distinctive response to the counted stimulus appears to be the consequence of the meaning associated with the stimulus. If the experiment were revised and the same subject was instructed to count the high-pitched tone and not the low-pitched tone, and if the target/nontarget presentation ratio remained 20/80, an essentially identical result would be obtained. The high-pitched or target tone would elicit the P300, whereas the low-pitched, uncounted, tone would produce little or no P300. Thus, a critical element in producing the P300 is not the physical properties of the stimulus but its mental categorization as a target or nontarget event. The average response is also dependent on the target/nontarget presentation ratio. This is illustrated in Fig. 5, which shows the progressive loss of the characteristic target response as the frequency of the target signal is increased.

The P300 is not limited to the auditory modality or to the discrimination of simple stimuli. It can be elicited by stimuli presented to the somatosensory and visual modalities. Moreover, P300s are elicited in complicated decision-making tasks that may require higher order abstractions. For example, male and female first names can be presented visually to the subject. In the experiment summarized in Fig. 6, the subject was asked to keep a running count of the number of times a male name appeared. Twenty presentations of male names were randomly distributed among 80 presentations of female first names. As in the preceding case, a large target effect is seen in the average signal. Thus, the categorization is an abstract class not bound by simple physical parameters of the stimuli.

In the examples shown here, the average target and nontarget responses are displayed. Typically, a single-trial response to the target stimulus and a single-trial nontarget response cannot be distinguished by direct visual inspection. In a classical ERP experiment, a large number of stimulus presentations are needed to generate average signals. The requirement for large numbers of stimulus presentations places severe limits on both the

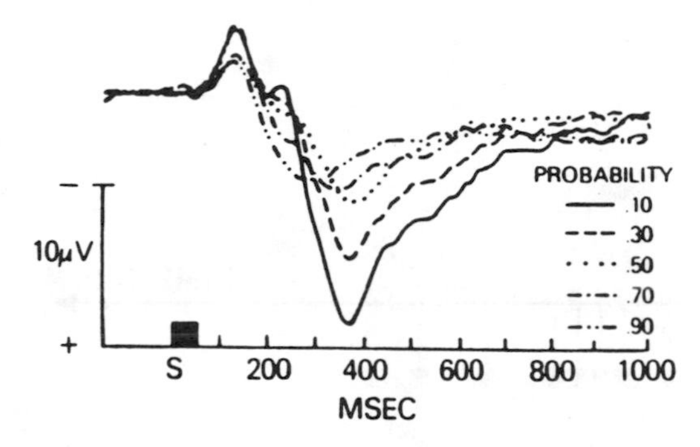

Fig. 5. The average target response in a sequence of auditory oddball experiments in which the frequency of the target stimulus is increased from 0.10 to 0.90. [Reproduced with permission from *(50)*, copyright 1986 by the Society for Psychophysiological Research.]

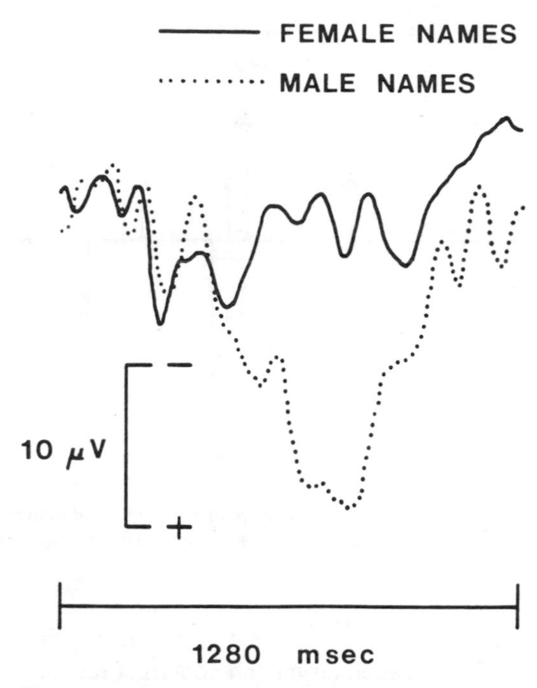

Fig. 6. Averaged response to female first names (nontarget stimulus, 80 presentations) and male first names (target stimulus, 20 presentations). [Reproduced with permission from *(56)*, copyright 1981 by the Society for Psychophysiological Research.]

design of scientific experiments and on the applications of ERP-based technologies. These limitations have motivated the search for analytical techniques that can distinguish between single-trial target and nontarget responses. A number of techniques have been investigated, including peak measurements, area measurements, discriminant analysis, and cross correlation techniques *(22)*. They have met with varying degrees of success. Results with discriminant analysis have been particularly impressive. Discriminant analysis requires the specification of a training set and identifies the contribution of each element of the training set, expressed as weighting coefficients, to the analyzed signal *(23)*. This procedure has been applied to single-trial ERP data obtained in auditory *(24)* and visual *(25)* oddball count tasks. Using stepwise discriminant analysis, approximately 80% of the single trial ERPs were classified correctly.

We have applied a variant of dimension calculations to the characterization of single-trial ERP signals. Because an ERP is a transient response followed by a return to an initial state, a true dimension cannot be calculated from this signal. However, it is possible to modify convergence criteria to produce a related measure, the correlation index [technical details are given in *(1)*]. An example of a successful application is shown in Fig. 7. In this figure we show the correlation indexes calculated from the averaged target and nontarget ERPs, as well as the calculations done with single-trial ERPs. For this subject the index values for single-trial target responses are always lower than those obtained from nontarget responses. Target and nontarget correlation index estimates fall into nonoverlapping ranges of values. While the correlation index of the average target response is indeed less than the index of the average nontarget response, the differential obtained with averaged signals is significantly less than that typically obtained from single trials. For some of the other subjects in the study, a clear separation in index values was not obtained *(1)*.

These mixed results must inevitably temper our enthusiasm for the dynamical analysis of single-trial ERPs with the correlation index. However, the pattern of failure is itself instructive. Preliminary results suggest that index calculations may complement single-trial analysis using amplitude calculations. Specifically, in a case where amplitude

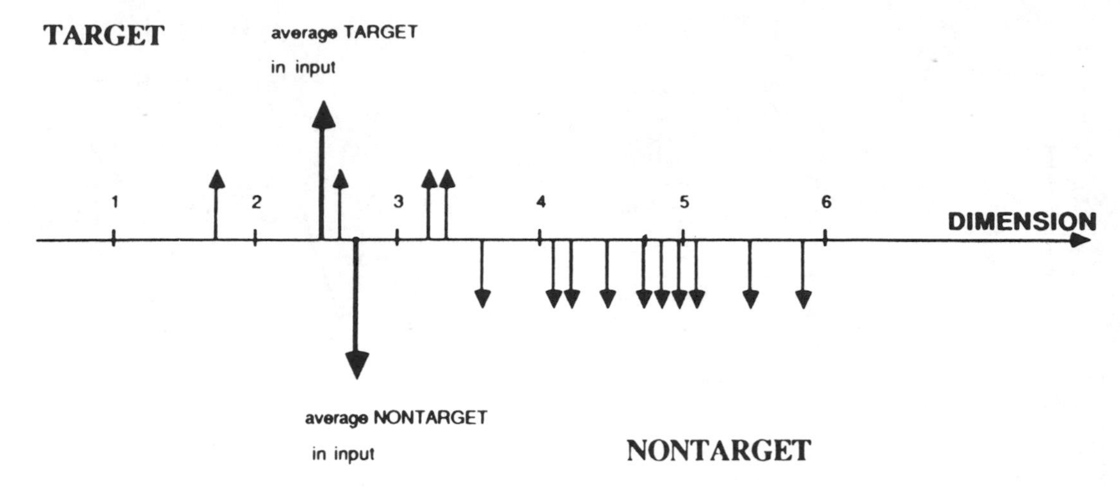

Fig. 7. Correlation index for responses to target and nontarget stimuli in an auditory oddball experiment. The horizontal axis gives the numerical value of the index. [Reproduced with permission from *(1)*, copyright 1989 by Human Sciences Press.]

measurements were unable to distinguish reliably between target and nontarget responses, there was a statistically significant separation in the target/nontarget index values. It should also be noted that there are a number of unresolved technical problems associated with dimension and index calculations. They are outlined in the last section of this chapter. As our estimation procedures improve, dynamical analysis may become an increasingly important complement to existing single-trial classification procedures.

The Unresolved Problems of Dynamical Analysis

The dynamical analysis of noisy biological signals introduces formidable technical difficulties. The following outline focuses on difficulties encountered in dimension estimation, but many of the issues raised are also applicable to other dynamical measures. As we have observed elsewhere *(26, 27)*, dimension estimation is not a numerically robust process. Unless reliability criteria are imposed, it is possible to generate spurious estimates. When these reliability conditions are in place, the calculation frequently fails to resolve a finite-dimension value. A number of operational questions concerning data requirements that must be answered as part of any systematic experimental design remain unanswered. Experimental design is made especially difficult by the fact that some of these questions do not have general answers.

(i) What digitizer resolution is required in EEG and ERP measurements? This is an example of a question that has no general theoretical answer, since acceptable resolution depends on the relative amplitude of noise and the fractal structures that are being investigated. Möller *et al. (28)* have shown that finite digitizer resolution can lead to overestimates of dimension. Empirically we know that, in the case of EEG signals, 8-bit resolution is often inadequate and 12-bit resolution is occasionally acceptable. Systematic studies have not been performed.

(ii) What embedding dimension should be used and what should be the sampling interval, T_s? Theoretically we know that embedding dimension N is big enough if it is greater than $2M+1$, where M is the dimension of the signal being investigated *(2)*. However, this is a conservative estimate of N, and it is often found that it is possible to use smaller values. It has also been shown that the choice of N and T_s are not independent. The important parameter in dimension calculations is the window [the product $(N-1)T_s$] *(29)*. The question is then, what is an appropriate window value? Previous research has compiled a list of signal parameters such as autocorrelation time and extrema of mutual information functions that fail to predict a successful window *(29, 30)*. Recent calculations of mathematical examples of chaotic systems (the Rössler attractor, the Lorenz attractor, and a three-torus) suggest that higher order correlation functions can predict successful windows *(54)*. This must be confirmed for EEG signals.

(iii) How long should the recording epoch be? Results described in Rapp *et al.* *(1)* suggest that, as would be expected, dimension estimates are sensitive to the nonstationarity of the EEG signal *(52)*. Segmentation techniques with their associated uncertainties thus become another element in the list of technical problems.

(iv) What effect does filtering have on dimension estimates? There are disagreements in the technical literature describing the effects of filtering on dimension estimates. In addition to a mathematical example, the Hénon attractor, Badii *et al.* *(31)* examined the effect of filtering on dimension estimates of data obtained from an NMR laser and from a Rayleigh-Benard convection experiment. Somewhat paradoxically, they found that filtering increases the dimension estimate. Specifically, dimension increases with decreasing bandwidth. Mitschke *et al.* *(32)* confirmed an increase in dimension after filtering. This may seem to be a counterintuitive result, since one would expect filtering to smooth a signal and therefore decrease its geometrical complexity.

These results offer a useful example of the general principle that dynamical analysis examines not the signal but the system that generates the signal. Adding a filter to the system increases its structural complexity. The dimension should therefore also increase. Badii *et al.* *(31)* constructed a formal mathematical argument based on this general idea to conclude that as long as the effects of a filter can be described by a differential equation, filtering should increase dimension. However, using EEG signals, Lo and Principe *(33)* obtained the opposite results. Several implementations of three classes of filter – linear phase finite impulse response, Butterworth, and Chebyshev – were used to filter EEG signals. Filtering caused a decrease in dimension. Dimension estimates of filtered signals ranged from 4.55 to 4.92. The dimension of the original signal was 6.56. Similarly, Martinerie (unpublished calculations) found that filtering with Parks-McClellan linear phase digital filters usually, but not always, decreased the correlation index of a single-trial ERP response. However, he also noted that the target/nontarget differences in dimension were largely unaffected by filtering.

It is difficult to think of any simple explanation that reconciles these two groups of observations, but the quality of the original data is probably a significant factor. In cases where the original signal includes a very large component of infinite-dimensional noise, the imposition of a filter, which is a finite-dimensional system, decreases the dimension estimate. Interestingly, Lange has results that indicate that filtering does not affect the metric entropy *(53)*. This empirical observation is consistent with theoretical arguments based on the definition of metric entropy and its relation to the correlation dimension *(55)*.

(v) How can the intrinsic uncertainty in a dimension estimate be determined? Happily, considerable mathematical progress has been made in this area [for examples, see *(14, 34–*36)]. Judd *(37)* has constructed a procedure for estimating dimension that is based on the Grassberger-Procaccia algorithm and thus shares with any correlation integral method the potential for biases introduced into dimension estimates by correlations in the data. Judd's procedure gives an estimate of the dimension with the confidence interval the data would have given if they had no correlation – that is, it gives the smallest confidence interval obtainable using any Grassberger-Procaccia–derived method. In tests with modest-sized data sets of intermediate quality, the method gave reliable results. It has not been tested with EEG data.

(vi) While an estimate of an attractor's dimension is "the first level of knowledge necessary to characterize its properties" *(38)*, it is only the first step in an analysis that would, if successful, lead to a deeper understanding of the dynamical structure of the system generating the measured signal. An important additional step is the construction of equations of motion from the time series. A number of procedures for doing this have been published *(39, 40)*. Recently, Hübler's procedure has been applied to simulated data and experimental data from human limb movements *(41)*. These investigators found that for systems with only one oscillator, there is excellent agreement between the limit cycling displayed by the experimental system and the reconstructed model, even if the data are very noisy. In systems formed by two coupled limit cycle oscillators, the reconstruction was successful only when applied to data with a sufficiently long transient trajectory and a relatively low noise level. This suggests that this method may not be able to provide successful reconstructions from electroencephalographic data. However, this possibility, which clearly merits investigation, has not yet been tested.

Technological and Clinical Applications of Dynamical Analysis

The discussion in the preceding section indicates that the magnitude of the technical difficulties encountered in the application of dynamical analysis to EEG and ERP signals should not be underestimated. With this warning in mind, it is still interesting, however, to speculate about the potential applications of this emerging technology. The discussion here, like the preceding material, is expressed in terms of dimension estimates. However, it should be remembered that the dimension is only one of several measures of behavior that have emerged from dynamical systems theory, included among which are Lyapunov exponents *(42, 43)*, Kolmogorov entropy *(44)*, and topological dimension *(45, 46)*.

The results in Fig. 3 suggest that the dimension of an EEG signal is sensitive to changes in cognitive activity. This in turn suggests that this measure might facilitate EEG monitoring of vigilance in individuals operating in demanding environments at tasks that require high levels of attention. Air traffic controllers are an example. Monitoring pilots of high-performance aircraft for gravity-induced loss of consciousness (GLOC) is another [see *(47, 48)* for reviews of alternative approaches to the study of GLOC]. Mayer-Kress and his colleagues *(7)* have found increases in dimension in response to fluroxene anesthesia and have suggested that this technology might be used to monitor the depth of anesthesia during surgical procedures.

Another potentially significant application area is epileptology. Babloyantz and Destexhe *(49)* observed a decrease in the dimension of an EEG signal recorded during an epileptic seizure. While these authors stress the preliminary nature of their observations, these results do give rise to a series of questions about the possible applications of dynamical analysis to the treatment of epilepsy.

(i) Is a bifurcation in dynamical behavior an early event in epileptogenesis?

(ii) Can changes in EEG dimension predict seizure onset prior to the clinical manifestations of the seizure?

(iii) Can the dynamical analysis of electroencephalograms assist in the localization of an epileptogenic focus? In this context, it might prove interesting to combine dynamical analysis with topographic maps. The inputs to the mapping algorithm need not be voltages. It would be possible to construct maps of dimension, Lyapunov exponents, or entropy.

(iv) Can changes in dynamical measures of EEG activity assess the efficacy of medication? In most instances, antiepilepsy medication is administered empirically. Visual inspection of EEGs provides an important, but imperfect, guide to clinicians. The reappearance of seizures, sometimes after seizure-free periods of months or years, is the definitive indication of the need to review medication schedules. Could long-term changes in dynamical measures indicate the need for modification of medication prior to the reappearance of seizure activity? Conversely, could these measures be used to monitor the supervised reduction of medication levels?

Much of what has been said in the specific

context of epilepsy would also be applicable to the administration of psychotropic medication – another domain of clinical practice in which matching patients with appropriate medication is an empirical process. Can dynamical measures of brain electrical activity assess the efficacy of psychotropic medication? Can changes in these measures indicate the need for changes in medication prior to an acute psychiatric crisis?

The potential applications of this analytical technology in the differential diagnosis of central nervous system (CNS)-centered disease merits consideration. A specific example can serve to clarify the general issues. In the case of the elderly patient, the differential diagnosis of dementia and depression can, in some instances, be difficult. The distinction is an important one because a successful discrimination leads to different courses of clinical action. Can dynamical analysis of electroencephalographic behavior facilitate this diagnosis?

Finally, it should be remembered that many drugs administered in response to disorders that are not primarily centered on the central nervous system – for example, medication for hypertension – can have CNS-based side effects that may be subject to investigation by sufficiently advanced forms of dynamical analysis of EEG activity.

In summary, while the magnitude of the unresolved technical problems should not be underestimated, it should be remembered that mathematical research into their resolution has shown some successes and is continuing. This, combined with the growing body of empirical results such as those presented earlier in this chapter, encourage us in the speculation that dynamical analysis will come to provide an important complement to classical forms of analysis.

Acknowledgments

T. R. B. and P. E. R. would like to acknowledge a grant from the Allegheny Singer Foundation. P. E. R. would like to acknowledge partial support from the Whitaker Foundation and NIH Grant NS19716. P. E. R. would also like to thank the Department of Mathematics and Kingswood College at the University of Western Australia for their hospitality. T. R. B. would like to acknowledge NIH Grants AG04581 and MH40267. J. M. M. wishes to thank INRIA, Paris. The continuing support of the College Computer Center of the Medical College of Pennsylvania is gratefully acknowledged. During the course of this research we have benefited from advice and instruction from many colleagues including E. Donchin, G. Goldberg, R. N. Harner, and J. A. S. Kelso. We wish to express our thanks to Unisys Corporation for the generous donation of computer equipment to this project. The technical assistance of Joseph Waldron is gratefully acknowledged.

References

1. Rapp, P. E., T. R. Bashore, J. M. Martinerie, A. M. Albano, I. D. Zimmerman, A. I. Mees, *Brain Topogr.* **2,** 99 (1989).
2. Takens, F., in *Detecting Strange Attractors in Turbulence. Lecture Notes in Mathematics* (vol. 898), D. A. Rand, L. S. Young, Eds. (Springer-Verlag, New York, 1980), pp. 365–381.
3. Mañé, R., in *Dynamical Systems and Turbulence. Lecture Notes in Mathematics* (vol. 898). D. A. Rand and L. S. Young, Eds. (Springer-Verlag, New York, 1980), pp. 230–242.
4. Albano, A., M. J. Muench, C. Schwartz, A. I. Mees, P. E. Rapp, *Phys. Rev.* **38A,** 3017 (1988).
5. Babloyantz, A., in *Measures of Complexity and Chaos,* N. B. Abraham, A. M. Albano, A. Passamante, P. E. Rapp, Eds. (Plenum Press, New York, 1989), pp. 51–62.
6. Lo, P.-C., and J. C. Principe, *Proc. Inter. Joint Conf. Neural Netw. Soc.* **1,** 693 (1989).
7. Mayer-Kress, G., *et al., Math. Biosci.* **90,** 155 (1988).
8. Roschke, J., and E. Basar, in *Dynamics of Sensory and Cognitive Processing in the Brain,* E. Basar, Ed. Springer Series in Brain Dynamics (Springer-Verlag, Berlin, 1988), pp. 203–216.
9. Hurewicz, W., and H. Wallman, *Dimension Theory* (Princeton University Press, Princeton, NJ, 1941).
10. Mandelbrot, B. B., *The Fractal Geometry of Nature* (W. H. Freeman, San Francisco, 1983).
11. Shaw, R., *Z. Naturforsch* **36A,** 80 (1980).
12. Farmer, J. D., *Z. Naturforsch* **37A,** 1304 (1982).

13. Eckmann, J.-P., and D. Ruelle, *Rev. Mod. Phys.* **57,** 617 (1985).
14. Theiler, J., in *Measures of Complexity and Chaos,* N. B. Abraham, A. M. Albano, A. Passamante, P. E. Rapp, Eds. (Plenum Publishing, New York, 1989), pp. 199 – 202; _____, *J. Opt. Soc. Am.,* in press.
15. Grassberger, P., and I. Procaccia, *Physica.* **9D,** 198 (1983).
16. Grassberger, P., and I. Procaccia, *Phys. Rev. Letts.* **50,** 346 (1983).
17. Grassberger, P., *Nature* **323,** 609 (1986).
18. Donchin, E., W. Ritter, W. C. McCallum, in *Event-Related Potentials in Man,* E. Callaway, P. Tueting, S. H. Koslow, Eds. (Academic Press, New York, 1978), pp. 349 – 412.
19. Donchin, E., D. Karis, T. R. Bashore, M. G. H. Coles, G. Gratton, in *Psychophysiology: Systems, Processes and Applications,* M.G.H. Coles, E. Donchin, S. Proges, Eds. (Guildford Press, New York, 1986), pp. 244 – 264.
20. Jasper, H.H., *Electroencephalog. Clin. Neurophysiol.* **10,** 371 (1958).
21. Gratton, G., M. G. H. Coles, E. Donchin, *Electroencephalog Clin. Neurophysiol.* **55,** 468 (1983).
22. Coles, M. G. H., G. Gratton, A. F. Kramer, G. A. Miller, in *Psychophysiology: Systems, Processes and Applications.* M. G. H. Coles, E. Donchin, S. Porges, Eds. (Guildford Press, New York, 1986), pp. 183 – 221.
23. Donchin, E., and R. I. Herning, *Electroencephalog. Clin. Neurophysiol.* **38,** 51 (1978).
24. Squires, K. C., and E. Donchin, *Electroencephalog. Clin. Neurophysiol.* **41,** 449 (1976).
25. Horst, R. L., and E. Donchin, *Electroencephalog. Clin. Neurophysiol.* **48,** 113 (1980).
26. Albano, A. M., A. I. Mees, G. C. deGuzman, P. E. Rapp, in *Chaos in Biological Systems,* H. Degn, A. V. Holden, L. F. Olsen, Eds. (Plenum Publishing, London, 1987), pp. 207 – 220.
27. Rapp, P. E., A. M. Albano, A. I. Mees, in *Dynamic Patterns in Complex Systems*, J. A. S. Kelso, A. J. Mandell, M. F. Schlesinger, Eds. (World Scientific Publishers, Singapore, 1988), pp. 191 – 208.
28. Möller, M., W. Lange, F. Mitschke, N. B. Abraham, U. Hübner, *Phys. Letts.,* in press.
29. Albano, A.M., J. Muench, C. Schwartz, A. I. Mees, P. E. Rapp, *Phys. Rev.* **38A,** 3017 (1988).
30. Martinerie, J. M., A. M. Albano, A. I. Mees, P. E. Rapp, *Phys. Rev.,* in press.
31. Badii, R., *et al., Phys. Rev. Letts.* **60,** 979 (1988).
32. Mitschke, F., M. Möller, W. Lange, *Phys. Rev.* **37A,** 4518 (1988).
33. Lo, P.-C., and J.-C. Principe, *Proc. IEEE Eng. Med. Biol. Soc.* **17,** 638, (1989).
34. Brock. W. A., W. D. Dechert, J. A. Scheinkman, *SSRI Working Paper No. 8702* (Dept. of Economics, University of Wisconsin, Madison).
35. Sayers, C.L., in *Measures of Complexity and Chaos,* N. B. Abraham, A. M. Albano, A. Passamante, P. E. Rapp, Eds. (Plenum Publishing, New York, 1989), pp. 183 – 186.
36. Ellner, S., *Phys. Lett.* **133A,** 128 (1988).
37. Judd, K., "Reliable Estimation of Dimension Error Estimation" (Preprint: Department of Mathematics, The University of Western Australia, Perth, 1989).
38. Farmer, J. D., E. Ott, J. A. Yorke, *Physica.* **7D,** 153 (1983).
39. Crutchfield, J. P., and B. S. McNamara, *Complex Systems* **1,** 417 (1987).
40. Cremers, J., and A. Hübler, *Z. Naturforsch.* **42A,** 797 (1987).
41. Eisenhammer, T., A Hübler, N. Packard, J. A. S. Kelso, "Modeling Experimental Time Series with Ordinary Differential Equations" (Preprint: Florida Atlantic University, Boca Raton, 1990).
42. Wolf, A., J. B. Swift, H. L. Swinney, J. A. Vestano, *Physica* **16D,** 285 (1985).
43. Eckmann, J.-P., S. O. Kamphorst, D. Ruelle, S. Ciliberti, *Phys. Rev.* **34A,** 4971 (1986).
44. Grassberger, P., and I. Procaccia, *Phys. Rev.* **28A,** 2591 (1983).
45. Mees, A.I. and P. E. Rapp, *Syst. Control Letts.,* submitted for publication.
46. Passamante, A., T. Hediger, M. Gollub, *Calculation of fractal dimension using local intrinsic dimension* (Naval Air Development Center Technical Memo #5032-TM-88-02, Warminster, PA, 1988).
47. Whinnery, J. E., *Aviat. Space Environ. Med.* **59,** 9 (1988).
48. Whinnery, J. E., and R. R. Burton, *Aviat. Space Environ. Med.* **58,** 468 (1987).
49. Babloyantz, A., and A. Destexhe, *Proc. Natl. Acad. Sci. U.S.A.* **83,** 3513 (1986).
50. Johnson, R., *Psychophysiology* **23,** 367 (1986).
51. Kutas, M., G. McCarthy, E. Donchin, *Science* **197,** 792 (1977).
52. Barlow, J.-S., *J. Clin. Neurophysiol.* **2,** 267 (1985).
53. Lange, C. G., personal communication.
54. Albano, A. M., A. Passamante, M. E. Farrel, "Using Higher Order Correlations to Define an Embedding Window" (Preprint: Bryn Mawr College, Bryn Mawr, PA, 1989).
55. Rapp, P. E., and A. M. Albano, unpublished results.
56. Donchin, E., *Psychophysiology* **18,** 493 (1981).

3

Sudden Death Is Not Chaos[1]

Ary L. Goldberger and David R. Rigney

Abstract

Sudden death due to cardiac arrhythmia is a major cause of mortality among Americans. A widely held view is that immediately before sudden death, the heartbeat is chaotic. Work from this laboratory and others, however, supports a different view, namely, that the heart's dynamics before sudden death are often relatively periodic, not chaotic, and the electrocardiographic signals are characterized by a narrow frequency spectrum. Patients at high risk of sudden death show periodic trajectories in the heart rate phase space with occasional abrupt changes that may be of diagnostic and prognostic value. In contrast, the electrical activity of the heart under healthy conditions appears to be fractal since a normal heart rate has a 1/f-like spectrum. In fact, two-dimensional phase-space plots of interbeat interval trajectories in healthy subjects suggest the presence of a strange attractor in a noisy environment.

The Chaos Debate in Physiology

Applications of nonlinear dynamics to the physiological sciences *(1)* initially consisted of publications demonstrating that nonlinear models are useful for understanding complex physiological phenomena *(2)*. These concepts are only now beginning to percolate into the medical sciences, which had overlooked typically nonlinear dynamics – such as abrupt transitions, sustained oscillations, and chaotic behavior – since they are not easily explained by traditional methods of analysis.

A controversy arose shortly after initial appreciation of the significance of nonlinear concepts in physiology, and this dispute is the subject of the present chapter (see Table 1) *(3–8)*. In general terms, the debate centers on the answer to the question, "Is chaos health or disease?" In cardiology, this question can be posed more specifically, "Is sudden death characterized by a chaotic or a relatively periodic type of dynamics?"

Sudden Death: Demographics and Dynamics

The question is of more than theoretical interest since over 400,000 Americans die suddenly each year: sudden cardiac death is the leading cause of mortality among American men aged 20–60 years, killing about one person every minute. Therefore, understanding the dynamics of this process is a major public health priority. The most common proximal cause of sudden death is a class of abnormal heart rhythms called ventricular arrhythmias. These electrical disturbances include ventricular tachycardias (Figs. 1–3), ventricular flutter (Fig. 4), and ventricular fibrillation (Fig. 1). The conventional model of sudden death, and in particular, of ventricular fibrillation, is that the heart has chaotic dynamics in the technical sense of the word *(5, 9)*. For example, in a frequently quoted article *(10)*, Ruelle called

1 This chapter is an expanded version of an article originally published in *Complex Patterns in Dynamic Systems,* J. A. S. Kelso, A. J. Mandell, M. F. Schlesinger, Eds. (World Scientific Publishers, Teaneck, NJ, 1988).

Table 1. The route to sudden death: two conflicting models.

Old model	periodic and steady healthy dynamics	→ chaotic arrhythmias
New model	chaotic healthy dynamics	→ periodic arrhythmias

fibrillation the prime physiological example of a strange attractor. The most explicit recent formulation of the chaos theory of sudden death has been by Smith and Cohen *(5)* who modeled fibrillation as a chaotic process that followed a Feigenbaum-type, period-doubling sequence. Their computer model was a descendant of an earlier analysis by Moe and coworkers *(9)* that described fibrillation as having turbulent dynamics. In a related paper, Smith *et al.* *(11)* suggested that ventricular fibrillation might represent a percolation process.

Support for this notion of sudden death as a chaotic process has come from several lines of evidence. The first was intuition and observation. The fibrillating heart seen during cardiac surgery or under experimental conditions certainly gives the appearance of chaos in the vernacular sense. This impression is further supported by the electrocardiographic waveforms recorded from the body surface of fibrillating animals and man; these oscillations also appear, at first glance, to represent completely erratic fluctuations. Furthermore, computer simulations of re-entrant-like path-

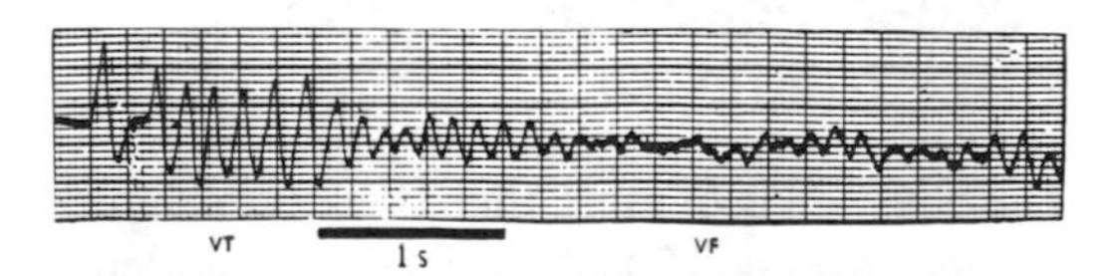

Fig. 1. Ventricular tachycardia (VT) abruptly changes to ventricular fibrillation (VF) in this electrocardiographic recording made during a cardiac arrest.

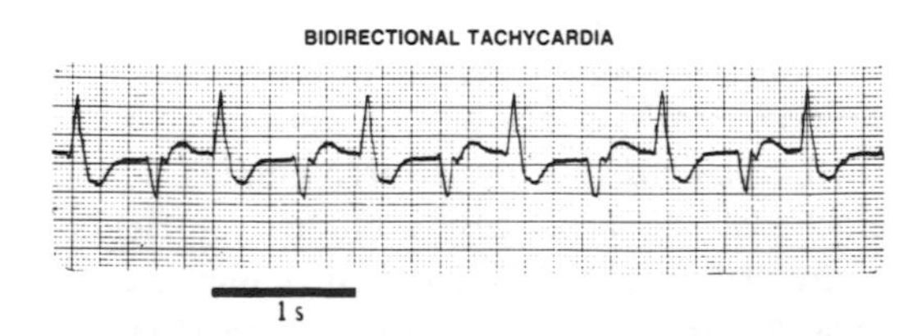

Fig. 2. Bidirectional ventricular tachycardia is characterized by ventricular premature beats that alternate in their vectorial orientation. This relatively rare bimorphic arrhythmia contrasts with the common type of monomorphic ventricular tachycardia shown in Fig. 1 and with the polymorphic arrhythmia called torsades de pointes (Fig. 3). Bidirectional ventricular tachycardia seems to represent a subharmonic bifurcation phenomenon. This patient's arrhythmia was due to digitalis toxicity and low serum potassium concentration.

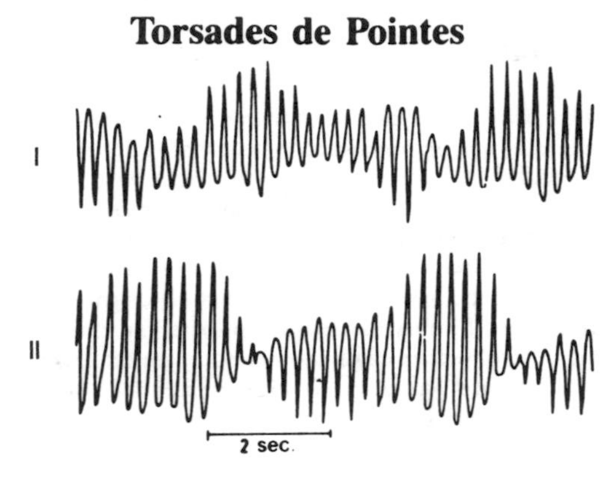

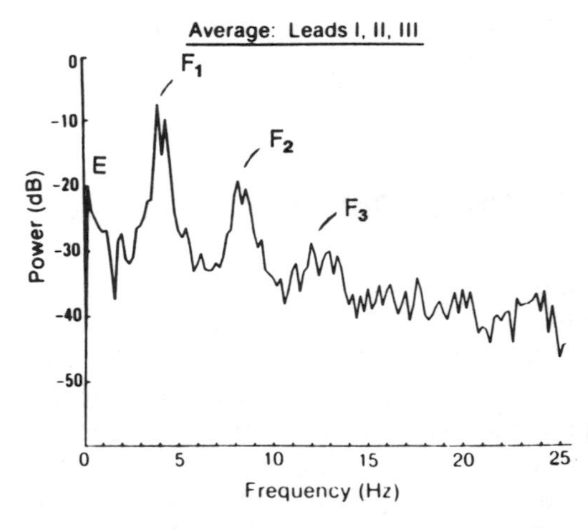

Fig. 3. Torsades de pointes type of ventricular tachycardia (upper panel) is characterized by relatively periodic oscillation of the ventricular depolarization vector, giving the surface electrocardiogram a spindle-like appearance. Spectrum (lower panel) shows spikes corresponding to the low "envelope" (E) frequency (repetition rate of the spindles) and higher frequencies corresponding to the actual heart rate (F1) and its harmonics (F2, F3, etc.). This arrhythmia is an important mechanism of sudden death *(12)*. [Adapted from *(19)*]

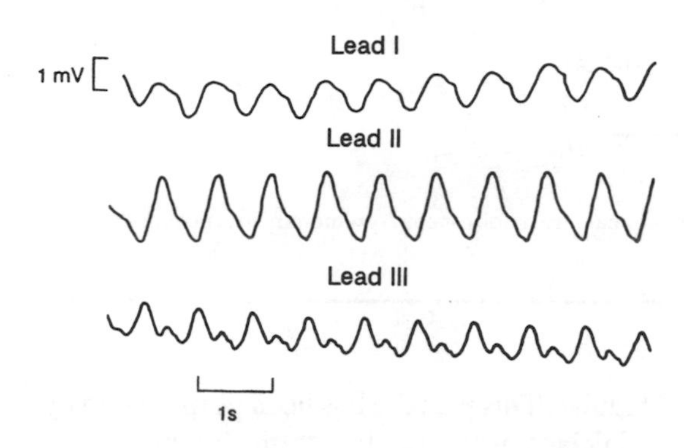

Fig. 4. Slow ventricular flutter-like rhythm due to extreme elevation of extracellular potassium (hyperkalemia). Note that the oscillations are regular but not purely sinusoidal, which is consistent with a periodic nonlinear wave mechanism.

ways – the mechanism thought to underlie fibrillation – yielded irregular oscillations that were thought to be analogous to actual fibrillation *(5, 9)*. Thus the theory of fibrillatory chaos has become an accepted model for the dynamics of cardiac death, but not without challenge.

The Dying Heart: A New Perspective

Over the past several years, we *(6, 7)* have contested the notion that the dying heart is best described as a chaotic attractor. Instead, we proposed an alternate view of cardiovascular dynamics, summarized by the following two propositions: (i) the onset of ventricular fibrillation and related tachyarrhythmias causing sudden death is usually a periodic process, not cardiac chaos (see Table 2) *(6, 12)*; and (ii) in contrast, healthy cardiac dynamics are chaotic (see Table 3) *(4, 6, 7, 13)*. *Sudden death, therefore, may be viewed as a bifurcation out of, not into chaos.*

Furthermore, period-doubling behavior, although not an uncommon feature of perturbed cardiovascular dynamics (Figs. 2 and 5; Table 4) *(14–19)*, does not appear to follow the classical route to chaos that was proposed by Smith and Cohen *(5)* on the basis of computer simulations. Instead, the term *inverse bifurcation (19)* may be more appropriate to describe this transition from healthy chaotic dynamics to the pathologic periodic dynamics that are associated with sudden cardiac death, thus distinguishing it from the classical bifurcation sequence that proceeds from periodicity to chaos in physical systems. In using the term inverse bifurcation we, of course, do not imply that all cardiac arrhythmias are periodic or that the pathologic states between health and sudden death are necessarily characterized by increasing regularity, only that such a progression is common.

These counterintuitive concepts – that health is a strange attractor and that sudden death is a periodic syndrome – are supported by the following lines of evidence (Tables 2 and 3). Spectral analysis *(6, 20–22)* of fibrillatory electrocardiographic waveforms from the body surface of animals and humans has consistently revealed spectra that are much narrower than normal (Fig. 6). Similarly, electrophysiological recordings from both the inner and outer surfaces of the fibrillating

Table 2. Evidence that sudden death is not a chaotic process.

Narrowband spectrum of body surface and epicardial electrocardiographic waveforms during ventricular fibrillation (Fig. 6) *(6, 20–22)*.
Organized waveforms of ventricular activation recorded from the outer surface of the heart at onset of fibrillation *(23)*.
Periodic electrical activity recorded from the innermost surface of ventricles during fibrillation *(24)*.
Periodic electrical activity seen during torsades de pointes (Figs. 3 and 12).
Periodic sinus rhythm and ectopic heart beat dynamics and loss of physiologic heart rate variability in patients at high risk of sudden death (Figs. 8 and 9) *(33, 35–39, 55)*.

Table 3. Evidence for cardiac chaos in healthy individuals.

Strange-like attractors seen in phase-space plots
Example: normal sinus rhythm in healthy subjects (Figs. 10, 11, and 14)

Broadband spectra
Example: spectrum of healthy heart rate variability (Fig. 7) *(25, 27, 28)*

Fractal-like structures *(13, 26, 27)*
Examples: coronary arteries, His-Purkinje system, chordae tendineae, tracheobronchial-pulmonary arteriovenous interface

heart also demonstrated periodic behavior *(23, 24)*. In contrast, spectral analysis of individual, normal QRS waveforms (representing ventricular depolarization) reveals a broad (1/f-like) spectrum consistent with activation of a fractal conduction network *(25)*, and the normal neurohumoral mechanism that regulates beat-to-beat heart rate dynamics is also quite variable (Fig. 7) *(26, 27)*.

Healthy Chaos, Chaotic Health

A common misconception among physicists and physiologists is that the normal heartbeat of a resting individual is metronomically regular. This mistake has been perpetuated by clinicians who use the term "regular sinus rhythm" to describe the normal pulse of their healthy patients. However, most cardiologists are surprised to find that the normal heartbeat is not strictly regular. In reality, even in resting subjects, it fluctuates in a highly erratic fashion and its spectrum is 1/f-like (Fig. 7), with superimposed spikes corresponding to physiological oscillations associated with respiration, baroreceptor control, thermoregulation, and the renin-angiotensin system *(26–29)*. This type of broadband spectrum with superimposed peaks is reminiscent of the "noisy periodicity" described by Lorenz in chaotic attractors *(30)*.

What is the mechanism of this 1/f-like spectrum of heart rate variability? At present, the details of the control system that is responsible for it are uncertain. We *(7, 25, 27, 31)* have proposed that physiological heart rate variability is caused by a regulatory process that exhibits self-similar fluctuations in interbeat intervals across multiple scales of time (Fig. 7). These fractal dynamics undoubtedly involve the complex interaction of parasympathetic and sympathetic branches of the autonomic nervous system impinging on the sinus node *(27)*.

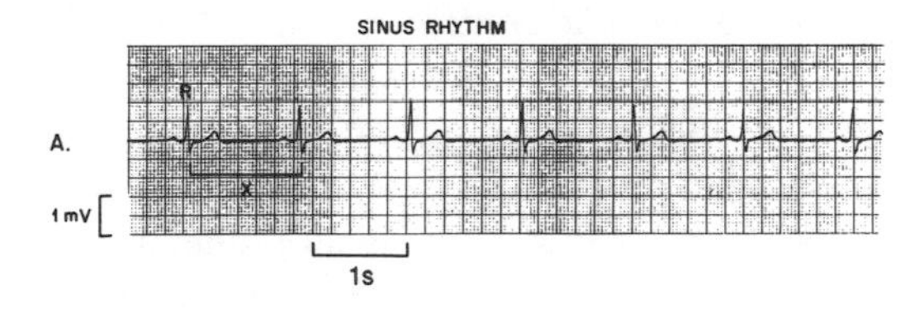

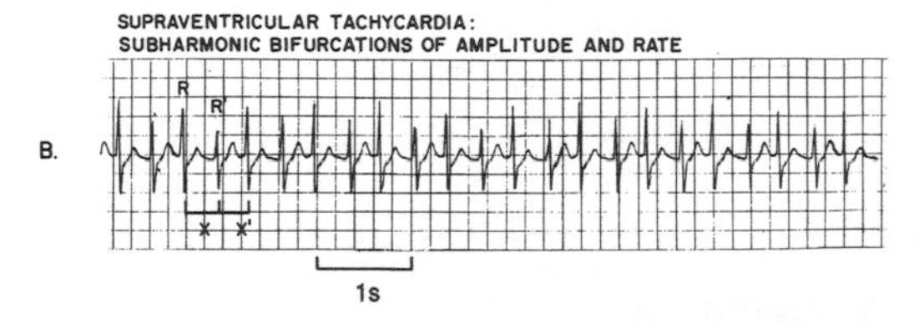

Fig. 5. Apparently subharmonic bifurcations of the QRS complex and interbeat interval in a patient with probable atrio-ventricular re-entrant tachycardia involving a concealed bypass tract (Wolff-Parkinson-White variant). Note periodic alternation of the interbeat interval between longer (x = 380 ms) and shorter (x' = 320 ms) values, as well as beat-to-beat oscillation of the amplitude of the QRS. This type of electrical alternans phenomenon is not normal (top panel). Subharmonic bifurcations are common in other examples of perturbed cardiac dynamics (Fig. 2; Table 4).

Losing Chaos: Pathologic Periodicities

The considerable variability of the normal heart rate is not present in several pathologic conditions, including diabetic neuropathy *(32)*, multiple sclerosis *(32)*, fetal distress *(33)*,

Table 4. Period-doubling–like behavior in pathologic cardiac dynamics.

Electrical alternans in pericardial tamponade *(17)*
Sick sinus syndrome *(18)*
Bidirectional ventricular tachycardia (Fig. 2)
ST segment-T wave alternans preceding ventricular fibrillation *(5, 6)*
T-U wave alternans preceding torsades de pointes *(13)*
QRS and R-R alternans with atrio-ventricular re-entrant tachycardia in Wolff-Parkinson-White syndrome (Fig. 4)
Pulsus alternans in heart failure *(17)*
Catecholamine-induced arrhythmias in animals *(16)*
Electrically perturbed chick heart embryo cells *(15)*

space adaptation syndrome *(34)*, and patients susceptible to sudden death *(31, 35–37)*. In fact, we recently identified two abnormally regular patterns in the heart rate dynamics of patients at increased risk of sudden cardiac death (Fig. 8) *(37)*. We have labeled one of them the *flat pattern* because it is characterized by a virtual absence of the normal beat-to-beat variability. The second is a *regular oscillatory pattern* characterized by relatively low frequency (0.01–0.06 Hz) heart rate oscillations, often accompanying oscillations in breathing rate and amplitude that are known as Cheyne-Stokes respiration *(4)*. These cardiopulmonary oscillations, which may persist for hours, are observed to start and stop abruptly (Figs. 8 and 9), a behavior that is diagnostic of perturbed nonlinear systems *(37)*. In addition to sinus rhythm heart rate oscillations, low frequency oscillation of ectopic beats is observed in some of these patients *(31, 38, 39)*.

Our preliminary data suggest that these pathologic heart rate dynamics – loss of variability and prominent low frequency oscillations – are sensitive markers of increased susceptibility to sudden death *(37)*. This is not to say that pathological heart rate dynamics are not chaotic, but that if chaos is present, they would appear to differ quantitatively from those present normally (e.g., having a lower dimension).

Problems with Diagnosing Chaos in Physiology

Can chaos be detected reliably in physiologic and clinical data? Time series plots showing erratic fluctuations such as those seen in healthy heart rate activity and their associated broadband spectra with 1/f-like (inverse

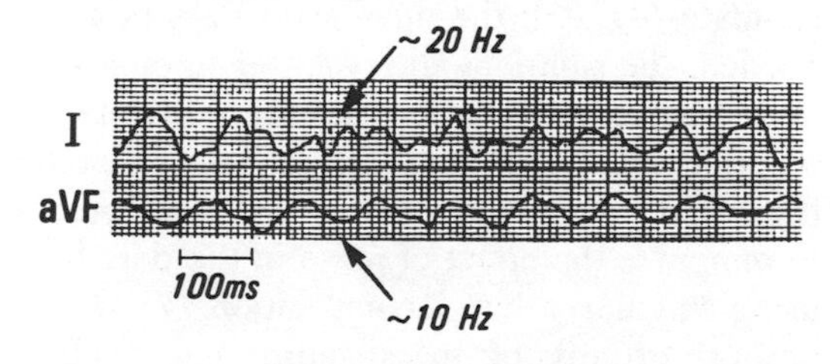

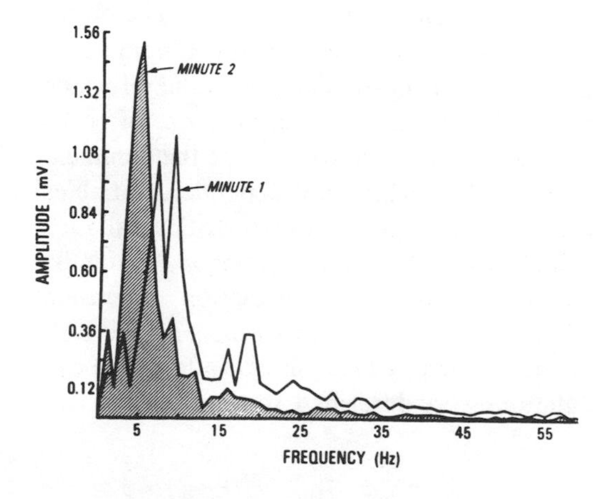

Fig. 6. Ventricular fibrillation (canine experiment) looks "chaotic" at first glance (see also Fig. 1). However, closer inspection reveals periodic oscillations (above), and the spectrum (right panel) of the epicardial or body surface electrocardiogram shows a relatively narrow band pattern during the first minute of fibrillation, with even further narrowing during the second minute. [Adapted from *(6)*]

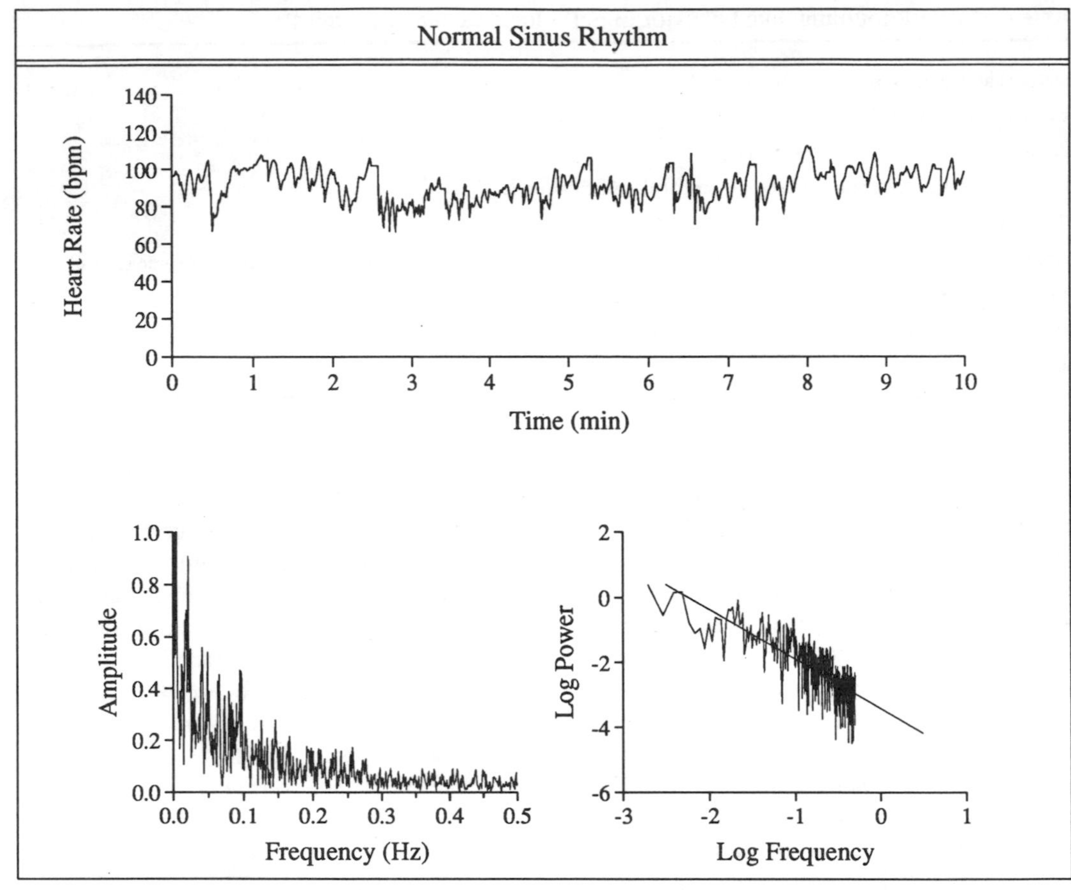

Fig. 7. The normal heart beat is not strictly regular (top panel). Instead, it displays fluctuations across many time scales. The frequency spectrum is broad (lower left panel) with an inverse power-law (1/f-like) distribution that is apparent when the same data are replotted on double log axes (lower right panel). The straight line is the linear regression fit to the data points.

power-law) distributions are consistent with chaotic dynamics. However, these features are not specific markers of chaos *(40)*.

As a result of such difficulties, several other tests have been used to diagnose the presence of chaos in a time series. The objective in each test is to compute a number (a dimension) from the data, the value of which may not only establish the presence of chaos but may also be used to estimate the minimum number of interacting, independent variables that may have given rise to it. Several such calculations have been proposed to identify chaos, including the correlation dimension *(41)*, Lyapunov exponents *(42)*, and the Kolmogorov entropy *(43)*. Shortly after these calculations were described, they were applied to a variety of time series, but the reliability of many of the dimensional estimates turned out to be questionable. Subsequent investigators proposed requirements that the data must exhibit in order to obtain a reliable dimension estimate *(44, 45)*: the number of data points needed, the sampling interval, the amount of noise that is acceptable, the effect of interpolation, etc. Other investigators have refined the original algorithms *(46)* or have attempted to minimize the effect of noise in the data by using singular value decomposition *(45, 47)*. Several groups of investigators have subsequently reported calculations of correlation dimensions from healthy heart rate data sets *(48, 49)*, and the results, while preliminary, are consistent with chaotic dynamics.

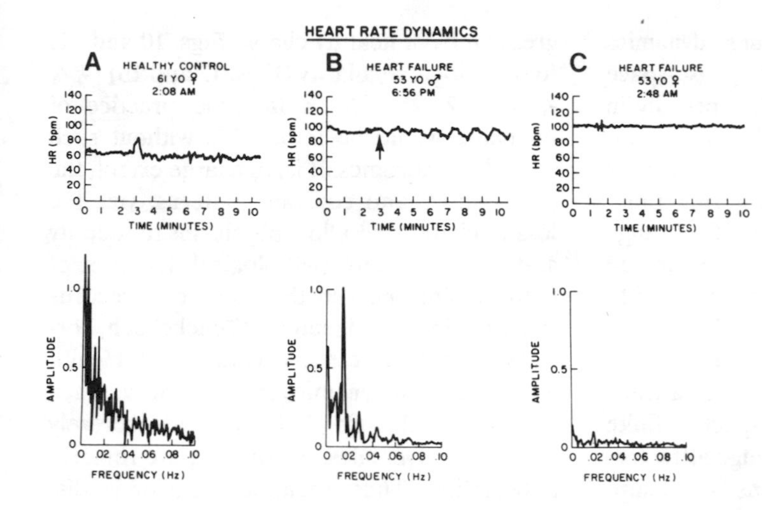

Fig. 8. Healthy subjects (A) show an erratic beat-to-beat heart-rate pattern, not a metronomically regular pulse rate. The spectrum is 1/f-like (see Fig. 7). Patients with heart failure, a group at particularly high risk of sudden death, characteristically show either low frequency (0.01–0.04 Hz) oscillations (B) or a marked overall reduction in beat-to-beat heart-rate variability with a flat spectrum up to 0.1 Hz (C). Note the abrupt onset of oscillations (arrow) in panel B. [Adapted from *(13)*]

Phase-Space Plots of Heart Rate

Another common test for chaos is to plot delay maps (phase-space plots) for the time series; if there is a discernible relation between the value of the time series at successive times, chaos may be present, even though the absence of such a relation does not rule out chaos *(50)*. The application of this technique to heart rate time series is now briefly described. The time series is represented as a vector, elements of which correspond to the data separated by equally spaced time delays. For example, Figs. 10–13 depict two-dimensional phase-space trajectories of the heartbeat in two healthy subjects and two patients who had cardiac arrests shortly after the recordings. The original time series and Fourier spectra of the heart rates for both subjects are also shown. Note the erratic beat-to-beat heart-rate variability in the healthy individuals, characterized by a 1/f-like spectrum. The phase-space trajectories of the normal interbeat interval follow a complex pattern that might be a strange attractor in the presence of noise. The trajectories appear to be driven toward a dense central region, leave the center along irregular paths, and then return to the center. The chaotic character of the trajectories is seen in the central region around the "average" heart rate where behavior becomes unpredictable and trajectories diverge.

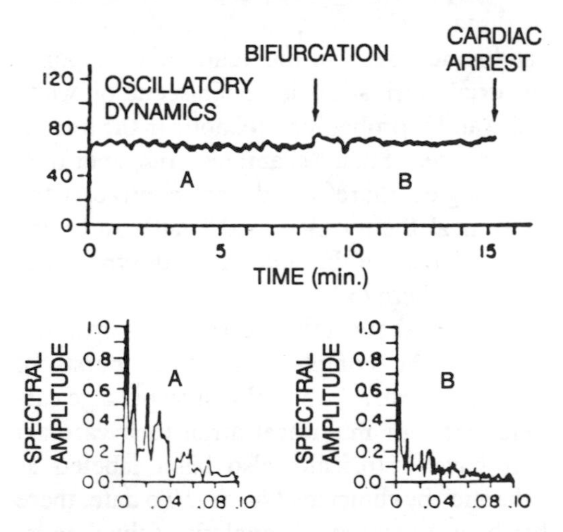

Fig 9. Heart-rate time series and corresponding amplitude spectra from the ambulatory electrocardiographic record of a patient who died suddenly. Low frequency interbeat interval oscillations (A) were followed by an abrupt change to a different frequency pattern (B) followed a few minutes later by ventricular tachyarrhythmias and cardiac arrest.

In contrast, in the patients with heart disease and impending cardiac arrest, the sinus rhythm beats appear to be more highly ordered. The time series shows a relatively regular oscillatory pattern having a narrow-band spectrum with a sharp peak at about 0.02 Hz. Unlike the healthy sinus rhythm phase plots in which trajectories converge toward a central node and then diverge outward, the pathologic sinus rhythm trajectories are center-avoiding and tend to follow relatively periodic elliptical or circular orbits.

Interpretation of a system's dynamics based on visual inspection of its phase-space representation may be difficult, especially in the presence of noise. The technique of singular value decomposition offers one particularly useful approach to filtering time-series data to extract the features of a strange attractor in a noisy environment. Figure 14 shows an example of this technique applied to a healthy heart rate data set.

The features of normal beat-to-beat cardiac behavior described above – erratically fluctuating time series, broad spectra, finite correlation dimensions, and strange attractor-like, phase-space representations – all support the formulation that healthy heart rate dynamics are chaotic. Can this hypothesis be more rigorously defended? Efforts are now underway to do so by formulating equations of motion for the cardiovascular feedback mechanisms that presumably generate the types of erratic fluctuations described here. If such a model is correct, it should also predict the pathologic periodicities when its parameters describing hemodynamics and the autonomic nervous system are varied over critical ranges. However, even without a detailed mechanistic explanation, recognition of the types of dynamic patterns reviewed here offers new approaches to quantitating healthy variability and to monitoring the risk of sudden death *(37)*. For example, subtle alterations in the beat-to-beat dynamics of the normal sinus rhythm heartbeats might precede the appearance of potentially fatal arrhythmias. The ability to detect and quantitate such changes may extend the diagnostic and prognostic capabilities of conventional cardiac monitors *(27, 37)*.

Why Are Healthy Dynamics Chaotic?

Chaotic dynamics appear to reflect the normal variability and adaptability necessary for responding to a fluctuating environment. Periodic behavior in response to a perturbation, in contrast, may represent an organism's progression from healthy chaos (Figs. 10 and 11) to pathologic regularity (Figs. 12 and 13) *(4, 6, 7, 13, 19, 25–27)*. In fact the practice of medicine would be impossible without such periodic syndromes. It is, to a large extent, the periodicities and constant patterns, viz., the loss of chaos, that allow physicians to identify and classify many pathological features of their patients, notably their heartbeat records (e.g., bigeminy, trigeminy, Wenckebach, torsades de pointes, electrical alternans). Health with its erratic dynamics is, necessarily, much harder to classify. Patients with severely pathologic dynamics, in contrast, are remarkably similar. Thus, periodic behavior in disease is likely to be more the rule *(13)* than the exception *(51)*. Indeed, the more extreme the pathology (e.g., impending sudden death), the more marked the periodicities often become (Figs. 1–6, 8, 12, 13).

Of course, this generalization is not intended to imply that all aspects of disease are periodic or that the progression from health to sudden death is always accompanied by a progressive increase in periodicity. The apparently random heart rate fluctuations during atrial fibrillation serve as an obvious counterexample to the notion of periodic dynamics in disease. However, the type of pathologic randomness seen in the ventricular response to atrial fibrillation is reminiscent of white noise and is probably not chaotic in the technical sense. Such examples of aperiodic pathologies, therefore, do not contradict the dynamical theory of fractal health and periodic behavior in the sudden death syndromes discussed here *(52)*.

What about other cardiac arrhythmias (such as the complex, multiform bursts of ventricular extrasystoles that may presage cardiac arrest or multifocal atrial tachycardias) that have historically also been labeled as "chaotic" by clinicians *(40, 53)*? To date, there has been no systematic analysis of the dynamics of these rhythm disturbances, so it is not clear whether they are in fact chaotic. Of interest, however, are observations that bursts of premature beats that appear to be "irregular" may demonstrate surprisingly periodic dy-

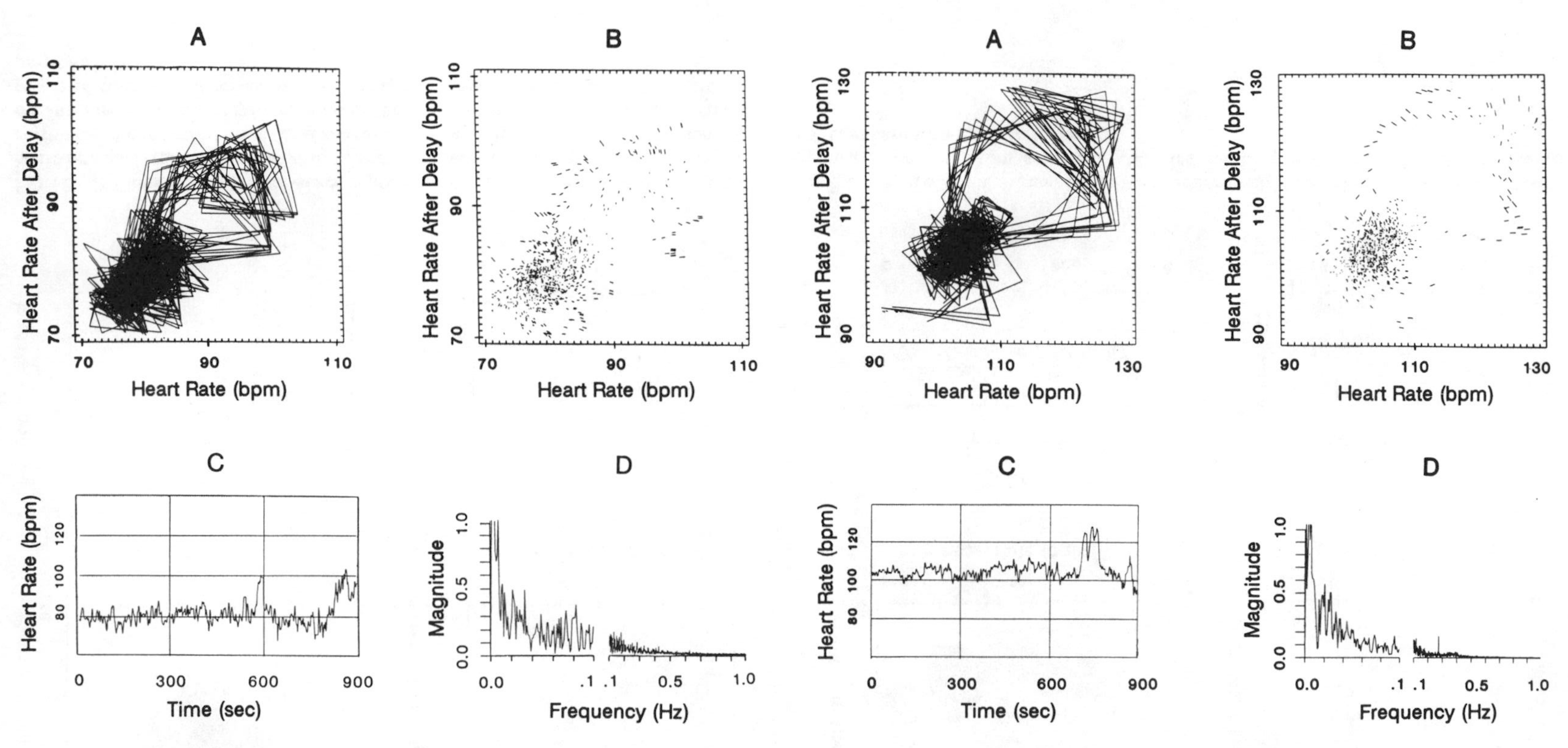

Fig 10. Two-dimensional phase-space representation of 900 seconds of a normal individual's heart rate (62-year-old female). Panel A shows the trajectory of the heart-rate vector where the first variable is the current heart rate, and the second variable is the heart rate after a fixed delay of 12 seconds. Panel B shows the same data as Panel A, except that for clarity, the trajectory at each point is shortened to be proportional to the distance that the trajectory travels after an evolution time of 12 seconds. Panel C shows the original heart-rate data from which the phase-space plots were calculated, and Panel D shows the amplitude spectrum of the time series.

Fig. 11. Two-dimensional phase-space representation of 900 seconds of another normal individual's sinus rhythm heart rate (35-year-old male). (See Fig. 10 caption for details.) The same parameters are used in Figs. 10–13. Note the similarity of the strange-like "arachnoid" attractors in the two normal subjects as compared with the orbits of the patients with sudden death syndrome (Figs. 12 and 13).

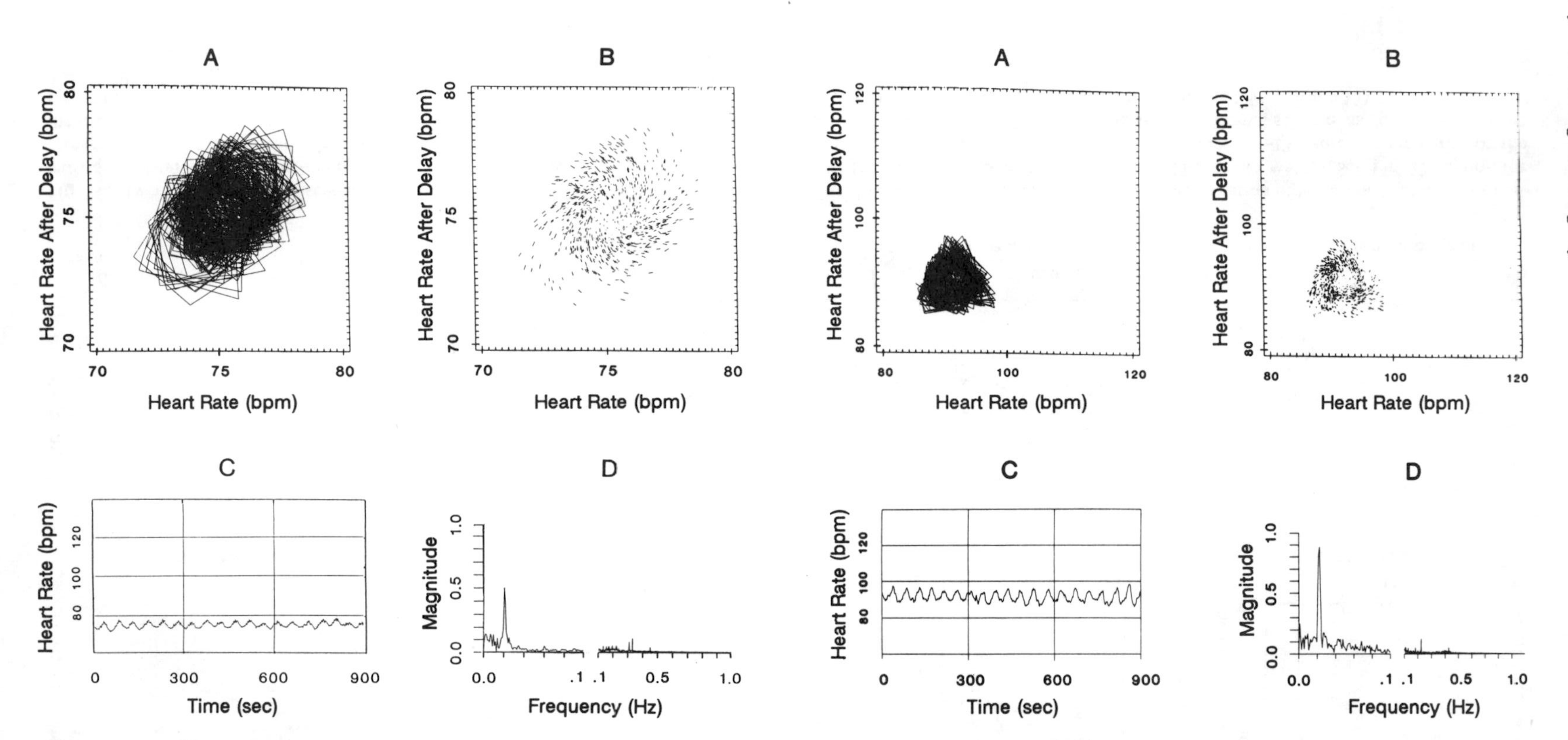

Fig.12. Two-dimensional phase-space representation of 900 seconds of sinus rhythm heart-rate data from a patient who sustained a cardiac arrest due to ventricular tachyarrhythmias about two hours later. Compare with Figs. 10, 11, and 13. Note that the dynamics here are relatively more periodic that those seen normally, consistent with the proposed concept that certain pathologic states represent a decrease in the degree of physiologic chaos.

Fig. 13. Two-dimensional phase-space representation of 900 seconds of sinus rhythm heart-rate data from a patient who died suddenly eight days after this recording. Compare with Figs. 10, 11, and 12.

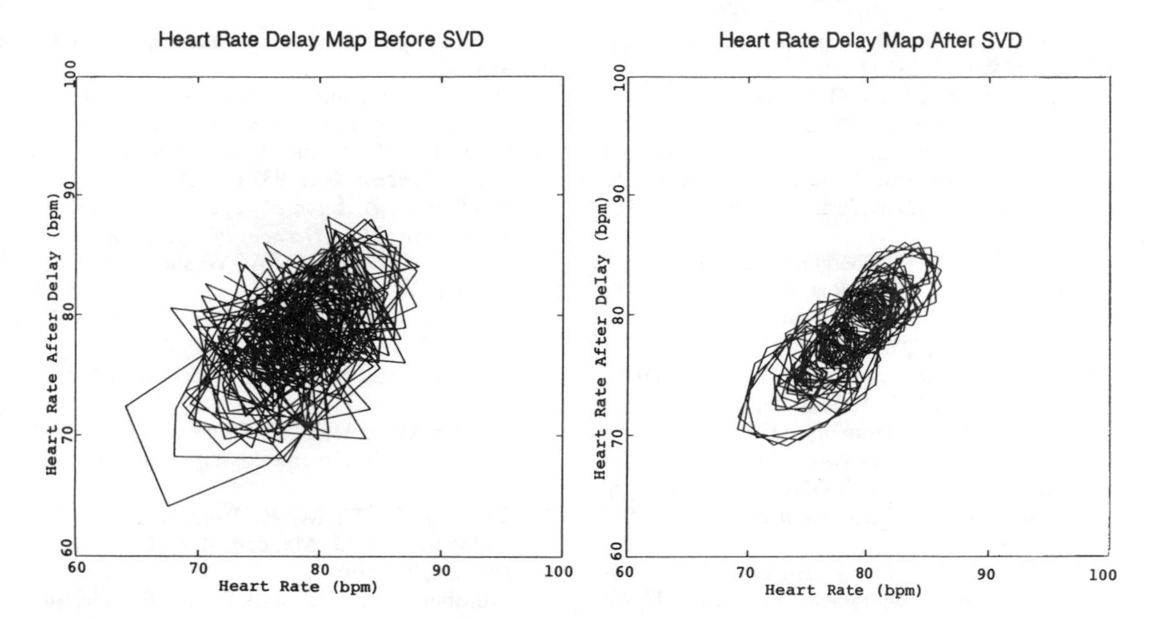

Fig. 14. Two-dimensional phase-space representations of 240 seconds of sinus-rhythm, heart rate data from a healthy subject. Left panel shows the unfiltered data; right panel shows the same data set after singular value decomposition (SVD), with filtering of all but the first principal component (eigenvector). Note that the filtering unmasks a strange-like attractor structure.

namics when more closely analyzed *(38, 39, 54)*.

Summary

We presented evidence supporting the counterintuitive theory that cardiac arrest follows a progression, often involving abrupt changes, from the apparently chaotic dynamics of the normal heartbeat to the pathologic periodicities and loss of variability of the dying heart. This scenario is opposite the conventional notion of sudden death as chaos and health as a periodic or constant state (homeostasis) *(39, 52)*. Loss of physiologic sinus rhythm heart rate variability and sometimes the emergence of highly periodic oscillations may be sensitive markers for increased risk of sudden death and may precede the appearance of potentially fatal cardiac arrhythmias (Figs. 8 and 9). An analogous loss of physiologic variability appears to be a common feature of pathologic dynamics in other organ systems as well *(27, 52)*.

Acknowledgments

This work was supported by grants from the Whitaker Health Sciences Fund, the G. Harold and Leila Y. Mathers Charitable Foundation, the National Aeronautics and Space Administration (NAG 2-514), and the National Heart Lung and Blood Institute (R01-HL42172). The authors acknowledge the technical assistance of Joseph Mietus and the secretarial skills of Susan Turner.

References

1. Gleick, J., *Chaos: Making a New Science* (New York, Viking Press, 1987).
2. Koslow, S. H., A. J. Mandell, M. F. Shlesinger, Eds. *Perspectives in Biological Dynamics and Theoretical Medicine. Ann. N.Y. Acad. Sci.* **504,** (1987).
3. Mackey, M. C., and L. Glass, *Science* **197,** 286 (1977).
4. Goldberger, A. L., L. J. Findley, M. R. Blackburn, A. J. Mandell, *Am. Heart J.* **107,** 612 (1984).

5. Smith, J. M., and R. J. Cohen, *Proc. Natl. Acad. Sci. U.S.A.* **81,** 233 (1984).
6. Goldberger, A. L., V. Bhargava, B. J. West, A. J. Mandell, *Physica* **D19,** 282 (1986).
7. Goldberger, A. L., B. J. West, in *Chaos in Biological Systems,* H. Degn, A. V. Holden, L. F. Olsen, Eds. (New York, Plenum Press, 1987), p. 1.
8. Glass, L., A. L. Goldberger, M. Courtemanche, A. Shrier, *Phil. Trans. Roy. Soc. A* **413,** 9 (1987).
9. Moe, G. R., W. C. Reinboldt, J. A. Abildskov, *Am. Heart J.* **67,** 200 (1964).
10. Ruelle, D., *Math. Intelligencer* **2,** 126 (1980).
11. Smith, J. M., A. L. Ritzenberg, R. J. Cohen, in *Computers in Cardiology* (IEEE Computer Society, Long Beach, CA, 1984), p. 175.
12. Bhargava, V., A. L. Goldberger, D. Ward, S. Ahnve, *IEEE Trans. Biomed. Eng.* **33,** 894 (1986).
13. Goldberger, A. L., in *Temporal Disorder in Human Oscillatory Systems,* L. Rensing, U. An der Heiden, M. C. Mackey, Eds. (Springer, New York, 1987), p. 118.
14. Schwartz, P. J., and A. Malliani, *Am. Heart J.* **89,** 45 (1975).
15. Guevara, M. R., L. Glass, A. Shrier, *Science* **214,** 1350 (1981).
16. Ritzenberg, A. L., D. R. Adam, R. J. Cohen, *Nature* **307,** 159 (1984).
17. Goldberger, A. L., R. Shabetai, V. Bhargava, B. J., A. J. Mandell, *Am. Heart J.* **107,** 1297 (1984).
18. Goldberger, A. L., V. Bhargava, B. J. West, A. J. Mandell, *Physica* **D17,** 207 (1985).
19. Goldberger, A. L., B. J. West, *Ann. N.Y. Acad. Sci.* **504,**195 (1987).
20. Battersby, E. J., *Circ. Res.* **17,** 296 (1965).
21. Herbschleb, J. N., R. M. Heethaar, I. van der Tweel, in *Computers in Cardiology* (IEEE Computer Society, Long Beach, CA, 1978), p. 245.
22. Kusuoka, H., W. E. Jacobus, E. Marban, *Circ. Res.* **62,** 609 (1988).
23. Ideker, R. E., *et al., Circulation* **63,** 1371 (1981).
24. Worley, S. J., *et al., Am. J. Cardiol.* **55,** 813 (1985).
25. Goldberger, A. L., V. Bhargava, B. J. West, A. J. Mandell, *Biophys. J.* **48,** 525 (1985).
26. West, B. J., and A. L. Goldberger, *Am. Sci.* **75,** 354 (1987).
27. Goldberger, A. L., and B. J. West, *Yale J. Biol. Med.* **60,** 421 (1987).
28. Kobayashi, M., and T. Musha, *IEEE Trans. Biomed. Eng.* **29,** 456 (1982).
29. Akselrod, S., *et al., Science* **213,** 220 (1981).
30. Lorenz, E. N., *Ann. N.Y. Acad. Sci.* **357,** 282 (1980).
31. Goldberger, A. L., and D. R. Rigney, *Proceedings of the Ninth Annual Conference of the IEEE Engineering in Medicine and Biology Society* **1,** 313 (1987).
32. Neubauer, B., and H. J. G. Gunderson, *J. Neurosurg. Psychiatry* **41,** 417 (1978).
33. Modanlou, H. D., and R. K. Freeman, *Am. J. Obstet. Gynecol.* **142,** 1033 (1982).
34. Goldberger, A. L., *et al., Space Life Sciences Symposium: Three Decades of Life Science Research in Space* (NASA: Washington, DC, 1987), p. 78.
35. Myers, G. A., G. J. Martin, N. M. Magid, *IEEE Trans. Biomed. Eng.* **33,** 1149 (1986).
36. Kleiger, R. E., *et al., Am. J. Cardiol.* **59,** 256 (1987).
37. Goldberger, A. L., D. R. Rigney, J. Mietus, E. M. Antman, S. Greenwald, *Experientia* **44** 983 (1988).
38. Findley, L. T., M. R. Blackburn, A. L. Goldberger, A. J. Mandell, *Am. Rev. Respir. Dis.* **130,** 937 (1984).
39. Goldberger, A. L., and D. R. Rigney, in *Theoretical Models for Cell-to-Cell Signaling,* A. Goldbeter, Ed. (New York, Academic Press, 1989), p. 541.
40. Glass, L., and M. C. Mackey, *From Clocks to Chaos: The Rhythms of Life* (Princeton University Press, Princeton, NJ, 1988), p. 54.
41. Grassberger, P., and I. Procaccia, *Physica D* **9,** 189 (1983).
42. Wolf, A., J. B. Swift, H. L. Swinney, J. A. Vastano, *Physica D* **16,** 285 (1985).
43. Grassberger, P., and I. Proccacia, *Phys. Rev. A* **28,** 591 (1983).
44. Mees, A. I., P. E. Rapp, L. S. Jennings, *Phys. Rev. A* **36,** 340 (1987).
45. Albano, A. M., A. I. Mees, G. L. de Guzman, P. E. Rapp, in *Chaos in Biological Systems,* H. Degn, A. V. Holden, A.V., L. F. Olsen, Eds. (Plenum Press, New York, 1987), pp. 207–220.
46. Fraser, A. M., and H. L. Swinney, *Phys. Rev.* **33A,** 1134 (1986).
47. Broomhead, D. S., and G. P. King, *Physica D* **20,** 217 (1986).
48. Destexhe, A., and A. Babloyantz, *Biol. Cybern.* **58,** 203 (1988).
49. Mayer-Kress, G., L. Benton, *et al., Math. Biosci.* **90,** 155 (1988).
50. Olsen, L. F., and H. Degn, *Quart. Rev. Biophys.* **18,** 165 (1985).
51. H. A. Reimann, *Periodic Diseases* (F. A .Davis, Philadelphia, 1963).
52. Goldberger, A. L., D. R. Rigney, and B. J. West, *Sci. Am.* **262,** 42 (1990).
53. Katz, L. N., *Electrocardiography,* 2nd ed. (Lea and Febiger, Philadelphia, 1946).
54. Smirk, F. H., and J. Ng, *Br. Heart J.* **31,** 426 (1969).

4

Lost Choices

Parallelism and Topological Entropy Decrements in Neurobiological Aging

Arnold J. Mandell and Michael F. Shlesinger

Abstract

Loss of fast frequencies and increasing dominance by slow power spectral modes characterize electroencephalographic, psychophysical, and motor behavior of aging mammals. We associate this finding with accelerating neuron losses due to the oxygen and nutrient deficiencies that are often secondary to the brain circulatory changes of most but not all of the elderly. This loss of complexity is described using measures of maximal (topological) entropy on the behavior of a distributionally conditioned, stochastically driven family of nonlinear deterministic interspike-interval discrete maps of the real line. These simple, single-maximum maps can be derived from a suitable reduction of the Hodgkin-Huxley nonlinear global membrane oscillator.

As return maps, the interspike-interval generating functions represent the behavior of the diffuse, norepinephrine-, dopamine-, and serotonin-mediated medullary, pontine, and mesencephalic neurons. They receive direct and transformed sensory information from throughout the brain and regulate states of consciousness, psychomotor behavior, and global limbic-neocortical dynamics, which range from obtunded lethargy through relaxed states of alertness to fearful excitation. The gradual loss of the property of "awakeness" in the aging process is the subject of the work. We then derive a theory that suggests that neuroanatomical losses of parallel input to "arousal" systems can be at least partially repaired by changing the parameter region in which the remaining biogenic amine neurons function through the use of psychotropic drugs. A partially successful example is described.

A deeper aspect of these studies involves the way neuronal activities can be seen to partition global neurobiological time. Their patterns of discharge range from bursting, as observed in short and long interevent intervals called "times" (see for example, the "runs" in the time series of Fig. 2), through more random firing to intermittent arrest with the information transport properties of a telegraphic code. One can imagine the way stroboscopic light frequencies and rhythms configure the perception of a ballet troupe dancing on a darkened stage. What we experience in part results from the temporal parsing of sensory and associative reality. We show how diffuse brain cell death (loss of parallelism of the input to the brain stem neurons) can "cause" and the expansion of the parameter space of individual neuronal possibilities may "cure" the uniformization and coarse graining of time that is characteristic of the aging process.

It appears that studies of the multiplicative interactions of stochastic processes and nonlinear deterministic maps have promise with respect to the development of theories of global neurobiological function.

Introduction

Two global dynamical theories of normal aging, one molecular, the other psychosocial, were derived from the implicit "second law" postulate that time-dependent convergence to equilibrium is associated with the maximization of atomic-molecular "disorder." Deterioration and death were said to occur in a state of maximum entropy. It may be that this application of thermodynamic law (even in the context of the elegant labors of de Donder, de Groot, Mazur, and Prigogine) epitomizes the inapplicability of the usual metastructure of theoretical physics to neurobiology.

The physicist's habitual move toward energies and potential functions and their conjugacies fails for patterned information transport in general and with the intrinsically unstable, nonstationary, healthy brain in particular. We are not helped in an assessment of the information transport processes of a computer with a current-voltage monitoring device on its power cord. Information transport through a semiconductor cannot be easily decoded using time-dependent changes in its heat capacity. Neurobiological processes make patterns out of the available entropy – its deficiency, not its surfeit, signifies functional impairment and impending death. It is only apparently paradoxical that some dimension algorithms reflect increases in relative disorder when electroencephalographic (EEG) signals are studied in humans during sensory discrimination tasks *(1)*.

In the spirit of this perhaps irrelevant downhill physics of statistical disorder, Orgel *(2)* suggested that an accelerating process of random mutations in the protein synthesizing machinery with resulting enzyme abnormalities supports a positive-feedback loop leading to "catastrophic" failure in macromolecular function in the aging organism. The psychosocial preventive theory of Cumming and Henry *(3)* suggested that gradual social disengagement avoiding perturbations would serve a protective function for the aged. More recent studies have been inconsistent with this point of view. For example, isolation from normal laboratory stimulation decreased the life span of laboratory rats from a mean of 938 to 835 days, whereas daily short periods of strong stimulation using a variety of sensory modalities increased it to a mean of 1120 days *(4)*.

We embed this and similar findings relating to other forms of central nervous system "desensitization" *(5)* in an elementary theory of neurobiological aging. We propose that the desirable dynamical complexity of normal brain function is associated with a higher number of independent, parallel inputs from a variety of sensory and sensory-association areas into midline, integrative, nonlinear neurons, the "nonspecific" biogenic amine neuronal system. Its actions include the regulation of arousal and electroencephalographic fluctuations (via the thalamic reticular system) related to states of consciousness *(6)*.

Parallelism is lost with neuron cell death caused by decreases in blood supply in arteriosclerosis associated with aging. This loss of parallelism is expressed through decreases in multiplicative driving (excitation) of quadratic neuronal firing mechanisms, the inverted "U" of graded excitation of neurons with autoreceptor or recurrent collateral-mediated self-inhibition as seen in the ubiquitous psychopharmacological "dose-response" curve. The results are decreases in (maximal) topological entropy, h: a decrease in the growth rate of new possibilities as a function of time.

More realistically, since parameter-space scenarios manifest the nonlinearities of their underlying equations by discontinuous changes in behavior with continous changes in the parameters, another form of pathological simplification might occur by inhibitory cell loss with increases in stochastic driving moving the system into periodic (mode-locked) windows and another kind of loss of h. In fact, increasing incidence of epilepsy (associated with periodic neuronal and global electrophysiological phenomena) in aging populations is well known.

In these studies, whereas neuronal firing rates are a function of both input amplitude and its delay, the stochastic driving term places the disorder of nonlinear and fractal time *(7)* into the former in a simpler one-dimensional model of a forced-dissipative neural mechanism with autoreceptor-mediated, self-inhibitory regulation. This work suggests that loss of parallelism by neuronal death and the resultant loss of "choices" – topological entropy – may be compensable by psychopharmacological manipulation of the codimensionality of the deterministic part of this stochastic-deterministic nonlinear dynamical system.

An underlying theoretical issue involves the influence of multiplicative noise on the behavior of discrete dynamical systems. Crutchfield and Huberman's studies of additive noise in this context *(8)* demonstrated an intensity-dependent loss of accessible states, resulting in simplification of the parametric evolutionary scenario with preferential loss of the shorter over longer periods. Other studies of additive noise in one- and two-dimensional nonlinear maps have demonstrated noise-induced new scalings, volume addition to attractors of measure zero, increased complexity of density distributions, and both ordering and disordering of near periodic behavior *(9)*.

Relatively little has been done with the interplay of multiplicative stochastic processes and the behavior of codimension one and two (here, codimensionality indicates number of control parameters) nonlinear maps. For example, the codimensionality of the nonlinear problem increases in that a noise forcing term makes the system nonautonomous. Does white noise driving make a low-dimensional nonlinear system infinite dimensional? Do different density distributions of the multiplicative noise (white, Gaussian, Levy) change the driven system's effective dimensionality? Do multiplicative noise-driven systems (like high-dimensional, forced-dissipative dynamics) demonstrate contractions onto finite-dimensional manifolds made smooth by the multiplicative stochastic term? We here add little with respect to a theoretical understanding of these difficult problems but rather demonstrate how such studies may contribute to an understanding of neurobiological systems where multiplicative, not additive interactions (including noise), are generic.

Background

The loss of a broad range of frequencies, including fast characteristic times in favor of dominance by broadband slower, Fourier modes, has been observed in the pattern of spontaneous neuronal discharge in desensitized neural preparations, autonomous motions of smooth muscle strips that "age" (lose sensitivity to polypeptide ligands in physiological solutions over time), cardiac interbeat intervals in some patients preceding sudden cardiac death, neuroendocrine cell hormone output following neoplastic mutation and loss of regulation, and the power spectral frequencies of the electroencephalogram in patients with epileptic disorders *(5, 10)*.

The hypotheses of "strange stability" and "chaos as health" have been suggested *(10, 13)*, using what appear to be the $f^{-\alpha}$, $1 \leq \alpha \leq 2$ power spectral patterns of fluctuations as observable characteristics of chaotic dynamics *(11)*, e.g., generated by the fractal point sets of deterministic nonlinear maps and flows in their parameter regions past that of homoclinic and heteroclinic intersection *(12)*.

Implicit in such global models is the assumption of a physically undefinable, emergent order parameter which controls the behavior of a low-dimensional set of central nervous system collective variables *(14)*. As yet, there has been no conjecture as to a general biological cause for the loss of the high frequency tail of the power spectra by these systems, nor have biological reasons been suggested for why the implicit nonlinear parameter(s) regulating the conjectured chaotic dynamical system should change in a way to induce "inverse bifurcations" from aperiodic complexity to more periodic order.

Fractal information transport mechanisms (such as the decorrelating influence of highly branched Purkinje cells on the distribution of excitatory arrival times at the ventricular pumps) have been suggested as anatomical support for the heart's normal hierarchy of characteristic interbeat-interval time scales *(15)*. However, no biological physics has yet been offered for the spontaneous and drug-induced reversibility of these health and disease related dynamical transitions, which may have less to do with anatomical change and more to do with more abstract topological dynamics of the nonlinear, self-excitatory oscillators of heart and brain.

Aging in the mammalian nervous system may be a propitious area to look for some ideas. It is well established that the electroencephalogram of unselected populations of aging patients shows a characteristic replacement of low voltage/fast (β waves ≅ 20–25 Hz) by both slowing in the characteristic waking relaxed record (α waves ≅ 10–12 Hz) and diffuse slow periodicity (3–7 Hz) *(16)*. We have shown that the loss of the p-300 EEG-expectancy wave predicting attentional and behavioral adaptability *(17)* and the autonomous fluctuations in the EEG *(18)* correlate well with dementing illness, including aging. Old rats manifest slowing and narrowing in the distribution of EEG frequencies recorded in the ventromedial, mammillary, and supraoptic nuclei of the hypothalamus associated with loss of response to pain, cold, and other forms of sensory perturbation *(4)*.

It is well known that the rate of neuron cell death increases in aging populations, and several in vivo measures are consistent with this neuropathological finding: cerebral blood flow and oxygen consumption in the brains of normal subjects demonstrated an exponential decline from 5.0 cc/100 gm/min at 10 years of age to 2.5 cc/100 gm/min at age 80 *(19)*, although an unusually well-preserved group of "supernormals" of 72 years did not demonstrate such a decline *(20)*. Sokoloff *(21)*, following an extensive series of studies applying glucose and oxygen utilization, concluded that circulatory disorder-induced loss of brain tissue is the cause of these electrophysiological, metabolic, and cognitive changes in the elderly.

We approach the problem of biological aging-associated loss of fast-frequency electrophysiological fluctuations and behavioral adaptability with the conjecture that loss of parallelism in activating input leads to topological entropy loss in the output of nonlinear neurobiological mechanisms. Unlike most other mammalian tissue, lost neurons don't regenerate. Our conjecture suggests an alternative treatment to current experimental surgery involving tissue transplantation to the brain: the manipulation of the parameters of the deterministic nonlinear part, introducing (suitably bounded) complexity by means of environmental and psychochemical intervention.

Loss of Parallelism

As the simplest representative case, we treat the neuron as a codimension two, discrete-time, noninvertible map of the unit interval representing the normalized range of interspike intervals. At one boundary is the minimum refractory period, at the other, the maximum delay of self-excitatory processes. It is quadratically driven by the distribution of parallel input times, $\Psi_n(t) \equiv$ *the probability density of an excitatory event occurring at time* t *through any of* n *independent channels of the input network.* Labos used a similar but piecewise linear quadratic automorphism of the real line as a "universal and chaotic" neuronal spike generator, which was capable of generating "any preassigned periodic or aperiodic interspike interval pattern" *(22)*.

The second codimension, ε, represents the average level of the neuron's indigenous "cotransmitter" polypeptide regulator, which changes the symmetry of the parabolic manifold and scales the parameter distance between bifurcations of the $\Psi_n(t)$-driven evolutionary scenario *(23)*. It is only in recent years that the "one neuron, one transmitter law" has been shown to be in error. Neuronal self-regulation dynamics is in at least codimension two (here, codimensionality is used as the number of parametric controls).

We treat $\Psi_n(t)$ as the sum of the rates of emission-arrival of the parallel neuronal sources of excitation (disinhibition), each with a Poisson distribution of (inverse) thresholds for excitation, $\psi_i(t) = \lambda_i e^{-\lambda i t}$, where λ_i represents the mean rate of emission-arrival of excitatory messages of the ith input neuron.

With respect to two parallel competing excitatory (disinhibitory) neuronal inputs, the first "suitable" one to arrive with respect to the quadratic relationship between arrival time and evoked membrane potential generates the potential that serves as the (intracellular electrode recorded) neuron observables that partition the interspike interval space. The probability, $\Psi_2(t)$, that an excitatory input arrives at the neuron at time t is

$$\Psi_2(t) = \psi_1(t) \int_t^\infty \psi_2(\tau)d\tau + \psi_2(t) \int_t^\infty \psi_1(\tau)d\tau, \quad (1)$$

i.e., the product of the probability that input neuron 1 arrived at time t and neuron 2 arrived after time t plus the product of the probability that input neuron 2 arrived at time t and that neuron 1 arrived later than t. For the (assumed) Poisson distribution of (inverse) membrane thresholds, emission-arrival rates for the two input neurons are

$$\Psi_2(t) = (\lambda_1 + \lambda_2)\, e^{-(\lambda_1 + \lambda_2)t}. \qquad (2)$$

The mean emission-arrival rate for n parallel-competing, parallel-input neurons (the first one there induces the membrane potential observable) is their sum:

$$\Psi_n(t) = (\lambda_1 + \lambda_2 + \ldots + \lambda_n)$$
$$e^{-(\lambda_1 + \lambda_2 + \ + \lambda_n)t}. \qquad (3)$$

For an aging system in which n is decreasing, the mean time, $\overline{T}$, for n competing parallel emission-arrival neuronal input shifts to longer times in a process that proceeds as follows:

$$\frac{1}{\overline{T}_n} = \lambda_1 + \lambda + \ldots + \lambda_2 = \Lambda_n \qquad (4)$$

where $1/\lambda_i$ is the mean time of arrival, $\overline{T}_i$, from neuronal input i.

For definiteness, the activity of the codimension-two neuron, ϕ^t, is that of a typical cell of the norepinephrine-, dopamine-, or serotonin-mediated brain stem systems, which receive axon collaterals from a wide variety of sensory-related "association" systems, each with emission-arrival time distributions of $\psi(t)$ and projects diffusely and bilaterally throughout subcortical and cortical brain regions. The midline reticular system, which has overlapping identities with the brain stem's biogenic amine pathways, is generally thought to support behavioral alertness in that its destruction produces long periods of sleep and coma associated with slow wave EEGs in animals and man. These neurons, some amine releasing, some not, have autoreceptor feedback mechanisms reinforcing the membrane's quadratic sensitivity to the input rates, $\Psi_n(t)$.

The microelectrode, intracellular membrane potential observables, v_i, partitioning the interspike-interval space is generated in an amine-reticular neuron as a codimension-two invertible map. They are driven by emission-arrival rates of the input neurons as regulated by its amine-peptide autoregulation term, ε,and are generated by iterating

$$v_{t+1} = \Lambda_n v_t - (\Lambda_n v_t)^{\varepsilon} \qquad (5)$$

such that the random variable driving term, $\Psi(t)$, is selected from a population of input neurons representing the sum of the inputs. $\varepsilon = 2$ for the symmetric quadratic map studied here.

This says, quite simply, that the stochastic driving term will have a distribution of higher rates (and shorter times) from which to randomly select the driving parameter with a larger population of input neurons than with a smaller population of emission-arrival units. Since neuronal death is a concomitant of the loss of blood supply associated with arteriosclerotic changes of aging, our theory of aging says that input neuronal death reduces the n of the driving distribution. This leads to decreasing average rates and increasing characteristic times (loss of bursting, for example) of amine-reticular neuronal driving, with inverse bifurcations from complexity to slow periodicity characteristic of parameter-space behavior of quadratic maps of the unit interval when examined over decreasing values of the driving term *(24)*.

Numerical Experiments

To exaggerate the effect for explication, we studied Eq. 5 using large $(m = 2000)$ and small $(m = 20)$ Poisson distributions as sources of input to represent the conditions of normal and lost parallelism (Fig. 1). The stochastic input was generated by

$$P(m) \equiv e^{-\lambda} \cdot \frac{\lambda^m}{m!} \qquad (6)$$

over ten input rates, reflecting ten discrete steps of interspike intervals (often between 50 and 500 msec), observed in many but not all populations of brain stem reticular-amine-polypeptide neurons such as dopamine-, norepinephrine-, and serotonin-containing cells with average values in the range of 150 msec, Poisson parameter $\lambda = 3$. The characteristic interspike intervals vary among biogenic amine neurotransmitter systems from 30 to 3500 msec.

The initial value of v_i was fixed at the derivative $= 0$ maximum of the symmetric map, $v_t = v_0 = 0.5$, since the forward orbit of this critical point converges most quickly to the underlying attractor *(24)*. The autoreceptor-peptide cotransmitter "scaling" parameter, ε, was set routinely at the parabolic symmetry parameter value of 2.0 but was explored between 1.3 and 2.6 as a technique modeling the "nonlinear treatment" of neuron loss syndromes (decreasing ε dilates the parameter range and, therefore, the number of resolution-dependent accessible states). Our previous study *(23)* demonstrated that ε could move this system into and out of fixed point regimes where they were driven by the nonlinear driving term, here the stochastic Λ_n, by changing the system's Feigenbaum δ.

We restricted the ten discrete steps of the low and high m Poisson distributions as sources of the random variable driving term to three finite ranges with decreasing potential for orbital complexity: $1.0 < \Lambda_n \leq 4.0$; $3.0 \leq \Lambda_n < 4.0$; and the parameter values for the first ten of the Metropolis, Stein, and Stein "U-sequence," $\Lambda_n \in MSS_{1-10}$ (measure zero singularities with parameters ranging from 3.236068 to 3.905707) *(24)*. The first case is the largest range of the forced-dissipative parameter with $\varepsilon = 2$; the second case involves a smaller connected parametric region of high complexity with respect to period doubling accumulation followed by homoclinic behavior. The third case represents another kind of loss of potential complexity to compare with the h decrements of neuron loss, a Cantor set of singularities constituting a further limitation on the effect of high and low parallelism in the stochastic driving term.

Choices: The Topological Entropy

The measure most central to our problem, reflecting the number of choices in a dynamical system, is the topological entropy *(26)*. The entropy of map ϕ_v^m is a number, $h(\phi_v^m)$, which approximates how many different orbits ϕ_v^m generates. If none, such as in limit cycle dynamics, $h = 0$. If the number of orbits $s(m)$ is counted over bounded time intervals, then as n grows in systems with $h > 0$, $s(m)$ grows

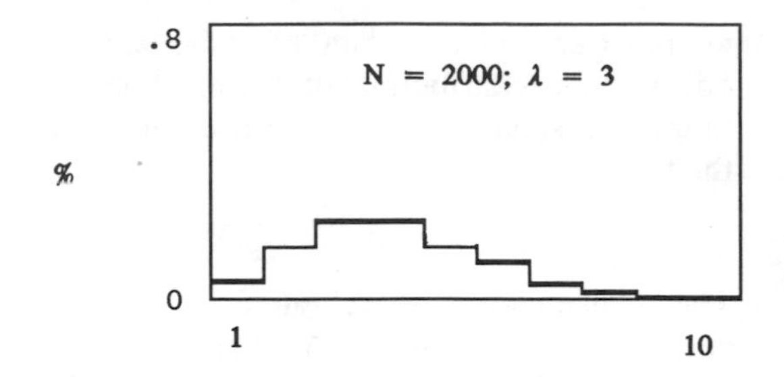

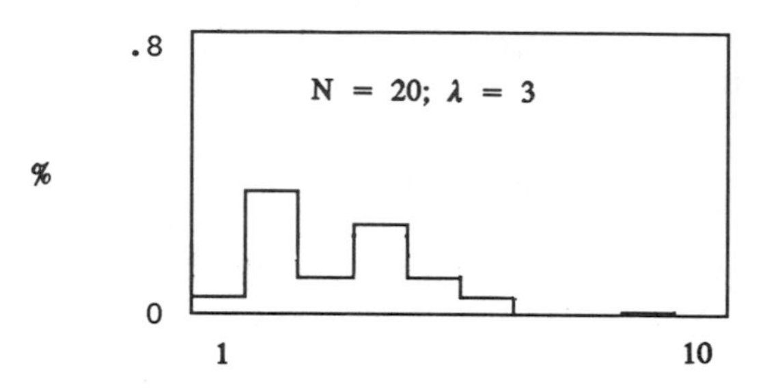

Fig. 1. Normalized Poisson process frequency distributions generated by higher and lower N sample lengths divided into ten intervals with Poisson parameter $\lambda = 3$ from which the stochastic forcing terms, Λ_n, of nonlinear map 5 (Eq. 5) (following assignment of values from three different parameter sets) were derived.

at a rate that we define to be the topological entropy,

$$h(\phi_v^m) = \lim_{m \to \infty} m^{-1} \log s(m) \tag{7}$$

If $s(m) \cong e^{hm}$, then $h(\phi_v^m) = h$, the asymptotic growth rate of the number of finite length orbits as their length goes to infinity. The obvious relationships between h, the Lyapunov exponent and spectral (return times) dimension has been developed by L.-S. Young, Anthony Manning, Dennis Sullivan, and others and will be addressed in the context of multiplicative-stochastic nonlinear maps in future work.

We calculate h as the logarithm of the largest positive eigenvalue which always exists in non-negative matrices (Frobenius-Perron), λ_{FP}, of the incidence matrices of Markoff transition matrices *(27, 28)* following the use of a binary tree of three generations with median values as nodes producing nine intervals of v_i values of equal occupancy as the arbitrary partition. We chose this number of divisions because it created an odd number of intervals closest to the ten Poisson intervals of the stochastic input distributions. We are then able to (roughly) compare the input versus the output h's for the low and high n stochastic driving conditions for nonlinear map 5 (Eq.5). Numerically, this involves the calculation of the asymptotic growth rate of the trace (periodic orbits maintain their ratio to all orbits) of the exponentiated incidence matrix reduced to its Jordan normal form by elimination of noncommunicating and unoccupied matrix subclasses *(27, 28)*.

For more general comparisons of the influence of the two conditions of parallelism on neuronal behavior, we have graphed the phase portraits and calculated the first four statistical moments, the frequency distribution, and the Fourier fast transform (FFT) Tukey-windowed power spectrum.

Lost Parallelism and Emergent Periodic Orbits Decrease *h*

Figure 2A(1,2), B(1,2), C(1,2) are graphs of the v_t/v_t+1 phase portraits and v_t/t time series of the output of Eq. 5 with high (left) and low m driving by $\Psi_n(t)$ under the three progressively restricted regimes described above. The clearest example of lost parallelism is in the comparison of Fig. 2A(1) with Fig. 2A(2) in which the ten discrete parametric steps achieve more occupancy of neuronal states and their dispersion under the former condition. We call this the *decreased Poisson spray of neurobiological aging.*

The loss of occupancy of states and their dispersion is less obvious in the Fig. 2B(1,2) reduced parametric smooth measure (1–4 to 3–4) conditions and the Cantor set of measure zero singularities Fig. 2C(1,2). Regions of up and down "two bursts" are seen in the time series of Fig. 2B and Fig. 2C, which reflect the prominence of 2^n bifurcations and "back and forth across the critical value" dynamics in both the smooth and measure zero 3–4 parameter space. We remind ourselves that even homoclinic point period three demonstrates $2n$ (left-right) orbits for much of its post period-doubling largest window. It is in the interaction of the parametrically deterministic window structure and the multiplicative random variable driving distributions that we anticipate finding a model of parametric reversibility of neuron-loss–mediated complexity decrement with aging.

Figure 3A(1,2), B(1,2), C(1,2) are frequency distributions of the output of the time series of Fig. 2 calculated using bin size B ≡ *max-min/10* and non-normalized densities. We see here a confirmation of the lost parallelism hypothesis in Fig. 3A(1,2) when studied over the full parametric range of the nonlinear map: the loss of available states and relative emphasis on small rates (long times) in the loss of input neurons, aging condition, Fig. 3A(1). Again we see that the prominence of singularities in the two reduced parametric measure conditions, Fig. 3B and Fig. 3C, make the influences of lost parallelism more subtle.

Figure 4A(1,2), B(1,2), C(1,2) are the FFT power spectral transformations of the time series of Fig. 2 and comparable to the frequency distributions of Fig. 3. We see more prominence of broadband "slows" in the full parallelism condition, Fig. 4A(1), than in the neuronal loss condition, Fig. 4A(2), as well as more occupancy of states and dispersion. The "smear" at the slow boundary in the Fig. 4A(2) condition represents long-range correlations (emerging statistically with a reduced number of possibilities iterated the same number of times) with time scales beyond that graphed in the power spectrum; consistent with the phase portraits and frequency distributions, it appears that "slowing" with neuronal loss is best seen when examined within the context of a full parametric range. The fast frequency "back and forth" dynamic dominates in both reduced parameter-space conditions (note that absolute magnitudes are not relevant since the three sets are normalized differently), and changes with loss of parallelism are more apparent in the singular value-driven spectra of 4C(1,2).

The Frobenius-Perron eigenvalues, λ_{FP}, and their logarithms, the topological entropies, h, (see Table 1) quantitate the loss of complexity resulting from loss of parallelism under all three regimes as well as the loss of complexity with Lebesgue measure of the

Fig. 2. The phase portraits, v_t versus v_t+1, and time series, v_t versus t of nonlinear map 5 (Eq. 5) using the higher N (left) and lower N (right) distributions of Fig. 1 and three parameter sets: (A) $\Lambda_n \in \{1,4\}$; (B) Λ_n $\{3,4\}$; and (C) $\Lambda_n \in$ {MSS1-10} representing parameter values for the first ten Metropolis-Stein-Stein "U-sequence" unstable periods.

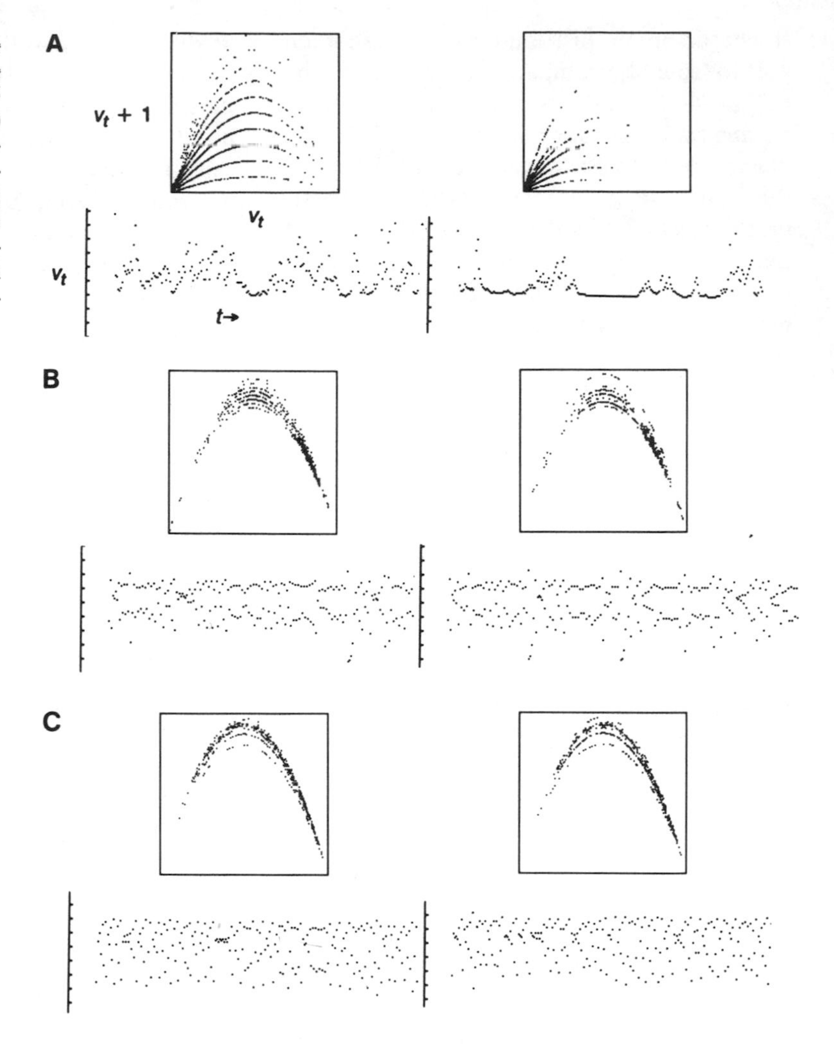

available parameter space. *These numerical experiments demonstrate two "causes" of decreased complexity with aging: neuronal loss and restricted dynamical possibilities.* It is obvious that in this interaction lies a rationale for the treatment of cell loss by parametric expansion of the number of available states. For example, we see that the 1–4 parametric condition slightly increases λ_{FP} and h of the stochastic input.

Figure 5 graphs an example of the difference between a deterministic parametric evolution and that driven by a random variable. The "period doubling to chaos" scenario demonstrates the smearing but not loss of manifold structure that was evident in Fig. 2. A systematic exploration of these differences with respect to multiplicative noise driving of automorphic $\Re^1$ and diffeomorphic $\Re^2$ maps is in progress.

Discussion

Viewing the global dynamics of neurobiological aging as a process of "inverse bifurcations" to increasing coherence, we anticipate a narrowing in the range of available time scales associated with selective loss in the fast frequencies. The classical neurological features seen in the normal aged patient *(29)* support this view. Slowed pupillary, ocular, and deep tendon reflexes are associated with a slow, near monotonic resting tremor. General motor activity manifests poverty of movement, rigidity, and perseveration. Simple and com-

A

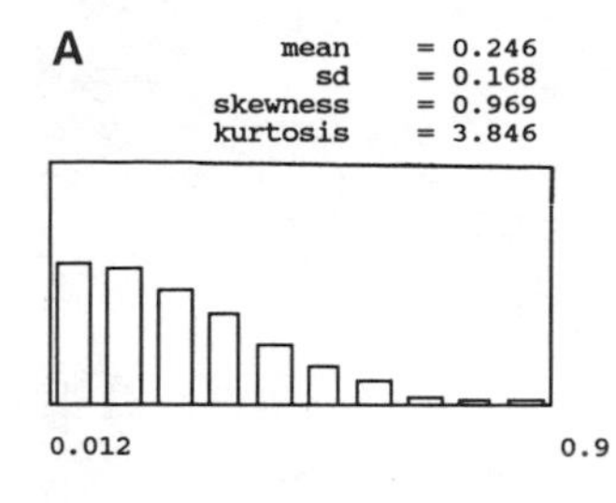

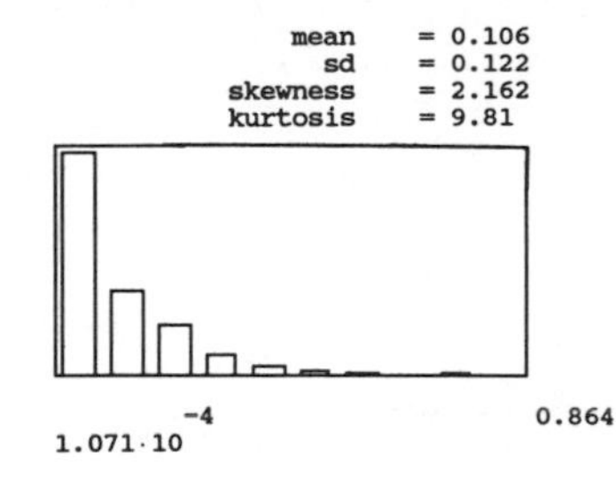

B

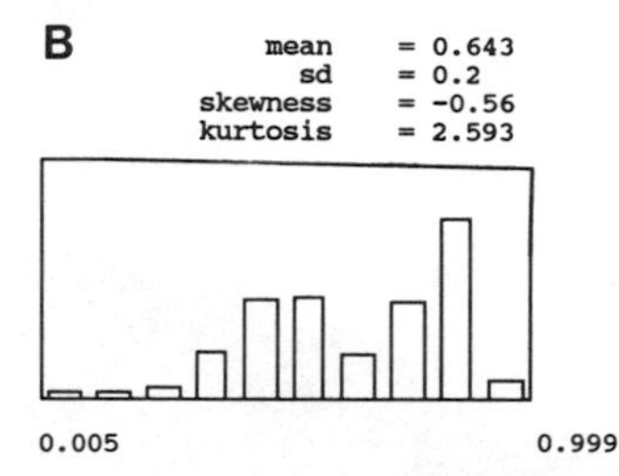

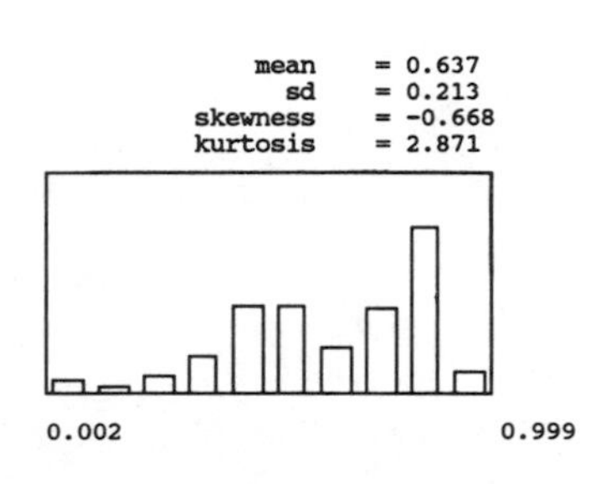

C

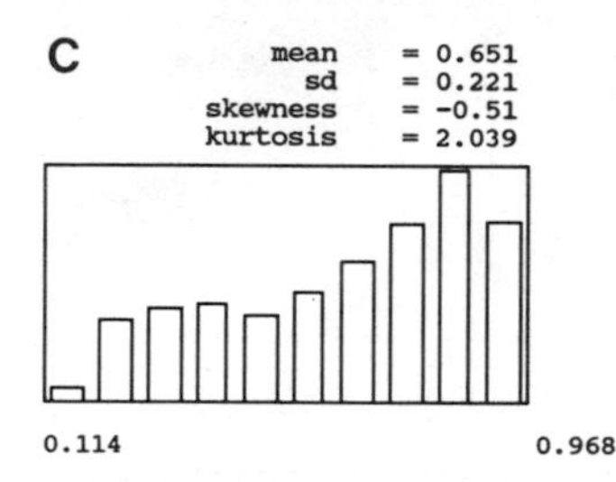

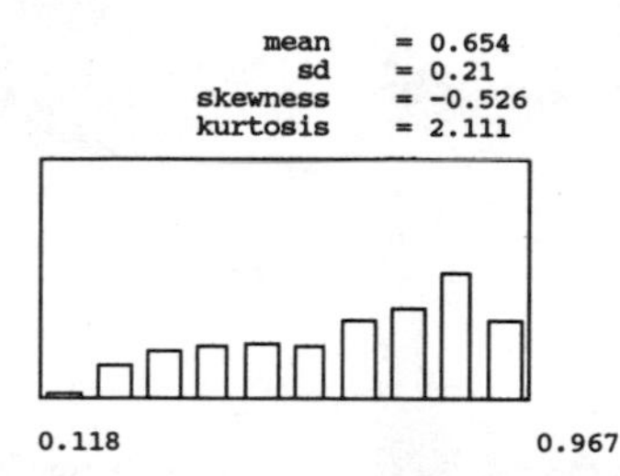

Fig. 3. Frequency distributions (not normalized) and the lower moments of the higher N (left) and lower N (right) distributions of Fig. 1, assigned the three sets of parameter values as described in text and Fig. 2, as generated by nonlinear map 5 (Eq. 5) and labeled here also as **(A)**, **(B)**, and **(C)**.

Table 1. Comparisons among the influences of higher N (full parallel) and lower N (reduced parallel) stochastic driving term, Λ_n, across the three parameter sets of the behavior of nonlinear map 5 (Eq. 5) on the largest eigenvalue of the transition probability matrices (as incidence matrices), the Frobenius-Perron eigenvalue, λ_{FP}, and its logarithm as the maximal (topological) entropy. We note that both the reduced N Poisson process of neuronal death and reduced parameter space for the individual stochastically driven reticular neurons result in decreases in the maximal entropy, h.

Parameter range	Full parallel		Reduced parallel	
	λ_{FP}	h	λ_{FP}	h
Input distribution $p(\Lambda_n)$	5.57	0.7459	2.72	0.4368
$1 < \Lambda_n \leq 4$	5.72	0.7574	4.06	0.6085
$3 < \Lambda_n \leq 4$	4.41	0.6444	3.26	0.5092
$\Lambda_n \in MSS_{1\text{-}10}$	3.63	0.5599	2.96	0.4713

plex response times increase *(30)* and decomposed information processing tasks reveal the same reduction in rates in peripheral and central feature detection, iconic storage, and short and long term memory *(31)*. There is a systematic decrease in critical flicker frequency with increasing age *(32)*. A decrease in the frequency range of the power distribution of the EEG with selective loss of fasts and a downward shift of the dominant, awake re-

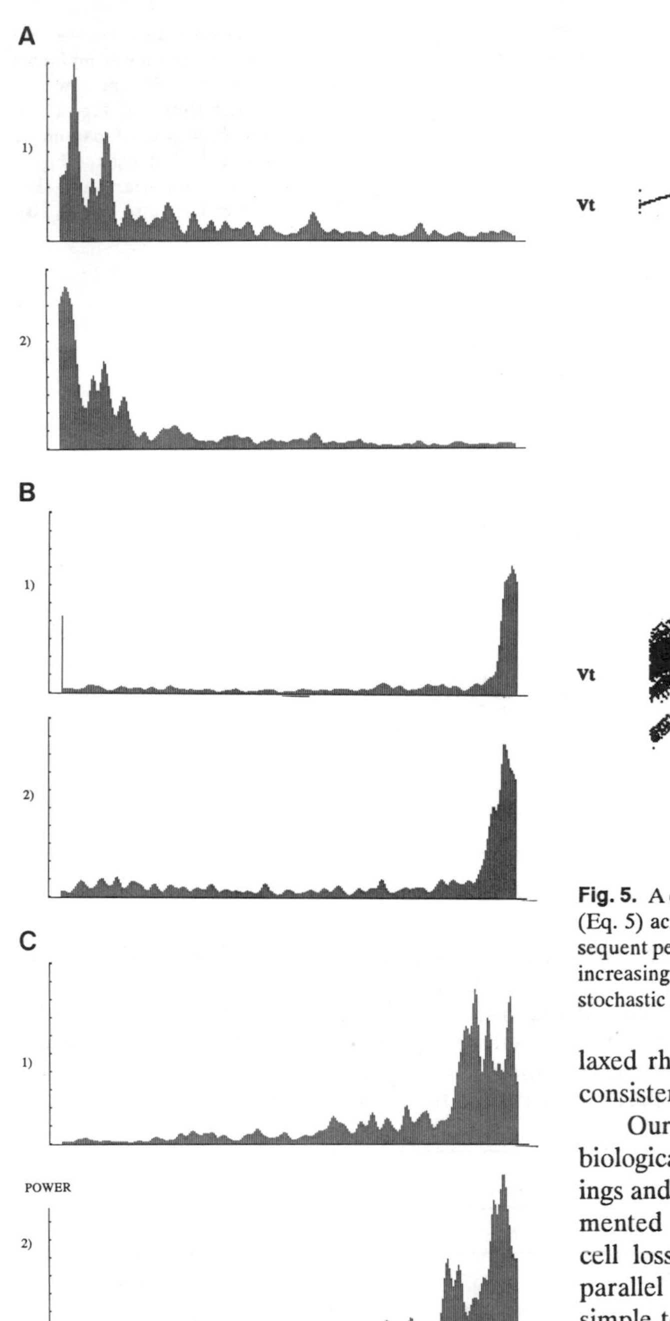

Fig. 4. Power spectral transformation of the time series of Fig. 2 (the square of the Fourier coefficients plotted against periods—period two at the extreme right, and periods ≅ {20–50} at the left) as calculated using a standard algorithm for the FFT and a Tukey window. We are here more interested in the "geometry" of the entire spectrum than in specific periodic broadbands. See text for descriptions and applications of these "portraits" to our problem.

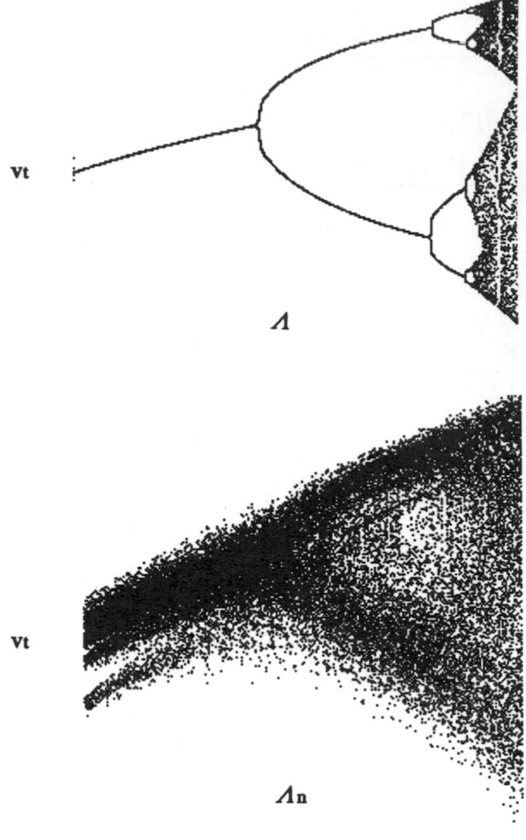

Fig. 5. A comparison of the behavior of nonlinear map 5 (Eq. 5) across a flip bifurcation to period two and subsequent period-doubling (tangent) bifurcations driven by increasing values of a deterministic (upper figure) or stochastic (lower figure) forcing term, Λ versus Λ_n.

laxed rhythm from 10.5 to 8 Hz has been a consistent finding in older people.

Our "loss of parallelism" model of neurobiological aging is consistent with these findings and starts with the simple and well-documented assumption of progressive neuronal cell loss with age. The additive feature of parallel input distributions makes it quite simple to conclude that, at least with respect to the driving term, less contributors lead to fewer possibilities. Our computer experiments also suggest that this topological entropy loss can be reversed by parametric manipulation of the measure of available states of the brain stem biogenic amine neuron dynamics. For example, replacing a Cantor set of measure zero available neuronal states with smooth measure of parameter space may serve as a compensatory source of complexity in syndromes of low *m* parallelism due to the neuronal losses of aging.

Consistent with the "hope for the aged" of the role of deterministic codimensionality in our system are the studies of serotonin autoreceptor agonists of the ergot alkaloid variety that result in disinhibition of downstream neurons, allowing them a wider range of available dynamical states. Following long term treatment (several months to years) with drugs like hydergine; the delay implicating the characteristic hysteresis of nonlinear systems, a small but significant percentage of aged populations regain the fast frequency tail of their EEG power spectra along with the reversal of much of their cognitive, affective, and behavioral disabilities *(33)*.

Acknowlegments

Appreciation is expressed to Office of Naval Research (Systems Neuroscience), the Army Research Office (Biophysics), and the National Institutes of Mental Health (Mathematical, Computational, and Theoretical Neuroscience) for support of this work. An earlier version of this work called "An Entropy-Loss Hypothesis of Aging" was presented at a conference sponsored by the National Institutes on Aging.

References

1. Rapp, P. E., personal communication.
2. Orgel, L. E., *Proc. Natl. Acad. Sci. U.S.A.* **49,** 517 (1963); _____, *Nature* **243,** 441 (1973).
3. Cumming, E., and W. Henry, *Growing Old: A Process of Disengagement* (Basic Books, New York, 1961).
4. Frolkis, V. V., *Aging and Life-Prolonging Processes* (Springer-Verlag-Wien, New York, 1982), pp. 92, 95, 205.
5. Mandell, A. J., *Math. Model* **7,** 809 (1986).
6. Magoun, H. W., in *Brain Mechanisms and Consciousness,* J. F. Delafresnaye, Ed. (Thomas, Springfield, Illinois, 1954), p. 1.
7. Shlesinger, M. F., *J. Stat. Phys.* **10,** 421 (1974); Hughes, B. D., M. F. Shlesinger, E. W. Montroll, *Proc. Natl. Acad. Sci. U.S.A.* **78,** 3287 (1981); Shlesinger, M. F., and B. D. Hughes, *Physica* **109A,** 597 (1981); Montroll, E. W., and M. F. Shlesinger, *Proc. Natl. Acad. Sci. U.S.A.* **79,** 338 (1982); Montroll, E. W., and M. F. Shlesinger, *J. Stat. Phys.* **32,** 209 (1983); Montroll, E. W., and M. F. Shlesinger, *Ann. Rev. Phys. Chem.* **39,** 269 (1988).
8. Crutchfield, J. P., and B. A. Huberman, *Phys. Lett.* **77A,** 407 (1980).
9. Ott, E., E. D. Yorke, J. A. Yorke, *Physica* **16D,** 62 (1985).
10. Mandell, A. J., in *Information Processing in the Nervous System,* H.M. Pinsker and W.D. Willis, Eds. (Raven Press, New York, 1980); Mandell, A. J., P. V. Russo, S. Knapp, in *Evolution of Chaos and Order in Physics, Chemistry, and Biology* (Springer-Verlag, New York, 1982); Goldberger, A. L., R. Shabetai, A. J. Mandell, *Am. Heart. J.* **107,** 1297 (1984); Findley, L. J., M. J. Blackburn, A. L. Goldberger, A. J. Mandell, *Am. Rev. Resp. Dis.* **130,** 937 (1984); Goldberger, A. L., V. Bhargava, B. J. West, A. J. Mandell, *Physica* **19D,** 282 (1986); Goldberger, A. L., V. Bhargava, A. J. West, A. J. Mandell, in *Fractals in Physics,* L. Pietronero and E. Tosatti, Eds. (Elsevier, New York, 1986); Babloyantz, A., and A. Destexhe, *Proc. Nat. Acad. Sci. U.S.A.* **83,** 3513 (1986); Tepper, J. Z., and A.J. Mandell, *Ann. New York Acad. Sci.* **504,** 301 (1987).
11. Farmer, D., J. Crutchfield, H. Froehling, N. Packard, R. Shaw, *Ann. New York Acad. Sci.* **357,** 453 (1980).
12. Smale, S., *Bull. Am. Math. Soc.* **73,** 747 (1967); Newhouse, S., *Topology* **12,** 9 (1974).
13. West, B. J., and A. L. Goldberger, *Am. Sci.* **75,** 354 (1987).
14. Kelso, J. A. S., G. Schoner, J. P. Scholz, H. Haken, *Physica Scripta* **35,** 79 (1987).
15. Goldberger, A. L., V. Bhargava, B. J. West, and A. J. Mandell, *Biophys. J.* **48,** 525 (1985).
16. Obrist, W. D., *Clin. Neurophys.* **6,** 235 (1954); McAdam, W., and R. A. Robinson *J. Ment. Sci.* **102,** 819 (1956); Muller, H. F., and V. A. Kral, *J. Am. Geriat. Soc.* **18,** 415 (1967); Orr, W. C., and P. Naitoh, *Int. J. Chronobiol.* **3,** 171 (1976); O'Conner, K. P., J. C. Shaw, J. C. Origley, *Brit. J. Psychiatr* **135,** 156 (1979); and reference 4.
17. Polich, J. M., C. L. Ehlers, S. Otis, A. J. Mandell, F. E. Bloom, *Electroencephalogr. Clin. Neurophys.* **63,** 138 (1986).
18. Basar, E., *EEG-Brain Dynamics: Relation Between EEG and Brain Evoked Responses* (Elsevier, Amsterdam, 1980).
19. Kety, S. S., *Res. Publ. Assoc. Res. Nerv. Ment. Dis.* **35,** 31 (1956).
20. Birren, J. E., *Am. Psychol.* **29,** 808 (1974).
21. Sokoloff, L., *Res. Publ. Assoc. Res. Nerv. Ment. Dis.* **41,** 237 (1966).

22. Labos, E., *Cyber. Sys. Res.* **2,** 237 (1984).
23. Mandell, A. J., *Physica* **27D,** 235 (1987).
24. Collet, P., and J-P. Eckmann, *Iterated Maps on the Interval as Dynamical Systems* (Birkhauser, Boston, 1980).
25. Metropolis, N., M. L. Stein, P. R. Stein, *J. Comb. Theory (A)* **15,** 25 (1973).
26. Adler, R. L., A. G. Konheim, M. H. McAndrew, *Tran. Amer. Math. Soc.* **114,** 309 (1965); Adler, R. L., and B. Marcus, *Mem. Am. Math. Soc.* **20,** no. 219, 133 (1979).
27. Seneta, E., *Non-negative Matrices and Markov Chains* (Springer-Verlag, New York, 1981).
28. Gantmacher, F. R., *Applications of the Theory of Matrices* (Interscience, New York, 1959).
29. Critchley, M., *Lancet* **i,** 1119 **(1931);______,** *J. Chron. Dis.* **3,** 459 (1956).
30. Koga, Y., and G. M. Morant, *Biometrika* **15,** 346 (1923).
31. Walsh, D. A., and L. W. Thompson, *J. Gerontol.* **33,** 383 (1978).
32. Corso, J. F., *J. Gerontol.* **26,** 90 (1971).
33. Heimann, H., *Brain Function in Old Age* (Spring-Verlag, New York, 1979).

5

Searching for Signal and Noise in the Chaos of Brain Waves

Walter J. Freeman

Abstract

Brain information is carried by patterns of neural activity that are manifested in electrical fields of potential known as action potentials and in electroencephalographic (EEG) waves. By recording and measuring these fields in the olfactory systems of rabbits trained to respond to odorants, we have found that perceptual information has the form of a burst of oscillatory activity like the vibration of a spoken word. The temporal pattern of the burst is brief, irregular, and nonreproducible, and it does not in itself contain the information. That is in the spatial pattern of the burst. It is like a frame in a motion picture, in which the light wave is the carrier, and the information is in the spatial pattern of light intensity.

Each burst forms rapidly by an abrupt transition like the flicker of a movie frame. Its pattern emerges from the pre-existing pattern under the influence of three factors: a familiar stimulus, past experiences of learning with the stimulus, and the chemical state of the brain constituting its degree and kind of arousal relevant to the stimulus.

From these observations and from our mathematical models of them, we conclude that the olfactory system is a chaotic generator. The synaptic interactions of its millions of neurons causes ceaseless, life-long cooperative activity. The temporal waveform looks like noise, but it is widely shared. Its spatial pattern changes by sudden jumps like small explosions, and it is in these reproducible patterns that we find the perceptual information.

We propose that chaotic neurodynamics solves the problem of information storage and retrieval in brains by mechanisms far removed from those of digital computers. An act of perception is not the readout of dead bulk. It is a fresh creation in a sensory cortex of a generic pattern of activity, under the influences of a familiar stimulus and the contexts of arousal and expectation mediated by other parts of the brain.

Introduction

There are many routes to chaos, and it has various faces in different chemical and biological systems. Several of the better-known tests are not useful in studies of brain dynamics. Sensitivity to initial conditions, for example, is meaningless because we cannot stop and restart brains at will. Scale invariance has no validity because nerve cell assemblies do not scale up from neurons or from their parts. Calculations of various measures of fractal dimensions have thus far failed to give reliable values due to the unresolved complexities of brain activities and our methods for observing them.

In order to characterize brain chaos, we begin instead with descriptions of brain states. Brains consist of collections of neurons, and their activities result from cooperation among neurons by synaptic interactions. If a brain is left alone without stimulation, and it goes to a rest state in which the neurons are silent, it is said to go to equilibrium. If it is stimulated into activity and then goes to rest each time the stimulation is halted, it is said to be "attracted" to rest. Therefore, it has an equilibrium "attractor." This occurs only in deep anesthesia, coma, or brain death.

If neurons, brains, or parts of brains generate periodic activity (typically at one frequency), they are said to be in a limit-cycle state. If each time they are driven by stimuli to some other state, only to return when the stimulus is removed, they are said to have a limit-cycle attractor. Such regular activity is uncommon. Brain activity is typically aperiodic and unpredictable in the absence of stimulation, and it returns to the same following termination of a stimulus. This may (or may not) manifest chaos and chaotic attractors. My aim in this chapter is to explain why some brain activities are apparently chaotic and what this may mean for brain function.

One principal, if not the most important, brain function is categorical perception. This is an act by which we combine sensory input with what we have learned in the past, with our present state of arousal, and with our expectations of future events. The crucial problem here is, What kind of past information is stored, and how? Our brains do not store memories in the manner that computers do. They have no components for laying down images or representations, retrieving them intact, cross-correlating to compare them with present or other past inputs, computing error functions, completing partial inputs, and so on. Each act of recognition and of recollection requires the creation of an appropriate pattern of neural activity in the brain. It arises from the flux of sensory information drawn into a sensory cortex. The questions I propose to address are, How might brains do this? and What might chaos have to do with it?

Patterns of Brain Activity

The data I will use to make my points come from observations on the olfactory system *(1, 2)*, which, in several important respects, is the simplest and most accessible sensory system in animals. It consists of an array of receptors in the nose (Fig. 1), which transmit odorant information directly to a simple form of cortex called the olfactory bulb. This, in turn, transmits to the nucleus that sends back into the bulb and to the olfactory cortex, which transmits broadly into the brain. Each of these structures generates characteristic patterns of neural activity that are observed through recordings of their electrical signs.

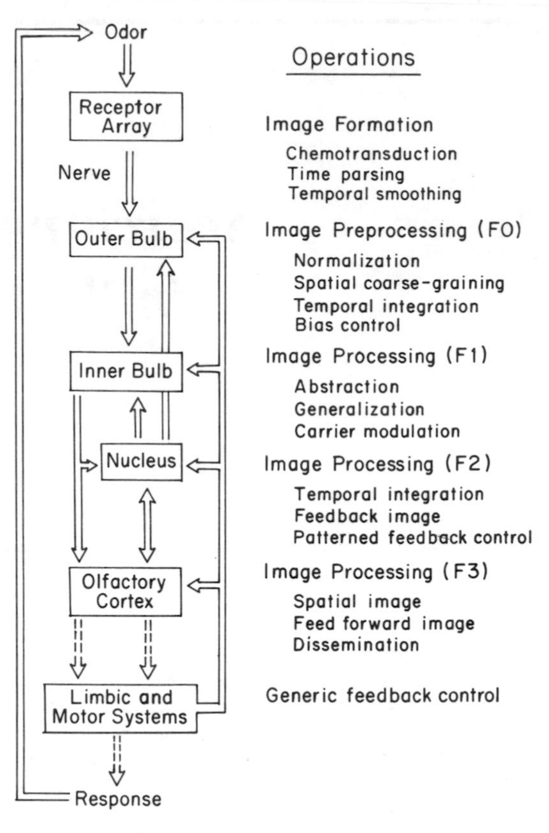

Fig. 1. The main stages are sketched for flow of odorant information through the olfactory system. Coupling of the bulb (F1), anterior olfactory nucleus (AON, F2), and olfactory (prepyriform) cortex (PPC, F3) by multiple forward and feedback neural pathways is essential for the genesis of olfactory chaos.

In the view of neurobiologists, the functions of our brains that we experience subjectively as "remembering" and "recognizing" are dynamic operations. These operations work on patterns of neural activity and transform them into other patterns of activity in our brains. We deduce these operations by making measurements on sequential neural activity patterns in the brain, identifying the patterns and their transformations with acts of recognition by animals, and constructing mathematical models of the dynamic operations so as to simulate our data.

Chaos comes immediately to the fore when we inspect our recordings of endogenous electrical activity, which has two forms. Trains of action potentials ("pulses") put out by neurons reflect the transmission of information from sensory receptors to the brain, from one part of the brain to another, and from brain to effectors. Fields of electromagnetic potential called "brain waves" or EEG waves reflect the integrative actions of local pools of neurons in brains, especially in areas of cerebral cortex. Such EEGs are recorded as "traces" from electrodes in the brain, on its surface, or on the scalp (Fig. 2).

Both forms of activity are aperiodic and locally unpredictable. And both have commonly been viewed as "noise" because distributions of intervals between pulses in trains are typically Poisson, histograms of EEG amplitudes are close to Gaussian, and spectra are broad with low and variable peaks. But EEG waves were not always perceived as noise. In the early decades of EEG research, brain waves were found to be useful signs of brain states such as waking and sleeping and to give useful clues to brain diseases such as epilepsy, coma, and cancer. However, in the 1950s signal detection theory was introduced into brain studies. Computer averaging of transient responses of brains to brief stimuli such as clicks and light flashes became and remains the most common procedure for experimental testing of brain function in human cognition. Its aim is to retrieve a sensory-induced "signal" from the brain noise in which it is embedded, thereby relegating the EEG to the status of noise, something to be eliminated by averaging.

Characterization of Brain Chaos

A realization that EEGs represent neither signal nor noise but chaotic carriers of brain information came from experiments by my students and myself in the study of olfaction in animals. Prior studies by Shepherd *(3)*, Kauer *(4)*, and others had established that (i) sensory input to the olfactory brain is by pulses from selected receptors that signal the presence of an odorant chemical; (ii) this information has the form of a spatial pattern of pulse activity among receptors; and (iii) with each sniff, it induces another spatial activity pattern in and among the neurons of the olfactory bulb. We reasoned that if we trained a subject to respond to an odorant, then when the animal sniffed the chemical, we should find a pattern of EEG activity in the bulb reliably reflecting the presence of the significant odor, just before the animal made a correct response.

Through an exhaustive search we devised methods to detect and extract that information from the EEG of the olfactory bulb *(5)*. This EEG is a mixture of electrical potentials from several sources, not all of which are known. We have found the information we sought in the spatial pattern of the amplitude

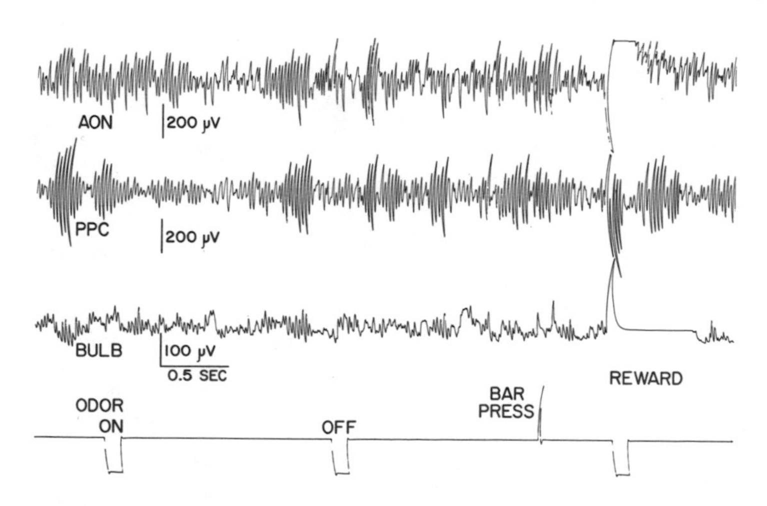

Fig. 2. Examples of EEG traces are taken from three sites in the olfactory system: the bulb, anterior olfactory nucleus (AON), and olfactory cortex (PPC) during the performance of a conditioned response to an odor. The rapid fluctuations are "bursts" that accompany inhalations. The spike at the reward is an artifact from movement, one of the many hazards that plague EEG research.

of a brief high-frequency fluctuation in the EEG called a "burst" (Fig. 2) that involves the entire bulb (Fig. 3).

This is a particular kind of chaos that is "spatially coherent." We discovered this in recordings made simultaneously from 64 electrodes placed on the olfactory bulb. No matter how irregular the waveform of the EEG from one electrode, the EEGs on all the other electrodes share that waveform to some degree. This holds both in and among bursts. The amplitude of the shared waveform differs between electrodes, so that there is a spatial pattern of waveform amplitude (Fig. 3) that we display by using contour plots.

The behavioral information is in these spatial patterns of the EEGs. We have shown this in the olfactory bulbs of rabbits trained to discriminate between odorants (as well as in the visual cortex of a Rhesus monkey trained to discriminate visual stimuli). The oscillatory carrier has been demonstrated to be spatially coherent in cats and to be closely related to pulse activity of neurons *(6)*. For each discriminated odorant (including the background with no deliberate odorant), there is a characteristic spatial pattern of the common burst waveform (Fig. 4). For this reason we call the common waveform the "carrier" of the information. It is analogous to the projection of a picture in which light is the carrier wave, and the information is in a two-dimensional pattern of light (high activity) and dark (low activity).

The carrier is shared by the entire bulb. It is not a response to being driven by an external generator of the carrier, because there is no global pathway to the bulb that can carry such high frequencies. It is not a resonance of the bulb to neural impact, like a struck bell, because it is too irregular. It is the manifestation of cooperative neural dynamics within the bulb.

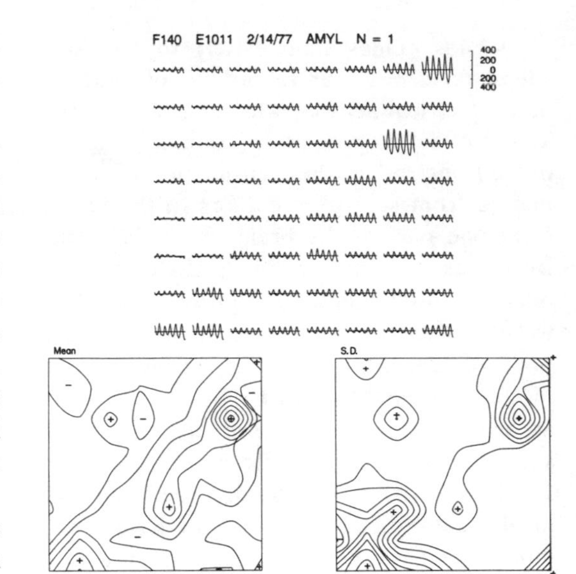

Fig. 3. A single, unaveraged burst is recorded simultaneously from 64 electrodes, showing the common waveform and its variation in amplitude. The time delays across the array are too short to be seen by the naked eye. The duration of the segment is 0.1 second. Amplitude is in microvolts. The spatial pattern of amplitude of the common waveform is expressed by contour plots. On the left is the mean of the amplitudes of 10 bursts, and on the right is the standard deviation.

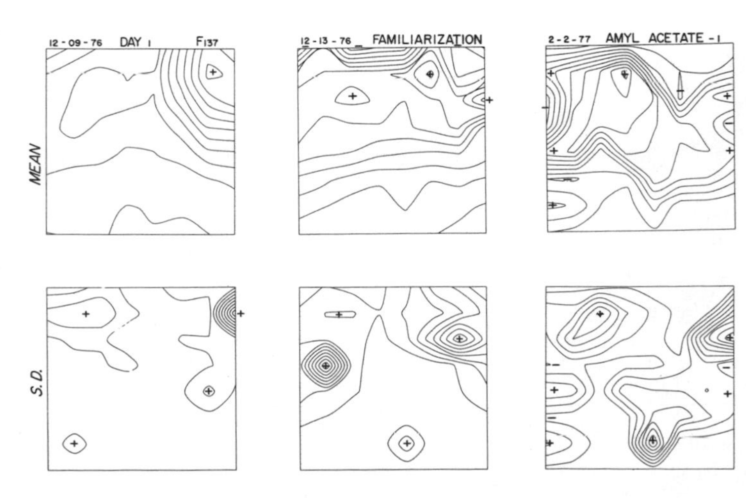

Fig. 4. The examples show the changes in form of the amplitude pattern for a set of 10 bursts (mean above and standard deviation below) from a subject on the first day of recording (left), after two weeks of familiarization to background odors of the testing apparatus (center), and after training to respond to a chemical stimulus (right). The array size of 8 × 8 electrodes is about 4 × 4 millimeters.

Evidence for Brain Chaos

We cannot prove mathematically that this brain activity is chaotic, nor can we prove that it is not. What makes the case convincing for us experimentally is our finding that the spatial pattern changes from one form to another almost instantly over the whole bulb. The transitions from interburst to burst and back again are sudden with each inhalation and exhalation. Each transition begins at some point anywhere on the bulb and spreads in all directions at high speed. We have measured the rate of spread by fitting sums of cosine waves to each of the 64 traces recorded during a burst and determining the phase values at the common frequencies. The values conform to the shape of a cone, as would be expected for the spread of waves from a pebble dropped into water (Fig. 4).

The location of the apex of the cone, which we call a site of nucleation, varies from one burst to the next without relation to behavior. The rate of spread is so fast that it is complete within a quarter of a cycle of the oscillation of the burst. The information introduced by pulses from sensory receptors into the bulb is disseminated as the burst spreads over the entire bulb to all bulbar neurons, and it takes place so rapidly that the areas of cortex to which the bulb transmits its information can carry out integration of bulbar activity from the entire bulb without degradation of the information by phase lags. In this way, the sensory information introduced by pulses from receptors is disseminated to all bulbar neurons in a few milliseconds.

This is an exceedingly important property for the management of neural information by the olfactory system, and we do not fully understand how it happens. It appears to be an instance of anomalous dispersion *(7)*, in which the velocity of spread of a state transition exceeds the velocity of a wave of excitation sweeping through the medium of the synaptic web.

This feature provides the most compelling evidence we have to date for the genesis of chaos by the olfactory system. No collection of noise generators could achieve the high speed of these abrupt state transitions and maintain the unity and order implicit in the reproducible emergence of spatial patterning that is dependent on the odorants (Fig. 5). Chaos provides the sensitization of the system for rapid controlled change under small perturbation *(8–10)*.

Physiological Mechanisms of Chaos

In order to analyze the complex neurodynamics of the olfactory system, we have used sets of coupled nonlinear ordinary differential equations. Their solutions serve to simulate the various chaotic patterns we have observed in EEGs from the several parts of the system (Fig. 6). This model has helped us understand the physiological mechanism by which chaos arises in the olfactory system. It has shown us what to look for experimentally when learning takes place, and it has given us insight into the role that chaotic attractors (Fig. 7) might play

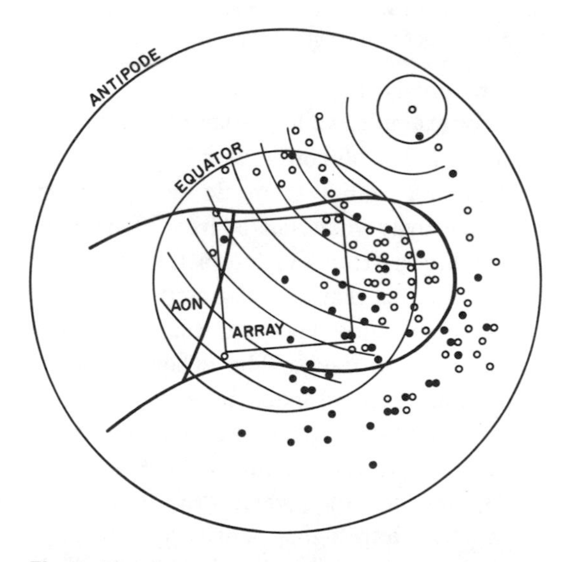

Fig. 5. The phase pattern of a burst is like that of a wave from a pebble dropped in still water. It is displayed by opening the near spherical surface of the bulb into a flat projection. The point of onset may be anywhere in the bulb, and it may start with maximal or minimal phase, independently of the odorant condition. The rectangle shows the size and location of the 8 × 8 array of electrodes. The solid dots show phase maxima; the open dots show phase minima of individual bursts. The isophase contour interval is 0.25 radians/millimeter.

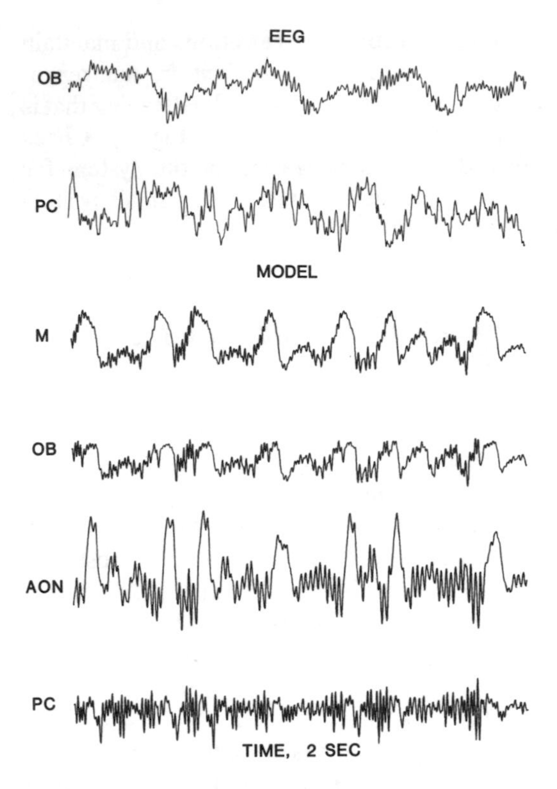

Fig. 6. A model consisting of nonlinear differential equations suffices to generate chaotic waveforms (lower four traces) that resemble traces of the basal interburst EEG from a resting animal (top two traces).

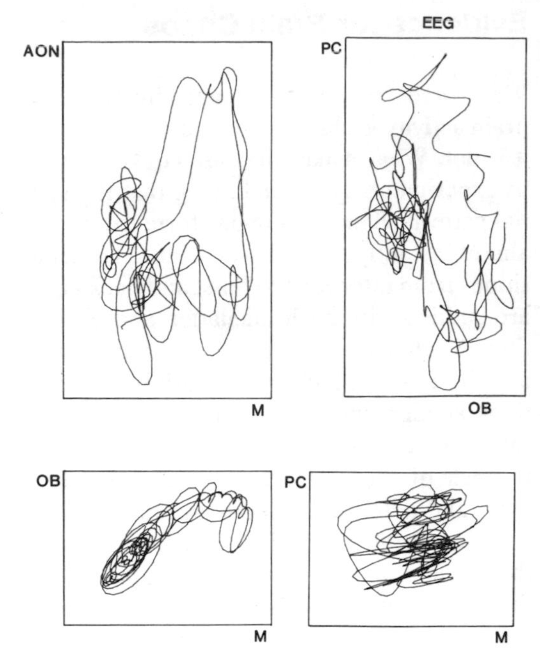

Fig. 7. Phase portraits of the chaotic attractor for the basal state are made by plotting pairs of traces against each other. The top right frame is from the real EEG, and the other three frames are from the model. With three or four traces we can plot these in 3-D or 4-D and rotate the images to get optimal views of their geometrical structure, using computer graphics like that in a flight simulator. This is known as "flying the attractor."

in perceptual categorization.

For example, we have found that relatively small areas of cortex with high rates of propagation within them do not generate chaotic activity when isolated from each other. Chaos arises from the interaction between areas of cortex that are separated by long or slowly conducting pathways that introduce communication delays. The delays are most significant in closed loops, in which one area of cortex acts on another and then, in turn, is acted upon. Moreover, the feedback is of two kinds, one excitatory, which constantly goads the cortices into higher activity by regenerative activation, and the other inhibitory, which constantly suppresses the cortical activity. It is the contradictory play of these two influences that keeps the activity going, and it is the inequality of delays in feedback that helps to disorganize and "chaoticize" the activity.

This kind of feedback connection between areas of cortex has been applied in artificial neural networks to implement algorithms of back propagation and of adaptive resonance, by which the networks modify their own connections and "learn" to categorize inputs in respect to standard patterns determined by the modelers *(11, 12)*. In real brains, there is no anatomical or physiological evidence that these feedback paths have the specificity of mapping from one area of cortex back to another that is necessary for error correction. On the contrary, in our model, the feedback is dispersed and divergent, in conformance with the properties of neural feedback, and the result is not error correction but chaos.

Perceptual Categorization by Chaotic Attractors

Our model has also helped us to get some idea

of what the chaotic attractors might look like in studies of brain dynamics by making "phase portraits" of their trajectories. We use at least three and preferably four dimensions to display the projections of the chaotic attractors revealed by EEGs. We do this by means of computer graphics similar to a flight simulator to show the structures in three dimensions (3D), and we use color to display amplitudes in the fourth dimension. (It is important here not to confuse the spatial patterns of activity on the cortex surface in brains with the geometric forms of dynamic attractors in state space.)

The best way for us to visualize the preferred activity of a part of the brain as it is revealed by the EEG is to record several EEG traces and then plot them against each other or against themselves, displaced in time rather than against time (Fig. 7). But the olfactory EEG is so complex that by this technique the portrait of its attractors look like cooked spaghetti on a plate. The complexity is caused by several factors, principal among them that the olfactory system is connected with the rest of the brain, and its activity in part reflects the participation of the brain in olfaction. Also, the EEG traces are a mix of activity from several sources, and the distinctive bursts do not last long enough to make phase portraits of them.

We developed our model to alleviate these problems. It does so because it is isolated, free from extraneous signals, and better controlled. For example, we can simulate a burst by giving input to the model while it is generating basal activity, and then deliver sustained input to hold the system in a prolonged burst (Fig. 8). The time series shows what looks like periodic oscillation, but the attractor looks more like a loose coil of wire than a circle (Fig. 9). It is irregular, chaotic. This experiment has led us to postulate that the burst does not manifest, as we had originally supposed, an aborted approach to a limit cycle attractor, but instead a trajectory into a spatially coherent chaotic attractor.

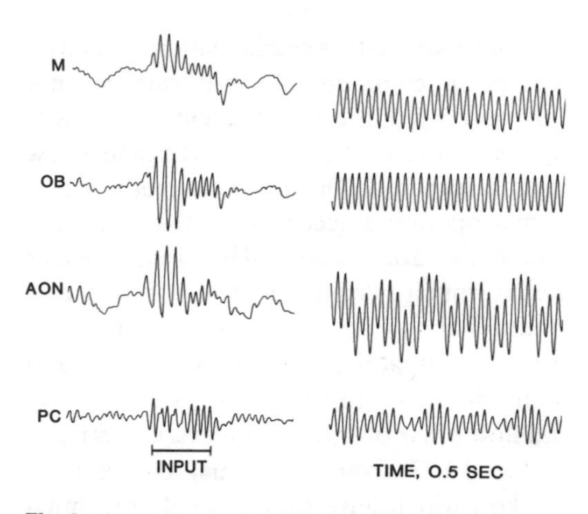

Fig. 8. An artificial burst is generated by the model when it is given input that simulates a surge of receptor input during an inhalation. When the input is kept on, the model goes into a continuous burst. This does not happen in the bulb unless it is poisoned by excitatory drugs.

At present, our model is helping us to explore the way in which chaotic dynamics categorizes olfactory stimuli, using this sense as the most accessible at present to model brain functions. In essence, the model learns to recognize classes of input by the same dynamical algorithms that we believe hold in the olfactory system. The work is still in its early stages, but we have already come to some preliminary generalizations, which we expect will help us to understand the brain mechanisms of categorical perception.

One finding is that when a patterned input is given to the model for it to categorize,

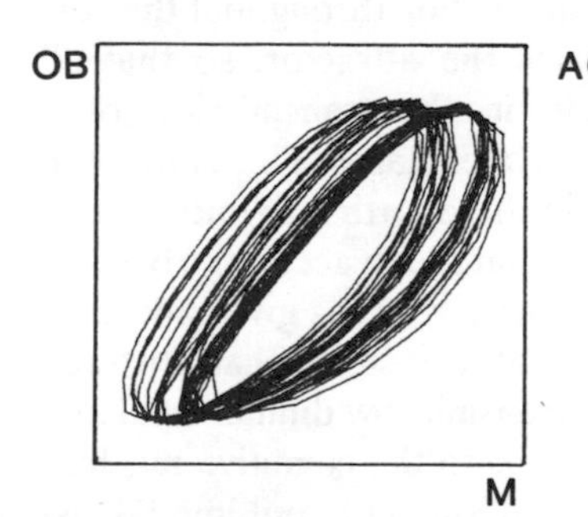

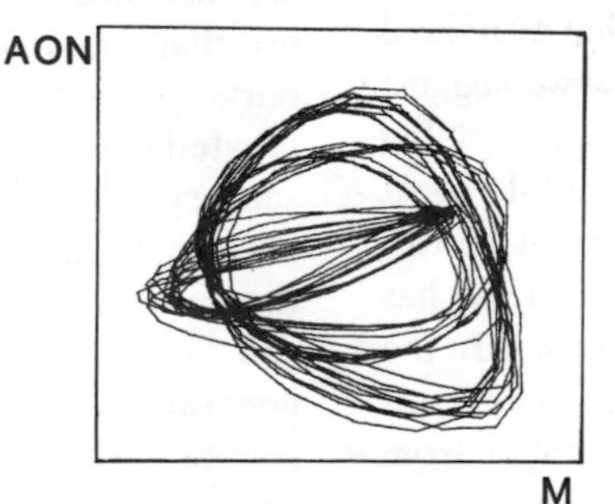

Fig. 9. The phase portrait of the artificially prolonged burst gives us an idea of the geometry of a part of a global olfactory chaotic attractor, which is accessed by delivery of an odorant, and which governs the perceptual categorization of the stimulus by the olfactory system. In various views of dynamic state space, it resembles a collapsed toroid (a squashed doughnut).

the convergence to a spatial pattern of amplitude of the common oscillatory carrier is not only very rapid, it is independent of the starting conditions of the system at the time a new input is imposed. This indicates that a neural system operating according to this algorithm need not return to a neutral basal state such as the interburst period before it is given a new input. That now appears to hold only in olfaction of air-breathing animals with nostrils and is an adaptation to the peculiarity of the ebb and flow of air over the receptors. Our studies of the EEG from the visual cortex of a monkey, which have shown visual perceptual information to be carried in the spatial amplitude pattern of a common chaotic carrier, have failed to show burst-interburst sequences. This suggests that there may be very little down time for state transitions in the visual cortex, for which the analogy of the frames in a movie may be apt.

Another finding is that, under the maintained input that is necessary to generate a burst both in the olfactory bulb and in our model of it, the model does not converge to an attractor and thereby lose all information about where in the basin it began its trajectory to the attractor. On the contrary, no information is lost about the input. The model performs a mapping from the stimulus to the output without loss of the information that was supplied by the stimulus, but it rearranges the information so that classification at a succeeding stage is facilitated.

These and related findings give us a tentative picture of the mechanism by which rapid and reliable categorization of inputs takes place in olfaction and, we believe, in other senses as well. In previous reports, we suggested that the bulb might have a learned limit cycle attractor for each odorant that an animal had learned to discriminate. Now we suggest that the olfactory system maintains a global chaotic attractor that has a complex shape. It may be analogized to the simpler but better-known Lorenz attractor *(10)*, which in 3-D has the shape of a butterfly with two wings. In the free-running chaotic state of the Lorenz system, the activity jumps unpredictably from one wing to the other and back again, corresponding to sudden shifts in the pattern of oscillation seen in the recording against time.

If the Lorenz system were biased by a sustained input, the system might preferentially stay in one wing or the other. We suppose that the olfactory attractor has multiple wings, one for each odorant that a subject has learned to discriminate, and that in the presence of the odorant, the neural mechanism preferentially orbits a repellor in a wing corresponding to that odorant. Categorization is the biased selection of a subspace of the global attractor and not a leap from one attractor to another. We postulate that the process of learning to identify a new odor causes the addition of a new wing to the global attractor, and that in the process, all the other wings are modified simultaneously. The emergence of a spatial pattern of amplitude for the common carrier reports which wing has been selected to the next stage for further processing. The emergence for the first time of a new spatial pattern and the concomitant modification of the preexisting patterns, as we have documented experimentally, reflects the addition of a new wing. This learning process introduces a structural change or bifurcation in the olfactory dynamics.

Conclusion

My conclusion is that chaotic olfactory dynamics supports a global attractor, which is a storehouse of the means for expressing olfactory experience, and that it affords rapid access to any of its wings without requiring a time-consuming, exhaustive search through all its wings. It provides for the very rapid dissemination of information throughout the cortex that maintains the attractor, so that all parts participate in the transmission of a created pattern to later stages. This unifies the sensory system in its reporting to and control of the next stages that are to act on the basis of the perception. Thereby chaos gives powerful vehicles to the central systems that abstract, generalize, and transmit low dimensional and relevant information to the cognitive machinery involving attention and multimodal association. But the chaotic voice is not the information, merely the vehicle that carries it. That is why we have had such difficulty in finding correct interpretations of EEGs. Our new approach, through nonlinear dynamics, affords an opportunity to exploit further these manifestations of brain activities and thereby better understand and emulate brain functions.

Acknowledgment

This research was supported by grant MH06686 from the National Institute of Mental Health, U. S. Public Health Service.

References

1. Freeman, W. J., *Mass Action in the Nervous System* (Academic Press, New York, 1975).
2. Skarda, C. A., and W. J. Freeman, *Brain Beh. Sci.* **10,** 165 (1987).
3. Shepherd, G. M., *Neurobiology* (Oxford University Press, New York, 1988).
4. Kauer, J., *Nature* **331,** 166 (1988).
5. Freeman, W. J., *Handbook of Electroencephalography and Clinical Neurophysiology,* vol. 3A, part 2 (Elsevier/North, Amsterdam, Holland, 1987), pp. 583 – 664.
6. Gray, C. M., P. Koenig, A. K. Engel, W. Singer, *Nature 338, 334 (1989).*
7. Jackson, J. D., *Classical Electrodynamics* (John Wiley, New York, 1987).
8. Prigogine, I., *From Being to Becoming* (W. H. Freeman, San Francisco, 1980).
9. Nicolis, J. S., *Hierarchical Systems* (Springer-Verlag, Berlin, 1985).
10. Gleick, J., *Chaos: Making a New Science* (Pergamon, New York, 1987).
11. Grossberg, S., *The Adaptive Brain. I. Cognition, Learning, Reinforcement, and Rhythm* (Elsevier/North, Amsterdam, Holland, 1988).
12. Rumelhart, D. E., J. L. McClelland, P. D. P. Research Group, *Parallel Distributed Processing: Exploration in the microstructure of cognition. Vol 1: Foundations* (Bradford/MIT, Cambridge, MA, 1986).

6

Applications of Chaos Theory to Shear Turbulence

Laurence Keefe, Parviz Moin, John Kim

Abstract

The Lyapunov dimension of the attractor underlying turbulent Poiseuille flow at a Reynolds number of 3200 has been calculated to be approximately 360. Because the resolution of the flow simulation used to calculate the Lyapunov exponents was coarse in homogeneous directions of the flow and the spatial domain small, we believe this dimension to be a lower bound on its true value at this Reynolds number. Assuming the general existence of strange attractors beneath turbulent flows, the existence and eduction of coherent structures is related to the geometrical characteristics of such attractors.

Introduction

Is the "strange attractor" the correct mathematical model for turbulent flows? Despite nearly a century of theoretical and experimental research into the origins and development of fluid turbulence, the explicit mathematical framework that connects the chaotic, time-dependent dynamics of real flows to the structure of the Navier-Stokes (NS) equations remains unclear. In recent years, nonlinear dynamical systems theory has laid claim to the candidacy for this framework, primarily as a result of its discovery and definition of that mathematical object called a "strange attractor." Such objects represent the solutions to differential equations and carry within their intrinsic structure a well-defined mechanism (usually called "sensitive dependence on initial conditions") that can produce chaotic and unpredictable behavior in all physical systems without requiring random forcing. In addition to conjecturing that the strange attractor connects the NS equations and turbulence, dynamical systems theory predicts that turbulence is asymptotically describable by a finite number of degrees of freedom, despite the fact that the NS equations are infinite dimensional. This has encouraged hope that the turbulence problem can be reduced in apparent complexity by projecting it onto some special basis. If there is a strange attractor underlying shear turbulence then, by implication, turbulent phenomenology must find some expression or source in the structure of these mathematical objects. The investigations described below provide strong evidence that strange attractors do underlie shear turbulence and demonstrate that the existence of "coherent structures" can be regarded as a manifestation of intrinsic structural details of such attractors.

Measuring the Complexity of Turbulence

The validity of dynamical systems theory as a descriptor of many supercritical fluid convection phenomena is well established *(1–5)*. Most importantly, the concept that the complicated dynamics of a thermal-fluid system are attributable to the interactions of a rather small number of degrees of freedom has been validated, at least in small aspect-ratio systems. The same can also be said for its utility in describing Taylor-Couette flow *(6)*. Chaos theory has found less acceptance as a framework for turbulence in channels, jets, boundary layers, and wakes, largely because experimental verification has proven to be so much more difficult than in convection.

Here we offer strong evidence that the turbulent solutions of the NS equations are confined to a strange attractor. At a single Reynolds number in turbulent channel flow we have determined the dimension, D_{λ}, of the underlying attractor, having measured sufficient of the Lyapunov exponent hierarchy, λ_i, to calculate this quantity from the Kaplan-Yorke *(7)* definition:

$$D\lambda = j + |\lambda_{j+1}|^{-1} \sum_{i=1}^{j} \lambda_i ,$$

where

$$\sum_{i=1}^{j} \lambda_i > 0, \quad \sum_{i=1}^{j+1} \lambda_i < 0 .$$

This evidence is indirect in the sense that, while the existence of an attractor guarantees the existence of its Lyapunov exponents *(8)*, calculation of a set of exponents using standard methods *(9)* may not be the guarantee of an underlying attractor (though we are at a loss to say what else there could be). Traditional methods of visualizing attractors (e.g., Poincaré sections, maps) fail utterly for high-dimensional objects, so we have only the existence of a converged Lyapunov spectrum to infer the existence of an attractor. Such inference has been common in the past *(10, 11)*, and we employ it now.

Because dynamical systems theory predicts that the asymptotic behavior of a dissipative dynamical system is confined to fewer degrees of freedom than needed to specify an initial condition, it has always been of interest to calculate the extent of this decrease. Calculating the dimension of the solution attractor supplies this information, for the dimension measures the number of degrees of freedom needed to characterize a point on the attractor and is a direct measure of the intrinsic complexity of the turbulence. Of the several definitions *(12, 13)* of dimension available, we have chosen to use the Kaplan-Yorke formula (which bounds the fractal dimension) *(14)*, since we have access to the dynamical equations (i.e., the NS equations) of the system, and can calculate the Lyapunov exponent hierarchy. This is in contrast to the one previous attempt *(15)* to calculate the dimension of attractors in Poiseuille flow, which employed methods most suited to data derived from experiments. This attempt failed, concluding that the dimension is greater than 40. Because our calculations indicate the dimension is almost an order of magnitude greater than 40, we argue that none of the "experimental" methods for measuring dimension can be expected to work on this problem because the data required exceed current computer storage capabilities. This is true whether the method is a variant of the "correlation" dimension *(13)*, or is one of the newer techniques *(16, 17)* to calculate the Lyapunov exponents experimentally.

The Lyapunov exponents of turbulent Poiseuille flow were calculated using standard *(9)* methods, except that the vectors separating solution trajectories were calculated in the phase space of the base trajectory (i.e., on the attractor) rather than in its tangent space. Thus, only the NS equations were integrated, not these equations and their variational form. The magnitudes of the separation vectors between members of the ensemble and the base trajectory were normalized to 10^{-2}. The core of the algorithm was a flow solver written to perform full numerical simulation of low Reynolds number channel flow *(18, 19)*. It has been tested and used extensively, demonstrating an ability not only to reproduce experimental results but to go beyond them in elucidating flow features not easily investigated in experiments.

The flow conditions for which the dimension was calculated correspond to a Reynolds number R_p, based on pressure gradient ($= |\nabla p| L^3/2\rho\nu^2, p$ = pressure, L = channel half-width, ρ = density, ν = kinematic viscosity) of 3200. This is below the value where the laminar flow becomes linearly unstable

(R_p = 5772), and in the region ($R_p \cong 2900$) where it becomes unstable to finite-amplitude, two-dimensional disturbances. The simulated flow is definitely chaotic and does a good job of predicting the skin friction on the channel walls *(20)*. However, it is a poorly resolved simulation in the sense that the computational grid in planes parallel to the channel walls is sparse (16 × 8), even though it is better resolved (33 points) in the direction perpendicular to the walls. Thus, it is not true turbulence. This Reynolds number is near the minimum for which turbulent channel flow can be sustained at these resolutions. Note further that the flow solver assumes periodicity in stream and cross-stream directions, so that this is not an open flow. As a result, the dimension is likely to be an increasing function of the size of the computational domain, as is found in studies of simple partial differential equations (*pde*'s) [Kuramoto-Sivashinsky *(11)*, Ginzburg-Landau *(21)*] on periodic domains.

Just such a change of size would be required to give the original flow simulation greater physical veracity, for its wave number spectrum extends neither high nor low enough. The restricted range of spatial scales resolved, as well as the small domain of the calculation, lead us to believe that our results are a lower bound on the true dimension of flow at this Reynolds number. However, proof of this contention awaits completion of calculations at both increased resolutions and domain size. Unlike results found in two-dimensional Bénard simulations *(22)*, we know that there is no "return to order" as resolution is increased at this Reynolds number.

Finally, the flow is absolutely, rather than convectively, unstable *(23)*. Thus, the calculated behavior is intrinsic, and may be subtly unlike that shown by an open flow in which the asymptotic state may or may not be dependent upon the stochastic forcing that originally excited it *(24)*. Since the flow simulation has already demonstrated its ability to reproduce both static and dynamic phenomena in experiments, it remains unclear what these subtleties might be.

The primary results of the calculation are displayed in Fig. 1, where the values of the Lyapunov exponents, λ_i, are plotted against their index, *i*. The first 450 exponents were calculated, at a cost of some 400 hours of CPU time, on a CRAY 2. Application of the Kaplan-Yorke formula to this distribution yields a dimension of $D_\lambda \cong 360$ and a metric entropy $h_\mu \cong 97$. The first 166 exponents are positive. The exponents with the higher indices converge most rapidly, and a strategy that averages the low index exponents to convergence before adding additional trajectories to the ensemble to calculate the high index exponents makes most efficient use of computational time. Though the computational grid is 16 × 33 × 8, and there are three velocity components at each node, particular features of the flow solver reduce the free nodes to 15 × 33 × 7, and the incompressibility of the flow means that only two of the velocity components are independent. Thus, there are 15 × 33 × 7 × 2 = 6930 degrees of freedom in the calculation, and the attractor dimension is roughly 6% of the dimension of the complete phase space.

The results described above have important implications for many current studies of fluid turbulence. Foremost, we believe that we have supplied strong evidence that turbulent solutions to the Navier-Stokes equations are confined to a finite-dimensional strange attractor, and thus temporal chaos in fluids results from the "sensitive dependence" mechanism intrinsic to such attractors. While the calculations have not been performed on "true" turbulence, the extension to such flows is strictly a mechanical problem, requiring more grid points and longer calculation times.

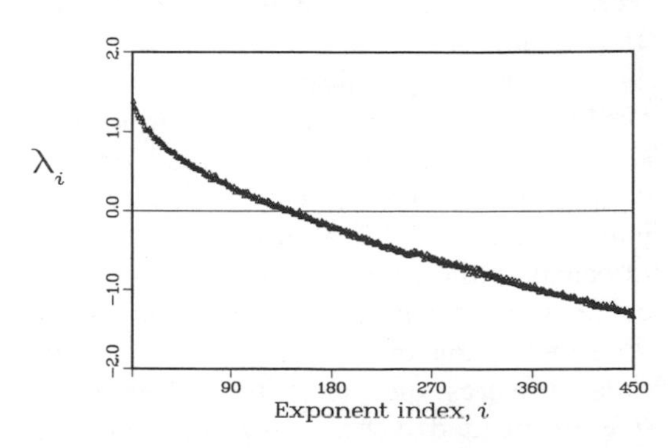

Fig. 1. Distribution of Lyapunov exponents, λ_i.

Such a study is currently under way at the same Reynolds number. Among other things, we hope this latter study will show how the dimension converges as more and more length scales are added.

Although the dimension of the attractor in the current study is finite, its magnitude places this dynamical system in an entirely different class from those analyzed experimentally in the Bénard or Taylor-Couette problems. For experimentalists the news is largely negative, for no available method of calculating dimension from measured data can handle such a high dimension. The number of points required for the "correlation" dimension *(13)*, if the scaling region is to extend over only a factor of 2, is 10^{108}! A billion points would give a scaling region within a radius variation of only 6%. Thus, there was no chance that the previous attempt *(15)* to calculate dimension by these methods could have succeeded. For those seeking a simplification of the turbulence problem by decomposing the flow onto special bases *(25, 26)*, the resultant dynamical system is still likely to be too complex for analysis. Simply extracting the fixed points of a 400th order system is a nontrivial task. And the subsequent analysis of phase space orbits in terms of these singularities is daunting even to contemplate. Thus, it seems unlikely, given the dimension of the attractor, that even a 40 mode truncation could get the basic, qualitative dynamics of the full system correct, though this remains to be demonstrated.

There is evidence *(27)* that postprocessing of correlation data in turbulent channel flow allows extraction of low order truncations that predict static distributions of second-order statistics. It is unclear that the dynamic exercise of these low order truncations *(26)* reproduces qualitative details of the complete system behavior. However, if the order of such a truncation is consistent with the known dimension of the attractor and the Whitney theorem *(28)*, the reduced system does reproduce global dynamics *(29)*. Thus, while truncated systems may lack dynamical veracity, they can still have a predictive power in engineering applications.

Connecting Coherent Structures and Strange Attractors

Historically, the view that turbulent flow contains structure has arisen from flow visualization studies. Two better known examples of this are the bubble pictures of Kim, Kline, and Reynolds *(30)* in the sublayer of a turbulent boundary layer, and the shadowgraph pictures of Brown and Roshko *(31)* in the mixing layer. In both cases, a recognizable sequence of events or physical structures, occurring irregularly, could be identified in the pictures. The belief that structures (or sequences) such as these were important to the flow dynamics led to the desire to extract their signature from the stochastic background for independent study. A problem in pattern recognition resulted, although the formalism and language of that field has rarely been applied *(32)* to the problem of defining a conditional sampling scheme.

In each case, investigators first construct a "feature" space from those characteristics they believe are unique to the event or structure. In this they are guided by their sense of what important physical processes are associated with a typical event and how best to identify such processes when they occur. Thus, choice of features includes an element of arbitrariness. In addition, this choice can be affected by the kind or amount of data *not* measured. (Readers are reminded of the story of the blind men examining an elephant. Each touches something the other does not; each conjectures the essential nature of the elephant to be that which he touched.) The result of this attempted characterization is a detection criterion that is a function of the features. When applied to available transducer signals, it announces occurrences of an event or structure.

Once an event's temporal origin has been determined by this scheme, it is aligned with transducer time histories from other events so that they can be ensemble averaged and the coherent signature extracted. Thus, development and application of a conditional sampling scheme proceeds according to the following sequence: choice of features that are functions of available transducer signals; development of a detection criterion based on the features; application of the criterion to determine event origins; ensemble averaging of transducer signals aligned on event origins.

In sampling schemes used to date, transducer signals have often been hotwire signals for the velocities U_{ij} (i denotes velocity component, j indexes the physical position of the measurement in an array). From these signals, M features f_n, have been constructed. Thus,

$$f_n = f_n(U_{ij}) . \tag{1}$$

Typical features have been time and space derivatives of the U_{ij}, as well as powers and short-term averages of these quantities. Choosing M features establishes an M-dimensional "feature" space.

The state of the flow field at the measurement array is characterized in time by a feature vector

$$\vec{f}(t) = \left\{ f_1(U_{ij}), f_2(U_{ij}), \ldots f_M(U_{ij}) \right\} \tag{2}$$

that sweeps out some trajectory in the feature space and whose components are the value of the features at each instant. The detection criterion, G, is a function of the feature variables. In simple cases, it can be written as

$$G(f_i) = 0. \tag{3}$$

Often this represents some surface in the feature space. When the feature trajectory $\vec{f}(t)$ crosses the surface $G(f_i) = 0$, an event or structure is signalled. The instant of crossing or some time referenced to the crossing becomes the event origin. It is used to align transducer signals or feature traces from different events for ensemble averaging. The result is a time trace "typical" of all qualitatively similar but randomly occurring events. Since averaging extracts a nonzero trace from a stochastic background, the average is considered to represent an underlying "coherent" structure, shorn of the incoherent fluctuation that obscured it

How do these notions relate to dynamical systems? Consider a system of ordinary differential equations (*ode*'s) (not necessarily finite)

$$\frac{d\vec{X}}{dt} = h(X) \tag{4}$$

$$\vec{X} = (X_1(t), X_2(t), X_3(t), \ldots) \tag{5}$$

With the initial condition

$$\vec{X}_0 = (X_1(0), X_2(0), X_3(0), \ldots),$$

the solution to the system (Eq. 4) can be written

$$\vec{X} = \Phi_{X_0}(\vec{X}) = \vec{\Phi}_{X_0}(t), \tag{6}$$

or

$$X_i = \Phi_i(t).$$

$\vec{X}$ is the phase space or "state" variable. Each point on the solution trajectory $\vec{X} = \Phi_{X_0}(t)$ completely characterizes the configuration of the system at an instant.

If the system of *ode*'s (Eq. 4) has only two members, the time asymptotic behavior of $\vec{X}$ has only three possibilities: steady state, periodic motion, divergence. If the system is dimension three or greater and is dissipative (the trace of the Jacobian matrix $[\partial h_i / \partial X_j] < 0$), then the asymptotic solution may be confined to a strange attractor in the phase space. Solutions confined to strange attractors are almost everywhere locally unstable but are globally stable to perturbations that do not cross the bounds of their basins of attraction. Thus, system states that are initially nearby diverge exponentially from each other for short times. Trajectories on strange attractors never repeat themselves exactly but ultimately will pass arbitrarily close to any previous portion of the trajectory.

Consider the character of a bundle of trajectory segments that all pass through some "window" in the space of the attractor. If the window is small enough, the system states occurring on each trajectory segment at the instant of passage through the window all will be similar. The sequence of states along each segment remains correlated with those along other segments for some time after (and before) passage through the window.

However, exponential divergence of nearby trajectories guarantees that this correlation time is finite. An ensemble average over trajectories passing through the window and aligned on the window passage time will yield a nonzero trace over finite time for any phase space component (or function of them) that has a zero mean value.

These concepts are illustrated in Figs. 2–4. The strange attractor that represents the asymptotic solution to the Lorenz system for a particular set of parameter values is pictured in Fig. 2a. The X-Z plane intersects the attractor transversely, and a large rectangular window has been outlined in this plane. Each time the solution trajectory passes through the window, the next 500 points in the time series of the Y component of the solution are traced in Fig. 2b. In Fig. 2c, the ensemble average of the 152 traces so obtained is displayed. Because the window is large, the selected traces are not substantially correlated with each other. Despite the fact their Y coordinates are identically zero at window passage, the range of X and Z values at intersection means that the system states (X, Y, Z) are "far" apart. In Fig. 3, the window area has been decreased by a factor of five, and the correlation between selected traces increases substantially. The ensemble average now clearly shows a nonzero value for finite time after passage. A further decrease in window size, now barely discernible in Fig. 4, again reduces the number of traces selected from a given record length but sharpens their correlation. The nonzero portion of the average increases both in duration and peak magnitude as a result.

It is easy to see that the "window" constitutes a conditional sampling criterion in the phase space of the Lorenz attractor. Here the "features" are the primitive variables (X, Y, Z) themselves, and the detection criterion, though not in the form of Eq. 3, is a simple function of the features, namely,

$$\begin{aligned} &Y = 0. \\ &a \le X \le b \\ &c \le Z \le d. \end{aligned} \tag{7}$$

The event criterion (Eq. 7) simply chooses the bundle of trajectories on the attractor that are ensemble averaged. If the detection criterion is too broad, selected trajectory segments are uncorrelated to begin with, and averaging fails to extract a structure. Tightening the selection criteria (i.e., decreasing the window size) increases the likelihood that trajectory segments are initially correlated, but exponential divergence of trajectories on the attractor guarantees that their mutual correlation time is finite. At times beyond the correlation limit, the ensemble average of the trajectories will be zero. If the size of a structure is defined as the length of nonzero trace that results from averaging, then the tighter the selection criterion, the greater the size of the educed event. Given the minimum distance, Δ, between trajectories at the event window, information theory *(33)* suggests that an upper bound on the event size or duration is given by a constant times the logarithm of $1/\Delta$. This separation between trajectories must be calculated in the full phase space of the dynamical system, not in some reduced dimension feature space.

In light of the previous paragraphs, eduction of coherent structures by conditional sampling can be viewed as the following set of operations on the full solution attractor of the Navier-Stokes equations. First, discretization of the flow field, by making a finite set of measurements, collapses the full attractor onto a reduced dimension subspace. Second, a feature space is constructed out of components from this subspace or functions of them. The number of features selected determines the dimension of the new space, which may actually be higher than that resulting from the discretization. This new space contains a finite size object to which the feature trajectory is confined, but this object is not an attractor in the rigorous sense. Third comes construction or specification of an event criterion in the feature space. The criterion is a subset of the feature space (e.g., curve, surface, surface patch) but may be infinite in extent. Fourth, the intersection points between the feature trajectory and the event criterion are determined. Finally, an ensemble average of primitive or feature traces aligned on these intersection points is obtained.

Specification of the event criterion in feature space creates a corresponding criterion in the full space of the attractor. It is there that the exponential divergence property has its effect, decorrelating system states both there and in the feature space. Carried to its logical extreme, it is clear that *any* sufficiently restrictive sampling criterion will extract a structure

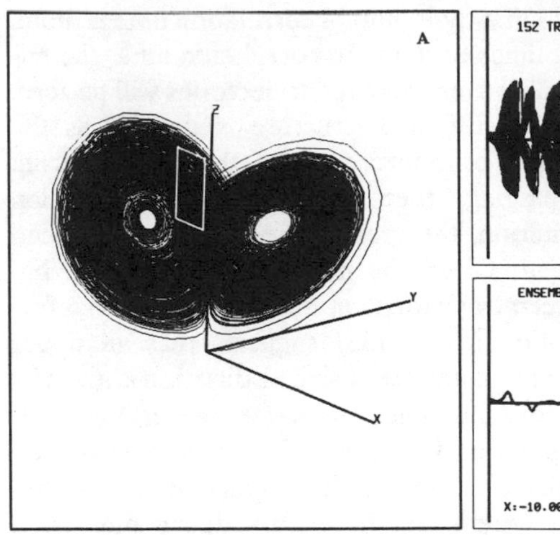

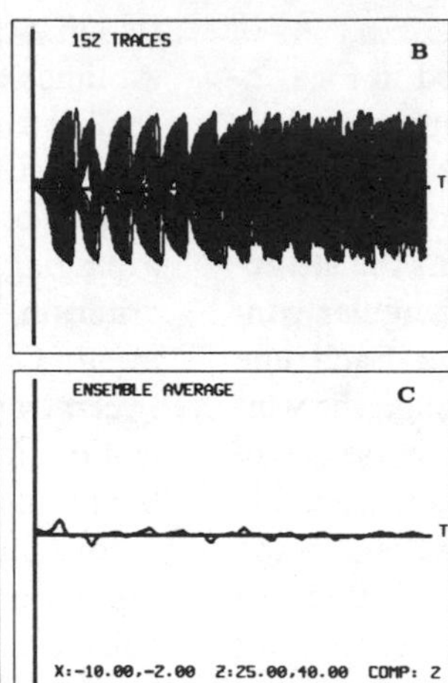

Fig. 2. (**A**) Lorenz attractor with window $-10 \leq X \leq -2$, $25 \leq Z \leq 40$. (**B**) Aligned Y traces. (**C**) Ensemble average of Y traces.

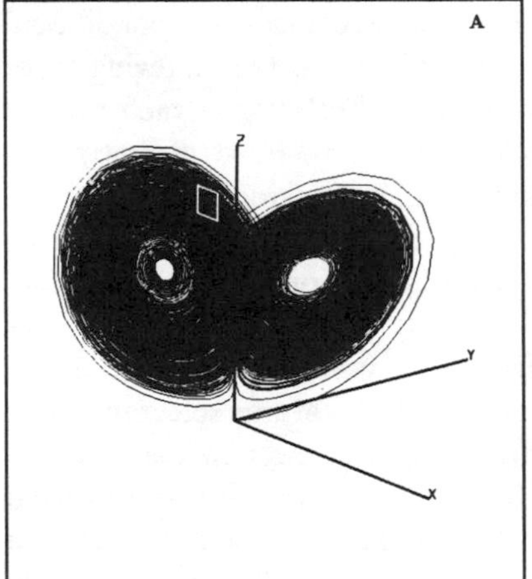

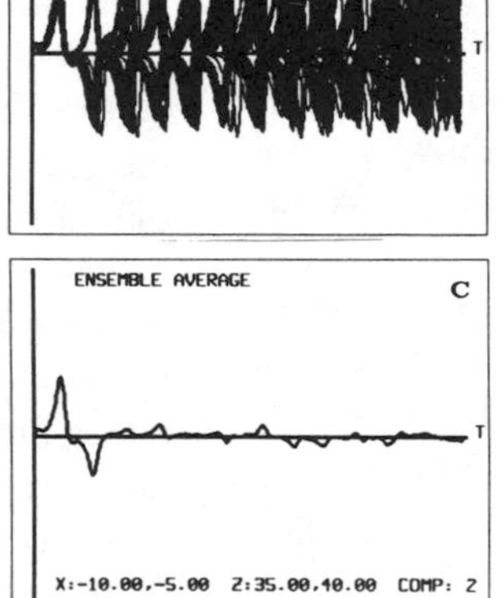

Fig. 3. (**A**) Lorenz attractor with window $-10 \leq X \leq -5$, $35 \leq Z \leq 40$. (**B**) Aligned Y traces. (**C**) Ensemble average of Y traces.

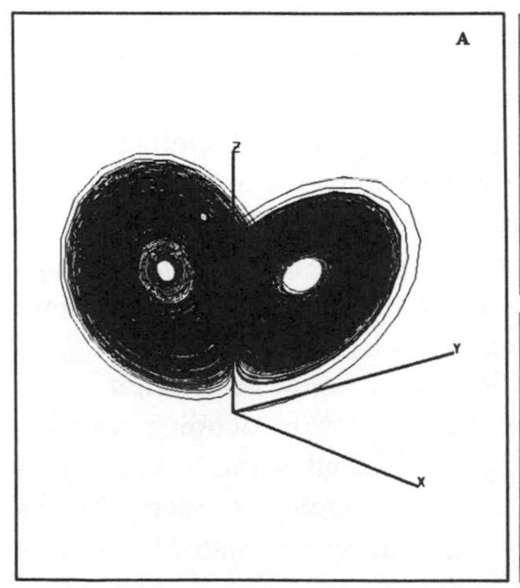

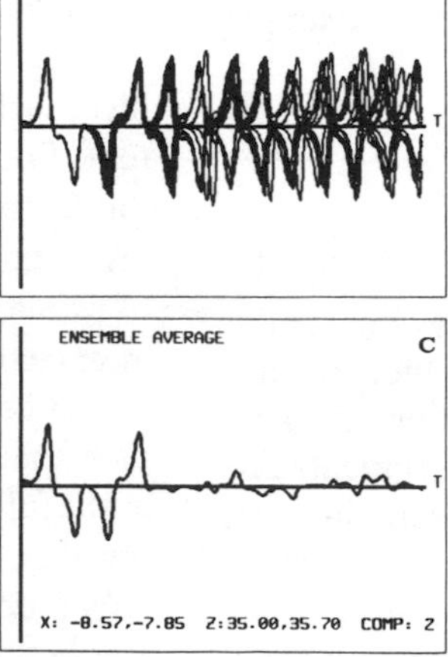

Fig. 4. (**A**) Lorenz attractor with window $-8.57 \leq X \leq -7.85$, $35 \leq Z \leq 35.7$. (**B**) Aligned Y traces. (**C**) Ensemble average of Y traces.

from a dynamical process confined to a strange attractor. Arbitrariness in the selection process begs the question of significance. There *is* structure in turbulence. However, it is up to each investigator to supply the physical arguments that raise their ensemble average to the status of a structure that contributes significantly to the dynamics of the flow.

Thus, the assumption of an underlying strange attractor is completely consistent with the observation and eduction of coherent structures by experimentalists. Indeed, it appears that the eduction of these structures depends strongly upon one of the fundamental properties of such attractors, the exponential divergence of nearby states for short times.

Acknowledgments

L. K. was supported, in part, by the National Research Council and by the Air Force Office of Scientific Research contract AFOSR-88-0056.

References

1. Gollub, J. P., and S. V. Benson, *J. Fluid Mech.* **100,** 449 (1980).
2. Libchaber, A., S. Fauve, and C. Laroche, *Physica* **7D,** 73 (1983).
3. Gorman, M., P. J. Widmann, K. A. Robbins, *Physica* **19D,** 253 (1986).
4. Moore, D. R., J. Toomre, E. Knobloch, N. O. Weiss, *Nature* **303,** 663 (1983).
5. Busse, F. H., in *Hydrodynamic Instabilities and the Transition to Turbulence* (Springer-Verlag, New York, 1981), p. 292.
6. Brandstater, A., and H. L. Swinney, *Phys. Rev. A* **35,** 2207 (1987).
7. Kaplan, J. L., and J. A. Yorke, in *Functional Differential Equations and Approximations of Fixed Points,* H. O. Pietgen and H. O. Walther, Eds., (Springer-Verlag, New York, 1979), p. 228.
8. Osledec, V. I., *Trans. Moscow Math. Soc.* **19,** 197 (1968).
9. Benettin, G., L. Galgani, A. Giorgilli, J.-M. Strelcyn, *Meccanica* **15,** 9 (1980).
10. Farmer, J. D., *Physica* **4D,** 366 (1982).
11. Manneville, P., *Proc. Conf. Macroscopic Modelling of Turb. Flows, Lecture Notes in Physics* **230** (Springer-Verlag, New York, 1985), p. 319.
12. Farmer, J. D., E. Ott, J. A. Yorke, *Physica* **7D,** 153 (1983).
13. Grassberger, P., and I. Procaccia, *Physica* **9D,** 189 (1983).
14. Constantin, P., and C. Foias, *Comm. Pure Appl. Math.* **38,** 1 (1985).
15. Brandstadter, A., H. L. Swinney, G. T. Chapman, in *Entropies and Dimensions in Chaotic Systems,* G. Mayer-Kress, Ed. (Springer-Verlag, New York, 1986), p. 150.
16. Eckmann, J.-P., and D. Ruelle, *Rev. Mod. Phys.* **57,** 617 (1985).
17. Broomhead, D. S., and G. P. King, *Physica* **20D,** 217 (1986).
18. Moin, P., and J. Kim, *J. Fluid Mech.* **118,** 341 (1982).
19. Kim, J., P. Moin, R. Moser, *J. Fluid Mech.* **177,** 133 (1987).
20. Rozhdestvensky, B. L., and I. N. Simakin, *J. Fluid Mech.* **147,** 261 (1984).
21. Keefe, L., *Phys. Lett. A* **140,** 317 (1989).
22. Curry, J. H., J. R. Herring, J. Loncaric, S. A. Orszag, *J. Fluid Mech.* **147,** 1 (1984).
23. Huerre, P., and P. Monkewitz, *J. Fluid Mech.* **159,** 151 (1985).
24. Deissler, R. J., *Physica* **25D,** 233 (1987).
25. Lumley, J. L., in *Transition and Turbulence,* R. E. Meyer, Ed. (Academic Press, New York, 1981), p. 25.
26. Aubry, N., P. Holmes, J. L. Lumley, E. Stone, *Sibley School of Mechanical and Aerospace Engineering Report FDA-86-15* (Cornell University, Ithaca, NY, 1986).
27. Moin, P., *AIAA-84-0174* (1984).
28. Whitney, H., *Ann. Math.* **37,** 645 (1936).
29. Sirovich, L., and J. D. Rodriguez, *Phys. Lett. A* **120,** 211 (1987).
30. Kim, H. T., S. J. Kline, W. J. Reynolds, *J. Fluid Mech.* **50,** 133 (1971).
31. Brown, G. L., and A. Roshko, *J. Fluid Mech.* **64,** 775 (1974).
32. Blackwelder, R. F., *Proc. Dynamic Flow Conf.* (Skovlunde, Denmark, 1979).
33. Shaw, R. S., *Z. Naturforsch.* **36a,** 80 (1981).

7

Fluid Dynamical Chaos in Vortex Wakes

Charles W. Van Atta

Abstract

Recent experiments show that temporal chaos and quasiperiodicity in vortex-shedding laminar wake flows behind two- and three-dimensional cylindrical and airfoil shapes is produced by competing vortex-shedding modes. These observations provide some clues as to how one might extend dynamical systems concepts to real fluid turbulence. The observed temporal chaos is produced by coherent large-scale vorticity patterns produced by the collective interaction of arrays of concentrated vortices due to the effects of vortex stretching, tilting, dislocation, and reconnection. The "mode competition" of different vortex patterns can be generated by unsteady boundary conditions (e.g., hydroelastic vibrations or intentional imposed forcing) or by spatial variations in body geometry and consequent spatially varying rates of wake vortex shedding.

The collective stretching, deformation, dislocations, and reconnections produced in the evolving vortex arrays sometimes arranges them into compact spatial patterns with a scale much larger than individual vortices. The convection of these vortex structures past a fixed probe gives rise to temporal chaos in the velocity field observed by a probe at that point. In complying with Helmholtz's vorticity principles, suitably modified by viscous effects, the vortices undergo stretching, tilting, dislocation, and reconnection, which generically produces pattern formation within the arrays with a scale an order-of-magnitude larger than the original spacing between vortices. The resulting vorticity fields are sufficiently complex that at some spatial locations in the fluid a continuous spectrum of frequencies is observed in the velocity field at a single point, a hallmark of temporal chaos.

In the same flow, in addition to chaotic motion in some regions, quasiperiodicity or mode locking may be observed simultaneously in other regions. This produces a spatial intermittency, where the intermittency is a function of the nature of the flow observed at a particular location. In the chaotic regions, excitations of the fluid are present over a wide range of scales, while in quasiperiodic or singly periodic regions only a few discrete scales may be excited. These chaotic laminar flows are not turbulent in the usual sense.

However, they bear similarity to the spatial intermittency encountered in turbulent flows, in which the fine scale vorticity, and hence the dissipation rate, exhibit a large spatial variability. The flows may thus constitute a useful paradigm for helping to define and to eventually understand the physics of spatial and temporal chaos in turbulent flows. The results reinforce the recurrent idea that a detailed and systematic description of the topology of vorticity fields in fully developed turbulent flows may reveal some essential physics and suggest that it may also shed some light on how to develop a chaos-like theory for fluid turbulence.

The Application of Chaotic Dynamical Systems Concepts to Fluid Dynamics

The analysis and interpretation techniques of dynamical systems approaches are currently being applied to geophysical fluid dynamics problems in oceanography and meteorology, and parallel efforts are being carried out in relation to laboratory fluid mechanical investigations. The wide range of possible applications can be appreciated by examining the references given at the end of this chapter. Vallis *(1, 2)* finds that simple models of the El Niño can exhibit chaotic behavior. Palmer *et al.* *(3)* find that bottom topography can produce chaotic behavior of sound waves transmitted in the ocean, and Samuelson and Allen *(4)* find limit cycles, period-doubling sequences, and chaos in a barotropic model for topographically generated mean flow over the continental margin. With respect to other basic problems in fluid mechanics, the main impact of dynamical systems scenarios has been confined to the routes to chaos in closed systems such as Bénard convection and Taylor-Couette flow. A succinct review of these successes has been given by Miles *(5)*.

Similar direct evidence for transition to turbulence in the open systems most often encountered in many engineering and geophysical applications has not been forthcoming, and the possible relation to fully developed turbulent flows is tenuous and not well defined even in principle. Recently, however, the ideas of temporal chaos have been applied to a fully turbulent open flow system, the wall region of a turbulent boundary layer *(6)*. This is a relatively low Reynolds-number region characterized by relatively limited spatial complexity and known to be dominated by coherent structures in the form of streamwise rolls. The velocity field in the wall region was expanded in empirical orthogonal eigenfunctions, obtained from measurements in the wall region of a pipe flow. In the wall region, one term is sufficient to reproduce some 80% of the fluctuating turbulent energy.

By Galerkin projection, a set of ordinary differential equations was obtained for the coefficients in the expansion, which are functions of time. The expansion was truncated with one eigenfunction and six cross-stream Fourier modes (only five of which are active; the zero mode decays to zero). No streamwise structure was included. It was necessary to model the energy loss to unresolved modes by an eddy viscosity. By applying the methods of dynamical systems theory, Aubry *et al.* found that as the value of the eddy viscosity dropped, first steady rolls were observed, then traveling waves (across stream), and finally intermittent "bursting," followed by an apparently chaotic regime.

Fluid Turbulence

Understanding and subsequent exploitation or avoidance of fluid turbulence is one of the major problems in many fields. To mention only one of these, in naval applications turbulence controls drag forces and noise on submersibles, and turbulent mixing determines the environment for propagation of underwater sound.

Turbulence is random and unpredictable on small scales, yet more organized and coherent on larger scales, and thus has traditionally been treated and analyzed by a combination of statistical and semi-deterministic methods. The chaos seen in many dynamical systems governed by ordinary differential equations or mappings is also unpredictable and can exhibit structure and scaling on many scales. Can we use it to help describe fluid turbulence?

We know that fluid turbulence is not simply temporal chaos in the sense of the definition used for dynamical systems. Turbulence obeys a set of partial, not ordinary, differential equations, with three space variables in addition to time. Also, an essential feature of real turbulence is a form (as yet not rigorously defined) of spatial chaos, perhaps in combination with a form of temporal chaos. However, the observation in real fluids of several classically unsuspected dynamical systems scenarios for transition to chaotic behavior as well as temporal chaos suggests that an extension of the dynamical systems ideas and techniques to include spatial variations may lead to vital insights into fluid turbulence. At present, it is not clear how one might proceed. It is clear that it may be a very difficult process, probably requiring additional discoveries, perhaps as equally revolutionary as those that led to the current interest in deterministic chaos.

The purpose of the present chapter is to

introduce the reader to some of the issues and to demonstrate a possible connection between the spatial intermittency in fully developed turbulent flows and chaotic vortex interactions produced by sufficiently complex vortex interactions in laminar wake flows.

Turbulence and Chaos in Fluid Dynamics

Fluid dynamicists want to know if adaptation and extension of the ideas that have arisen from the observation of chaotic solutions of dynamical systems problems can help to "solve" the "turbulence problem," i.e., understand particular turbulent flows in a more incisive way [see *(7)*, for example]. We know that *fluid turbulence* is not simply *chaos* in the sense of dynamical systems. Chaotic solutions have clearly given great insight into a variety of simpler (nonfluid) mechanics problems, describable in terms of ordinary differential equations, as discussed, for example, in the books by Moon *(8)* and Thompson *et al. (9)*. But fluid turbulence obeys the partial differential equations of Navier and Stokes, and, not surprisingly, progress in utilizing the concepts of chaos in dynamical systems to illuminate the behavior of fully turbulent flows has been slow and uncertain, suggesting that the wrong questions have probably been asked so far.

Definitions of Chaos

To discuss chaos in fluids, we first need to define what kind of chaos we are talking about. Here I want to discuss the occurrence of temporal or Eulerian chaos, and not Lagrangian chaos. Lagrangian chaos occurs when particle trajectories in a fluid are chaotic. This can occur even when the velocity and vorticity fields are not themselves chaotic. As shown by Aref *(10)*, such sensitive dependence on initial conditions, or "chaotic advection," can occur in two-dimensional unsteady flow, but not in steady two-dimensional flow.

Chaotic advection had already been analyzed in the context of steady three-dimensional flows by Arnold *(39)* and Hénon *(11)*. Oceanographers have frequently observed that two drifter floats put in at nearly the same points in the ocean could take very different paths. If two floats were launched a short distance apart at Santa Barbara, after drifting for a month one could end up near Seattle and the other off Baja California. Temporal chaos occurs by definition only in unsteady flows, which can be laminar as in deterministic chaos or, presumably, turbulent in the traditional sense of this term. An essential feature of turbulence is the order-of-magnitude increase in transport of momentum and heat compared with a corresponding laminar flow at a comparable Reynolds number, perhaps just below transition. Whether or not a chaotic laminar flow can have comparable transport is an interesting question which has not yet received the attention it deserves, although Ottino *(12)* indicates that chaotic transport is much smaller than turbulent transport.

Temporal Chaos

The notion of temporally chaotic fluid flow was suggested by Lorenz's observation of chaotic solutions of systems of first-order ordinary differential equations, or "dynamical systems" *(13)*. Some other theoretical milestones in this development are the Ruelle and Takens generic route to chaos *(14)*, the Feigenbaum period-doubling route *(15)*, and the intermittent behavior of Manneville and Pomeau *(16)*. Some work initially stressed the contrast of the transition scenarios with Landau's conjecture of an infinite cascade of successive incommensurate instability frequencies leading to turbulence *(17)*. The fact that Landau's conjecture was long known not to correspond with actual transition in many situations (e.g., boundary layer instability, pipe flow, mixing layers, and jets) appears to have been overlooked by many contemporary authors.

One intrinsic property of all real turbulent flows is their spatial intermittency. There are regions of very strong vorticity and large dissipation spatially adjacent to much quieter regions. Probability densities of the rate of dissipation of turbulent kinetic energy (a positive definite quantity heavily weighted by the smallest scales of the turbulence) have long tails and are roughly lognormal. Since fluid turbulence is spatially chaotic or "intermittent" as well as temporally chaotic, turbulence is obviously not chaos in the restricted sense of dynamical systems. To extend the concept, we need to find a way to characterize spatial intermittency. Spatial intermittency in dissipation is closely connected with spatial variations in the vorticity of the fluid, and this may be visualized, following many earlier workers, as associated with the evolution of interacting vortices embedded in the fluid. Before discussing how the recent experimental evidence might guide us in characterizing the intermittency, let us finish the discussion on purely temporal chaos.

The amazing thing about the dynamical systems scenarios of temporal chaos was that experimentalists studying the transition of laminar to turbulent flow actually observed some of the scenarios for increasing complexity or transition to chaotic behavior in their experiments. These include the observation of a Ruelle and Takens-type transition in circular Couette flow by Gollub and Swinney *(18)*; Gollub and Benson's observations of periodic doubling, quasiperiodic, and intermittent routes to transition in Bénard convection, Libchaber and Maurer's observation in Bénard convection of period doubling obeying Feigenbaum universality, with quantitative numerical agreement in both the accumulation point and spectral values *(20)*, and the observation by Brandstater *et al.* *(40)* of a "strange attractor" in the phase space of circular Couette flow. These scenarios for transition occurred only in a very limited parameter space of geometrical or fluid property parameters, with other classical routes to transition occurring outside these ranges as found by many other earlier investigators.

Bénard convection and circular Couette flow both occur in a fully bounded region, a "closed" flow. The fluid being studied always consists of the same fluid particles, which recirculate within the volume, carrying information about their entire previous time history with them. In the open fluid mechanical systems to which we are so accustomed in aerodynamics, meteorology, and oceanography, new fluid particles with no previous memory of the flow continually enter the system, while others leave it. Dynamical systems theory does not differentiate between open and closed fluid systems and so predicts the same scenarios for both. Experimental attempts to find Ruelle-Takens and period-doubling transition scenarios in unforced open systems have been unsuccessful to date. However, the recent computer simulation observation of period doubling via a three-dimensional oscillation in an untapered cylinder wake at Reynolds numbers near 300 by Karniadakis and Triantafyllou *(37)* may revive interest in this subject.

What are the intrinsic differences between the stability properties of fluid motion in closed and open systems? One clue may be that closed systems are always "absolutely" unstable but open systems can be either "convectively" or "absolutely" unstable. An instability in the closed system is felt everywhere in the fluid, while a convective instability in an open system may be confined to a temporally and spatially restricted region of time and space. A convective instability may grow continually as it propagates to a new region, or it may die out as it moves. For a full discussion of these ideas, the reader may consult Huerre and Monkewitz *(21)*.

The Search for Temporal Chaos in Open Fluid Systems

Where should one look for the manifestations of chaos in fluids? As mentioned earlier, the essential physics of many fluid problems is well described by looking not simply at the velocity field but at the vorticity, which is the curl of the vector velocity, or at the behavior of well-defined vortices when they comprise a major component of the flow field.

Chaos can be observed in the very simple idealized two-dimensional laminar fluid flows comprised of collections of irrotational point vortices. The flow produced by up to three vortices cannot be chaotic, but flows produced by four or more vortices can be chaotic. Such two-dimensional flows do not possess the vortex stretching essential to turbulent flows. However, if idealized vortices in two dimension can produce chaotic flow, it seems that an even richer structure should be anticipated for interactions of three-dimensional vortices. The following examples show that this is no idle conjecture.

Three-Dimensional Chaos and Spatial Intermittency in Vortex Wakes

The temporally periodic "vortex street" wake flows behind bluff bodies or airfoils seem to be a natural place to look for transitions to chaotic behavior. First, the flow is temporally periodic due to the periodic formation and shedding of well-organized and nominally two-dimensional vortex lines in the wake. The formation of these vortices is intimately connected with the fact that the mean wake flow has a considerable region of absolute instability in the near wake, which, by analogy with closed system behavior, might be expected to enhance the possibility of observing some of the canonical dynamical systems routes to chaos.

While scenarios for chaotic behavior of such three-dimensional collections of vortex lines have not yet been theoretically constructed, several recent experiments on vortex shedding in wakes of two- or three-dimensional cylinders, cones, and airfoils have uncovered quasiperiodic and chaotic behavior produced by mode competition between different patterns of vortex shedding. It appears that understanding the reason for chaotic behavior of these flows might be an important or, perhaps, even necessary step to approaching similar questions for fully developed turbulence. Before discussing these experiments, for conceptual purposes it is useful to recall some basic properties of idealized inviscid three-dimensional vortices, due to Helmholtz.

Vortex shedding from a single circular cylinder is a flow which has a long history [see the review by Blevins *(22)*]. For flow with velocity, U, past a cylinder of diameter d, the nature of the flow is controlled by the value of the (dimensionless) Reynolds number, Re = Ud/ν, where ν is the kinematic viscosity of the fluid. When the Reynolds number is greater than about 40, vortices are periodically shed from the cylinder and the wake flow is unsteady.

When the cylinder is motionless, i.e., does not vibrate or is not otherwise driven in any way, then only a single frequency of fluid oscillation is observed. As shown recently by Williamson *(23)*, the frequency observed depends on whether the consecutive parallel rows of vortices are shed parallel or nonparallel to the cylinder. The vortex shedding can produce self-excited oscillations of the cylinder at its natural vibration frequencies. The vortex-shedding frequency can then become locked to the vibration frequency over a certain range of velocity, which depends on the hydroelastic structural parameters as well as the flow parameters.

Generation Mechanisms for Quasiperiodicity and Chaos

Aeroelastic-response forcing

From his hot-wire measurements in cylinder wakes, Sreenivasan *(24)* proposed that his observations of quasiperiodicity and chaotic behavior were a manifestation of the Ruelle-Takens-Newhouse (RTN) transition to chaos. Van Atta and Gharib *(25)* then showed experimentally that RTN transition was not the cause, but that the quasiperiodicity and chaos were produced by three-dimensional aeroelastic interactions due to high-order–harmonic cylinder vibrations. The flow visualization results of Van Atta *et al.* *(26)* showed that, in certain cases, the chaotic regions were compact, soliton-like disturbances that were periodic in space and time. The full topology of complex vortex interactions leading to the observed pattern formation is under study.

External forcing of body oscillations

Olinger and Sreenivasan oscillated a cylinder in its fundamental mode of vibration, thus modulating the wake at a second frequency differing from that of the natural vortex-shedding frequency *(27)*. At rational values of the frequency ratio, the system displayed Arnold tongues due to phase locking, leading to a behavior approximating the devil's staircase along the critical line. Associated with quasiperiodic transition at the golden mean frequency ratio, spectral peaks appear at various Fibonacci sequences. These, as well as the "singularity spectrum" of the Poincaré section of the attractor at the critical golden mean point, are largely similar to the universal characteristic of the sine circle map, a behavior which has been observed by Jensen *et al.* in Rayleigh-Bénard convection and in a number of other systems *(28)*. No broadened spectra of the type found in the aeroelastically excited cases were reported, and the relative simplicity of the spectra suggests that the topology of the vortex interactions, which was not investigated, is probably much simpler than in the self-excited cases of Van Atta and Gharib *(25)* and Van Atta *et al. (26)*.

External forcing of wake via boundary layer instabilities

Williams-Stuber and Gharib *(29)* used the strip heater technique in a water tunnel to externally force the wake of an airfoil operated at a low Reynolds number in the laminar vortex-shedding range. Strip heaters on the top and bottom of a symmetrical airfoil were used to introduce Tollmein-Schlichting waves into the top and bottom boundary layers on an airfoil. The waves were subsequently amplified downstream and served to force the vortex-street wake. Within a certain range of frequencies, the wake velocity signal could be locked to a single heater frequency or made quasiperiodic, using two different frequencies with two forcing heaters.

The possibility of a RTN-like transition was investigated using three strip heaters. The wake frequency was locked with one heater frequency, and then two other frequencies were imposed, lying just outside the locking range. In one case, this procedure produced a sharp increase in the continuous part of the velocity spectrum, possibly indicating a RTN-like transition to a chaotic state. An attempt to characterize the chaotic and quasiperiodic states in terms of dimension and Lyapunov exponents was unsuccessful, a result that was attributed to the inherently higher noise level of such an open system as compared with the closed ones in which such diagnostics have been successfully carried out.

Topological defects, quasiperiodicity, and chaos produced by spatial variations in body geometry

Quasiperiodic and chaotic flow can also be produced by the variations in local vortex shedding induced by the variable geometry of the body from which wake vortices are shed. The simplest paradigm for continuous spatial variation of vortex-shedding frequency is a tapered cylinder or cone.

Exploratory hot-wire measurements by Van Atta and Piccirillo *(30)* show that vortex shedding behind uniformly tapered cylinders with taper ratios in the range 32:1–13:1 is two-frequency quasiperiodic or chaotic, depending on the value of the local spanwise-dependent Reynolds number. For quasiperiodic shedding, the modulation frequency of the vortex shedding is found to be equal to the frequency difference between equally spaced multiple-shedding–frequency spectral peaks. The relative energies, but not the frequencies, of these discrete peaks evolve in a continuous fashion with spanwise location, implying a long range spanwise correlation over most of the cylinder. Spanwise interactions are thus much stronger than in the cellular shedding observed by Gaster for mildly tapered cylinders, for which adjacent cells had their own individually different, but cellwise-constant, shedding frequencies. Comparison with flow visualization pictures show that chaotic regimes may be associated with topological defects in the vortex pattern. These results will now be discussed in more detail to illustrate how three dimensionality of steady boundary conditions leads to interesting new quasiperiodic and chaotic flow phenomena.

Vortex Puzzles Posed by Spatially Varying Vortex-Shedding Rate

Vortex shedding from a two-dimensional circular cylinder at low Reynolds numbers (40–200) produces a periodic "vortex street" array of parallel vortex filaments whose velocity signature at a fixed point is characterized by a spectrum consisting of a single sharp peak and its harmonics. When, as in the present case, the cylinder is tapered, the vortex-shedding frequency f will naturally vary as a function of spanwise coordinate z so that the vortex lines cannot be continuous along the full span as in the two-dimensional case, and dislocations in the vortex structure can be anticipated. This leads to complex three-dimensional vortex patterns in which curved vortex lines may be shed from distinct cellular spanwise regions. Mode competition is thus generic in the tapered cylinder wake, as is the possibility of RTN transition.

The general flow patterns and vortex shedding from conical objects depend strongly on the value of the included cone angle. When the angle becomes quite large, as in flows around volcanoes and seamounts, the patterns, as investigated by Barnett *(31)* and Brighton *(32, 33)*, are much different than those for the relatively slender shapes of interest here. The only previous studies of vortex shedding from slender cones are those of Gaster *(34, 35)*.

From Gaster's *(34)* water tunnel experiments on two cones of taper ratio 36:1 and 18:1, the local Roshko number fd^2/ν, where $d(z)$ is the local diameter and ν the fluid kinematic viscosity, appeared to be a universal function of the local Reynolds number Ud/ν, where U is the free-stream velocity. Gaster concluded that the frequency of shedding was controlled by the local diameter and had a slightly lower value than that for a nontapered cylinder of the same diameter. Gaster found that the shedding process was modulated by a low-frequency oscillation, which he concluded was dependent only on U^2/ν and independent of any physical dimension of the cylinder. For a slightly tapered (120:1) cylinder, Gaster *(35)* found that, at the lowest Reynolds numbers, the velocity signals were singly periodic in distinct cells along the span except at cell boundaries, where the signals had low-frequency modulations characteristic of the more highly tapered cylinders, with a modulation frequency precisely equal to that which would be produced from a summation of the two velocity signals of differing frequency in neighboring cells. At higher speeds, his hot-wire signals were similar to those observed on steeper cones, with modulated signals occurring over the entire span. The present experiments were undertaken to investigate the origin of the quasiperiodic behavior associated with the low-frequency modulation observed by Gaster and possible connections with dynamical systems theory for open fluid mechanical systems.

Experimental Arrangement for Cone Studies

The experiments were carried out with tapered cylinders placed in the exit plane of a small open circuit wind tunnel. The tapered cylinders used consisted of three different uniformly tapered stainless steel cones, 2.5 cm in length, with taper ratios of 32:1, 22:1, and 13:1. The geometrical parameters of the cylinders were measured both with micrometer tools and with the aid of a microscope and ruled steel scale. The tip and base diameters ranged from 0.015 to 0.03 cm, and 0.07 to 0.18 cm, respectively. Each of these slender cones was mounted on an airfoil cross-sectioned strut supported outside the tunnel, with a 1-cm diameter circular endplate at the base of the cone, separating it from the supporting strut. The fluid velocity was measured with a hot-wire probe positioned by a micromanipulator traverse located outside the tunnel. The hot-wire was placed between 5 and 10 diameters downstream of the cone and in the center of the row of vortices shed from one side of the cone. A hot-wire is a small (0.0001-inch diameter) velocity sensor that is heated electrically and cooled aerodynamically. The hot wire was operated with a DISA 55M10 anemometer circuit. The freestream velocity U was measured with a small pitot-static tube and, as a check, from the vortex-shedding frequency of a second nontapered cylinder also mounted in the test section exit plane. Power spectra of the velocity signals were measured on-line with a Spectral Dynamics SD 380 Signal Analyzer.

Van Atta and Gharib *(25)* found that even very small cylinder vibrations can have a profound effect on the vortex-wake structure. A vibration detector of the type used in their study was used in some preliminary experiments to check for such effects in the present work. No vibrations of the tapered cylinders were detected under any of the conditions employed.

For flow visualization, a smoke wire located three base diameters downstream of the cone passed through the strut and cone endplate and was held tautly in place by two circular endplates. A controller sent a pulse to the wire and a delayed pulse to a strobe backlighting the flowing smoke, which was then photographed with a camera.

Results

Quasiperiodic velocity spectra

The visual appearance of the hot-wire signals was generally very similar to that reported by Gaster for cones with similar taper ratios, but the present spectral analysis revealed a number of essential features not observed by him. For a fixed freestream velocity, the nature of the signals evolved with *z*, the spanwise coordinate with origin at the tip of the cone. At and near the tip, no vortex shedding was observed. With increasing *z*, the appearance of the hot-wire signal indicated low frequency modulated vortex shedding as reported by Gaster.

Spectral analysis of the velocity signals revealed some unexpected features. The spectra typically did not simply contain a single vortex-shedding peak, as would be inferred from Gaster's description of his results, but, instead, the spectra contained multiple high-frequency peaks near the expected shedding frequency. An example in which the two central "shedding peaks" have equal amplitudes is shown in Fig. 1. The low-frequency modulation frequency is seen to be equal to the frequency difference between two adjacent high-frequency peaks. Typical evolution of the velocity power spectra with spanwise location is illustrated by the example in Fig. 2. The spectra at all locations contain a discrete series of strong peaks at a group of discrete frequencies that are independent of spanwise location. The most energetic peaks are a single dominant or two neighboring dominant high-frequency, vortex-shedding peaks and a less energetic but, nevertheless, dominant peak at a much lower (modulation) frequency.

The frequencies of the vortex-shedding peaks are all simply related to one another by adding or subtracting an integer multiple of the modulation frequency. Moving from the tip of the cone toward the base, as *z* and, therefore, *d(z)* increases, each previously

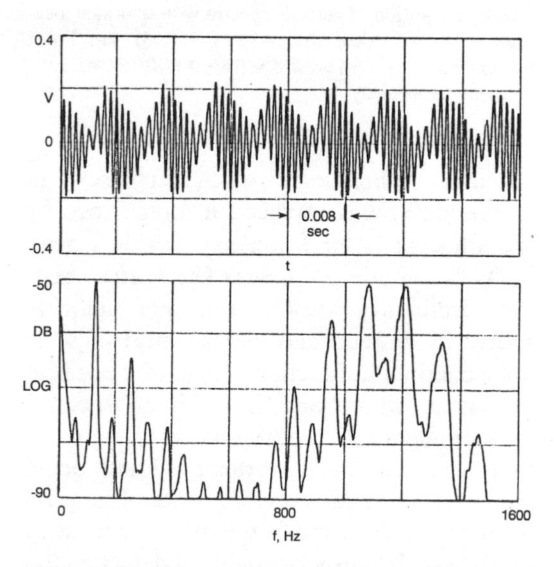

Fig. 1. Example of quasiperiodic hot-wire velocity signal and corresponding spectrum. U = 3.64 m/sec, Re = 118, *d(z)* = 0.03 + 0.031*z* cm.

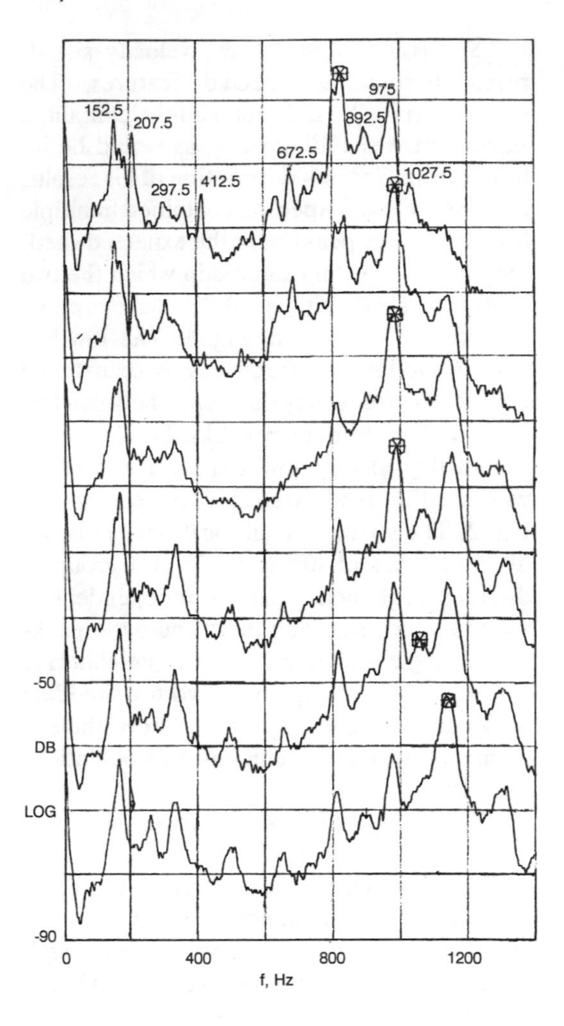

Fig. 2. Evolution of velocity spectra with spanwise location. U = 3.44 m/sec, $d(z) = 0.03 + 0.031z$ cm. From bottom to top, spectra are at $z = 0.66, 0.74, 0.76, 0.81, 1.02, 1.12$ cm, respectively.

dominant shedding peak decreases in amplitude and is replaced in dominance by the adjacent lower frequency shedding peak as its amplitude increases. The vortex-shedding frequency itself is thus not uniquely spectrally defined and in a spectral sense is not a continuous function of spanwise position z. Gaster, who made very limited spectral measurements, did not report observing more than one dominant vortex-shedding peak. However, the peak-counting method he used to measure shedding frequencies for most of his data is obviously incapable of detecting the presence of multiple spectral peaks.

It is remarkable that the vortex-shedding peaks are evenly spaced at multiples of the modulation frequency. The constant frequency between adjacent peaks suggests that the modulation frequency is a function of only the taper ratio and not the local scale or Reynolds number. The frequency that one would measure by counting peaks as Gaster did is thus composed of contributions from several discrete frequencies that do not change with z, but whose relative contributions to the observed frequency do change with z. This behavior suggests a kind of cell-like spanwise structure in which each cell is dominated by a single shedding frequency, but the continuous evolution of the vortex-shedding peaks precludes the presence of the very simple independent cell structure observed by Gaster *(35)* for much larger taper ratios.

The frequency of the dominant low-frequency peak is equal to the difference in frequency between adjacent dominant vortex-shedding peaks and is independent of z, and the magnitude of the low-frequency peak varies relatively little with z. Spectral peaks at higher-order harmonics of the low frequency and at high frequencies separated from the dominant peaks by integer multiples of the modulation frequency are also present. The signals thus contain two basic incommensurate high frequencies associated with vortex shedding, which by nonlinear interaction produce a much lower frequency modulation frequency. At a given z location, the velocity signal consists of a sharply weighted mixture of contributions from different spanwise locations with different vortex-shedding frequencies, with a weighting factor that decreases with spanwise distance from the measurement point. This suggests a "leaky" cell structure of vortex shedding, unlike the independent cell structures observed by Gaster *(35)* for much smaller taper ratios.

Because of the continuous evolution with z of neighboring vortex shedding peaks, at certain spanwise locations the velocity spectra contain two dominant peaks of equal amplitude. If the shedding frequency is taken as equal to the higher of these two frequencies, the corresponding values of the Roshko parameter and local Reynolds number (fd^2/ν, $Ud(z)/\nu$) fall right on the Roshko curve for a nontapered cylinder. Thus, whenever the competition between two dominant shedding frequencies of two neighboring cells is "equal," the upper frequency is the same as would be obtained from the Roshko relation for an untapered cylinder. The average frequency in this case is equal to the frequency obtained by simply counting peaks. When the largest shedding peaks are of substantially unequal magnitude and the shedding frequency is taken as that of the dominant peak, the fd^2/ν versus *Re* points all lie below the nontapered cylinder curve, as noted by Gaster *(34)*. The departure from the Roshko curve increases with increasing taper angle α. The spanwise distance between the locations at which equal dual peaks occur can be used to define a "cell" wavelength. For the case shown in Fig. 2, the spanwise wavelength is 3.0 mm.

Chaotic velocity spectra

Broadened "chaotic" velocity spectra were often observed as the Reynolds number was increased, either by increasing U or at sufficiently large $d(z)$. The spectra broadened gradually as the Reynolds number increased, with the peaks becoming less pronounced until they were finally overcome by the increasing continuous background. Some examples of chaotic spectra are shown in Fig. 3.

The broad peak at high frequency and the low frequency rise of these velocity spectra look similar to those of the chaotic spectra reported by Van Atta and Gharib *(25)* and Van Atta *et al.* *(26)*. In those cases, the chaotic spectra were produced by high-order harmonic cylinder vibrations. Spanwise differences in local shedding frequency associated with differences between natural and forced shedding frequencies produced dislocations in the vortex patterns, which, in turn, produced chaotic velocity spectra as they convected past the hot wire. In the present case, the mechanism for producing chaos is not aeroelastic but purely fluid mechanical. Preliminary comparison with flow visualization pictures *(30)* suggests that chaotic regions are associated with visual disruptions of regular vortex line patterns. More precisely, more recent experiments by Piccirillo and Van Atta *(36)* show that for tapered cylinders the low-frequency modulation is produced by a splitting or "bifurcation" of vortex lines which occurs at cell boundaries once every modulation period.

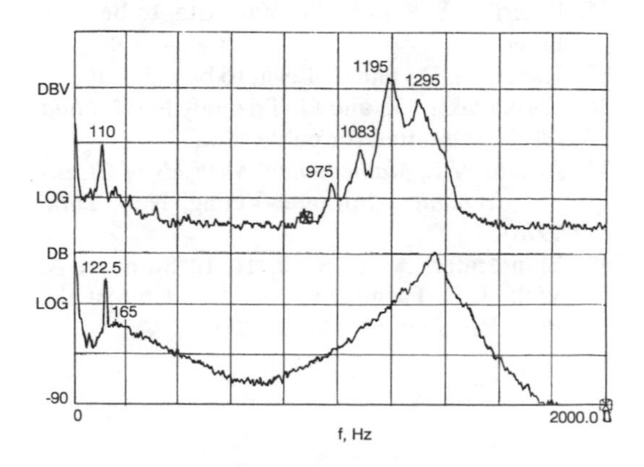

Fig. 3. Examples of (**A**) nearly chaotic, Re ≈ 130, and (**B**) fully chaotic, Re ≈ 150 hot-wire velocity spectra.

Numerical Simulations

Three-dimensional numerical solutions of the Navier-Stokes equations for vortex shedding from tapered cylinders are currently being computed by Jesperson and Levit *(37)* at the NASA Ames Laboratory. These simulations exhibit behavior similar to the experimental data of Figs. 1–3 and the vortex line splitting observed by Piccirillo and Van Atta *(36)*, indicating that direct numerical simulations are becoming a valuable tool for analyzing three-dimensional chaotic behavior in wake flows.

Summary

Three-dimensional wake vorticity dynamics lead to chaotic fluid motion in a great variety of experimental situations. While the presently described flows are not turbulent and occur in the low Reynolds-number, laminar flow regime, the vortex interactions are often sufficiently complex to produce continuous velocity spectra, a feature that fluid dynamical chaos has in common with truly turbulent flows. Chaos is produced by three-dimensional viscous vortex interactions, in which vortex lines are not totally constrained by Helmholtz's inviscid laws, but may bifurcate and reconnect as required by competition of different wake vortex-shedding modes.

Acknowledgments

This work was supported by DARPA grant N00014-86-K-0758 and ONR contracts N-00014-86-K-0690 and N-00014-88-K-0522.

References

1. Vallis, G. K., *Science* **232**, 243 (1986).
2. Vallis, G. K., *J. Geophys. Res.* **93**, 13937 (1988).
3. Palmer, D. R., M. G. Brown, F. D. Tappert, H. F. Bezdek, *Geophys. Res. Lett.* **15**, 569 (1988).
4. Samuelson, R. M., and J. S. Allen, *J. Phys. Oceanography* **17**, 2043 (1987).
5. Miles, J., *Adv. App. Mech.* **24**, 189 (1984).
6. Aubrey, N., P. Holmes, J. L. Lumley, E. Stone, *J. Fluid Mech.* **192**, 115 (1988).
7. Liepmann, H. W., *Amer. Sci.* **67**, 223 (1979).
8. Moon, F. C., *Chaotic Vibrations: An Introduction for Applied Scientists and Engineers* (John Wiley, New York, 1987).
9. Thompson, J. M. T., and H. B. Stewart, *Nonlinear Dynamics and Chaos: Geometrical Methods for Engineers and Scientists* (John Wiley, New York, 1986).
10. Aref, H., *J. Fluid Mech.* **143**, 1 (1984).
11. Hénon, M., *C.R. Acad. Sci. Paris A* **262**, 312 (1966).
12. Ottino, J. M., *Sci. Am.* **260**, 56 (1989).
13. Lorenz, E., *J. Atmos. Sci.* **20**, 130 (1963).
14. Ruelle, D., and F. Takens, *Comm. Math. Phys.* **74**, 189 (1971).
15. Feigenbaum, M. J., *J. Stat. Phys.* **19**, 25 (1978).
16. Manneville, P., and Y. Pomeau, *Physica D* **1**, 219 (1980).
17. Landau, L. D., and C. R. Doklad, *Sci. USSR* **44**, 311 (1944).
18. Gollub, J. P., and H. L. Swinney, *Phys. Rev. Lett.* **47**, 243 (1975).
19. Gollub, J. P., and S. V. Benson, *J. Fluid Mech.* **100**, 449 (1980).
20. Libchaber, A., and J. Maurer, *J. Phys. (Paris) Coll.* **41**, C3 (1980).
21. Huerre, P., and Monkewitz, P., *Annu. Rev. Fluid Mech.* **22**, (1989).
22. Blevins, R. D., *J. Sound and Vibration* **92**, 455 (1984).
23. Williamson, C., *Phys. Fluids* **31**, no. 10, 2742 (1988).
24. Sreenivasan, K. R., in *Transition and Turbulence in Fluid Flows and Low-Dimensional Chaos, Frontiers in Fluid Mechanics*, S. H. Davis and J. L. Lumley, Eds. (Springer Verlag, New York, 1985), pp. 41 – 67.
25. Van Atta, C. W., and M. Gharib, *J. Fluid Mech.* **174**, 113 (1987).
26. Van Atta, C. W., M. Gharib, M. Hammache, *Fluid Dyn. Res.* **3**, 127 (1989).
27. Olinger, D. J., and K. R. Sreenivasan, *Phys. Rev. Letters* **60**, no. 9, 797 (1988).
28. Jensen, M. H., L. P. Kadanoff, A. Libchaber, I. Procaccia, J. Stavans, *Phys. Rev. Lett.* **55**, 2798 (1985).
29. Williams-Stuber, K., and Gharib, M., *J. Fluid Mech.* **213**, 29 (1990).
30. Van Atta, C. W., and P. Piccirillo, *Topological Defects in Vortex Streets behind Tapered Circular Cylinders at Low Reynolds Numbers.* in *New Trends in Nonlinear Dynamics and Pattern Forming Phenomena: The Geometry of Non-Equilibrium*, P. Huerre and P. Coullet, Eds. (Plenum Press, New York, 1990), in press.
31. Barnett, K. M., *Boundary-Layer Meteorol.* **2**, 427 (1972).
32. Brighton, P. W. M., Ph.D. thesis, University of Cambridge (1977).
33. Brighton, P. W. M., *Q.J.R. Meteorol. Soc.* **104**, 289 (1978).
34. Gaster, M., *J. Fluid Mech.* **38**, 565 (1969).
35. Gaster, M., *J. Fluid Mech.* **46**, 749 (1971).
36. Piccirillo, P., and C. W. Van Atta, to be published.
37. Jesperson, D., and C. Levit, to be published.
38. Karniadakis, G., and G. Triantafyllou, *J. Fluid Mech.*, submitted for publication.
39. Arnold, V.I., *Mathematical Methods of Classical Mechanics* (Springer-Verlag, New York, 1978).
40. Brandstater, A., J. Swift, H. L. Swinney, A. Wolf, J. D. Farmer, E. Jen, J. P. Crutchfield, *Phys. Rev. Lett.* **51**, 1442 (1983).

8

Microwave Excitation and Ionization of Excited Hydrogen Atoms[1]

Peter M. Koch

Abstract

An active and broad area of atomic physics research involves *atoms in strong fields.* Atoms ionized by a strong oscillatory electric field present a particular challenge to theory when the field amplitude becomes non-negligible compared to the Coulomb field binding the electron. Moreover, the continuum final states involved are difficult to treat theoretically and simulate numerically. Usual methods based on time-dependent quantum mechanical perturbation theory can be used only with extreme caution, and they are hopeless if many absorbed photons strongly couple a large number of atomic states.

This chapter provides an overview of the multiphoton ionization of highly excited hydrogen atoms by microwave electric fields. The experimental results have confronted quantal theory with even greater challenges than those just mentioned. Because the atoms in such experiments are in states with large principal quantum numbers, they have been modeled in terms of classical dynamics theory. Ionization can take place classically through unstable, chaotic orbits.

The experimental apparatus includes no "chaos meter" and, therefore, does not measure anything chaotic. The notion of chaos has entered only through the interpretation of experimental data with concepts from classical dynamics. These concepts have given important insights into the physics of this manifestly quantal system that is driven periodically, leading to classical calculations of surprising quantitative accuracy over some ranges of experimental parameters. In other ranges, classical dynamics fails. The desire to understand how and why quantal dynamics mimics the behavior of chaotic classical dynamics in some ranges, but not in others, will continue to drive both theory and experiment and to teach us important physics.

Introduction

Motions of heavenly bodies

The apparent regularity and stability of the motions of heavenly bodies, particularly those in our solar system, has always attracted human attention. The qualifier *apparent* warns that there is more to their motions than meets the naked eye of an earthbound observer. Closer examination shows that chaos is rife in extraterrestrial physics. Specific instances of this behavior are discussed elsewhere in this volume, and the interested reader is urged to browse through chapter 16 by Thieberger, Spiegel, and Smith and chapters 17 and 18 by Buchler and Regev for detailed discussions. Chapter 9 by Jensen is even more directly relevant to the topics covered here. See also publications by Swinney and Wisdom listed in the appendix to this volume.

In the early 17th century, Kepler expressed as laws some observed regularities of planetary motions. Newton later derived them for two gravitationally interacting bodies, following his discovery of the universal law of gravitation. The problem simplifies to that of the relative motion of a single particle moving in a gravitational potential that falls off as r^{-1}.

---

1 A version of this chapter was also presented at the Soviet-American Conference on Chaos, National Academy of Sciences Study Center, Woods Hole, Massachusetts, 24–28 July 1989.

Beginning with the work of Poincaré at the end of the 19th century, we know that for more than two bodies bound together gravitationally, e.g., our solar system, irregular or *chaotic* behavior completely unlike that for just two bodies becomes possible – deterministic trajectories that are exponentially sensitive to initial conditions. One cannot compute and, therefore, not predict their exact behavior far into the future.

A fictitious but simple model system. Let us see how easy it is to make a simple but contrived deterministic dynamical system unpredictable. Imagine one point-like planet of mass m in an elliptic orbit about a point-like, infintely massive sun. Let its orbit be so elongated that the planet moves back and forth in one spatial dimension (1D), along the positive z-axis. Once per orbit, it reflects elastically off the sun at $z = 0$, instantaneously reversing its momentum vector $\vec{p_z}$. We have constructed a conservative, integrable dynamical system in N = 1 degree of freedom. Expressed in terms of the 2N variables z and p_z, the constant total energy is

$$E = \frac{p_z^2}{2m} - \frac{\beta}{z}, \tag{1}$$

where β is a constant. Its bounded trajectories with $E < 0$ are regular, periodic, and predictable. After a canonical transformation from (z,p_z) to (scaled) action-angle variables (I,θ), the energy of bound orbits is expressed even more simply,

$$E = -\frac{1}{2I^2}, \tag{2}$$

The bounce frequency ω_b is easily calculated from

$$\omega_b = \frac{\partial E}{\partial I} = \frac{1}{I^3} = (-2E)^{3/2}. \tag{3}$$

Imagination now adds more. Suppose at a fixed position very far out on the z-axis is a fantastic heavenly body whose mass varies sinusoidally in time,[2] $M(t) = M \sin(\omega t)$. The periodic forcing of the planet by the gravitational field of this imagined object adds two new parameters (a driving frequency, ω, and a driving force, F, proportional to M) and one new variable, the time, t. It is well known *(1)* that a time-dependent Hamiltonian system having N degrees of freedom can be transformed to another time-independent system having N + 1 degrees of freedom. Hence, the time-dependent system we have constructed is equivalent to a time-independent ("autonomous") Hamiltonian system in two degrees of freedom. N = 2 is the minimum number required for chaotic and hence unpredictable behavior to be possible in a nonintegrable Hamiltonian system.[3]

It is clear how to generalize this particular system at the expense of increasing N. The sun can have a finite mass. The planet (which could as well be an asteroid or a comet) could be free to move in two or even all three spatial dimensions, depending on how the periodic forcing is oriented with respect to the orbit. The forcing does not have to come from an imagined, fantastic heavenly body. Approximately periodic forcing could be supplied by the orbital motion around the sun of some

---

2 Never mind that we cannot actually do this. It is all in our minds!

3 One must be careful when using the word *chaotic* to describe the system we have constructed. As Berry *(2, 3)* and others have cautioned, classical chaos involves both exponential sensitivity to initial conditions and strictly bounded motion. The latter condition ensures that the irregular behavior brought about by the former condition persists for infinite times. Mathematicians studying chaos are wont to take infinite time limits. Physicists are a bit more cavalier, perhaps because we cannot study things for infinite times. In our imagined system, however, there is another problem. If F is not too far above the critical force at which escape becomes possible for some initial conditions, other initial conditions at the same initial energy, $E < 0$, lead to driven orbits along which the planet remains bound for very long times, perhaps forever. Thus, the phase space is divided into regions of initial conditions that do not escape and those that do. As it recedes from its sun, the planetary motion on any escaping orbit settles into the regular motion of a jiggling at frequency ω superimposed on a drift motion. The irregular part of the driven motion occurs when the planet is driven for long enough times with a force comparable to the binding force. "Long enough" means long compared to important time scales in the system: the inverse of the bounce frequency or the inverse of the driving frequency. If the motion lingers long enough in the irregular region, it will imitate the motion of a classically chaotic Hamiltonian system well enough to satisfy all but the mathematical purists.

other planet. One hopes, however, that the simpler system will have much of the physical behavior of the more complicated, but more realistic systems.

What about atoms?

Although this chapter is about real atoms, not about imagined solar systems, this historically motivated introduction sets the stage for what will follow.[4] Over some but not all ranges of the parameters, we shall find strong similarities between behavior calculated for periodically driven, classical Kepler orbits and that observed in experiments studying excitation and ionization of excited hydrogen atoms driven by a microwave electric field. The former is a classical system governed by nonlinear equations of motion that allow the kind of local instability that can lead to the onset of global chaos *(1)*. The latter is a quantal system governed by the linear time-dependent Schrödinger equation that does not have this kind of local instability. Understanding how and when the behavior of one can mimic that of the other is one of the worthwhile goals of a branch of research that most call "quantum chaos," although Berry prefers "quantum chaology" *(2, 7)*.

A classical model for the atom . . . fails. Atoms are real systems consisting of charged particles – heavy, positively charged nuclei and much lighter, negatively charged electrons – bound together by electromagnetic forces. The strongest of the electromagnetic forces in atoms is Coulomb's electrostatic force. It is identical in mathematical form to Newton's gravitational force, but, with two signs for the electric charge, it is attractive between the nucleus and each electron and repulsive among the electrons. Considering the similarity between the Coulomb and gravitational forces, it is not surprising that early in this century a solar system model was proposed for the atom, motivated by some key experimental observations by Rutherford and co-workers. It helped to explain what has come to be called Rutherford scattering, or how heavy ions are deflected when they pass near and "through" atoms.[5]

The atomic analogue of a sun and one planet is the hydrogen atom, an electron orbiting a proton that is 1836 times more massive. A wise professor once commented during the author's student days that half of what we know about quantum physics started from careful studies of the hydrogen atom. Let us appreciate that wisdom with a little more history. The solar system model of the atom is based on Newtonian concepts that do not include a crucial bit of physics that comes out of Maxwell's mid-19th century unification of electricity and magnetism. Calculations showed that the electron accelerating around its orbit should emit electromagnetic radiation over a broadband, continous spectrum, conserving total energy by spiraling in toward the proton and lowering the sum of its kinetic and Coulomb potential energies. This inherent instability of atoms contradicted observations. Atoms certainly existed in stable forms having very long, perhaps infinite, lifetimes.

Atoms could absorb and, suitably excited in an electrical discharge, flame or even emit electromagnetic radiation, but their spectra were dominated by "lines" – patterns of discrete wavelengths that were different for each atom or ion. This was also incompatible with the Newtonian/Maxwellian atom.

Bohr's quantal atom. A solution to some of the puzzles concerning the structure and line

4 The periodically forced, one-planet, 1D solar system model we have constructed above actually has a close analogue in the quantal world. A single electron above the surface of liquid helium can be bound by the small Coulomb force between it and its image charge below the surface. Such "surface state electrons" are very nearly a realization of a "1D hydrogen atom"; their line spectrum has been observed *(4)*. Jensen *(5, 6)* first proposed driving this system with microwaves to realize a quantal, periodically driven 1D system. His early papers presented a classical theoretical analysis of this driven 1D system.

5 Anyone who forgets the lasting impact the solar system model has had on atomic and molecular physics should recall that even though orbital trajectories simply do not exist in quantum mechanics, quantal treatments of atomic structure still often refer to basis functions for "atomic orbitals."

spectra of hydrogen grew out of Bohr's application in 1913 of Planck's quantum hypothesis to a solar system model of the atom. Only those bound (circular) orbits in hydrogen would be allowed for which the angular momentum p_θ of the orbiting electron took on positive integral multiples of Planck's constant h divided by 2π, $p_\theta = n_0 \hbar$, where $n_0 = 1,2,...$ is a quantum number. This is equivalent to quantizing the action I_0 of the periodic orbit, leading to an energy expression equivalent in form to that of Eq. 2,

$$E = -\frac{1}{2n_0^2}. \tag{4}$$

(In Eq. 4 the energy is measured in Hartree's atomic units; see the section entitled "Simple properties of the hydrogen atom," page 79.) The energy for $n_0 = 1$ would be an absolute minimum, meaning that there was an atomic state – the ground state – of lowest energy. This neatly allowed for a stable form of atomic matter. If the atom were somehow put into an excited state with $n_0 > 1$, it was hypothesized to be capable of sudden jumps downward in energy, with the energy difference emitted as a quantum of the electromagnetic radiation field, a photon. This neatly explained the origin of the spectral lines.

Absorption was also allowed and tied up nicely with Einstein's earlier explanation of the photoelectric effect. Electromagnetic radiation of frequency ω could be absorbed by the atom if addition of the photon energy $\hbar\omega$ to the energy, Eq. 4, of an initially populated state equalled the energy of a higher, final state. Ionization – the photoelectric effect – would result if the photon energy was sufficient to reach a final state in the continuum, $E > 0$.

***Multiphoton ionization*.** Contrary to what is usually taught or implied in introductory physics courses, this does not prevent ionization from taking place if $\hbar\omega < \frac{1}{2n_0^2}$, i.e., below the threshold for one-photon ionization. In this case, conservation of energy means that the continuum is reached only when more than one photon is absorbed. This is a nonlinear process that may be described by the linear theory of quantum mechanics, a situation that should *not* be confused with the nonlinear dynamics of the driven classical Kepler system.

It is useful to differentiate between three situations in "multiphoton ionization." They are conveniently explained via three examples for the hydrogen atom initially in an $n_0 = 2$ level, which lies 3.4 eV below the ionization continuum. (i) If the photon energy is 1.8 eV, "simultaneous" absorption of two photons from an intense source, e.g., a pulsed laser, will raise the electron $(2 \times 1.8) - 3.4 = 0.2$ eV into the continuum. This is normal, nonresonant multiphoton ionization. (ii) A slightly higher photon energy of about 1.89 eV still requires at least two photons for ionization, but this photon energy equals the $n_0 = 2 \rightarrow n = 3$ energy splitting. It is possible for the first photon to make a "stepwise excitation" of the $n_0 = 3$ level, and then for the second photon to ionize this level. This is closely related to multiphoton ionization involving an intermediate resonant state. (iii) With respect to example (i), analysis of the emitted photoelectron shows it is possible for it to have kinetic energies higher than 0.2 eV. For an intense enough 1.8 eV pulsed-laser source, one might record photoelectrons with kinetic energies near 0.2, 2.0, 3.8 eV. The first corresponds to ionization by the minimum number of photons. The last two correspond, respectively, to so-called "above-threshold ionization" (ATI) by three and four photons (one and two "above threshold," that is).

All three kinds of behavior have been observed in pulsed-laser ionization of tightly bound atoms and molecules. ATI, especially, is still under active study because the actual positions, shapes, and relative intensities of the photoelectron "lines" are affected by the peak intensity, polarization, duration, and shape of the laser pulses *(8)*.

All three multiphoton ionization processes just discussed play a role in microwave ionization of highly excited atoms; see especially the discussion of Regime-II–Regime-V, pages 85–94.

***Quantization procedures*.** Bohr's quantization procedure described on page 77 is the simplest of those used to infer quantal behavior from regular classical behavior. For N > 1 degree of freedom, the procedure used to quantize the classical actions is usually called "EBK," after Einstein, Brillouin, and Keller. Because classical actions are not calculable in the strongly chaotic regime, how to effect quantization in this case is unclear. However, in regions of phase space where chaos is relatively mild, the existence of approximate actions suggests a natural extrapolation of the EBK methods to near-integrable systems *(9, 10, 86)*.

Correspondence principles. Understanding within what limits and how quantal solutions tend toward classical solutions is the other important relationship between classical and quantum mechanics; it goes under the name "correspondence principle." Although the concept is usually associated with Bohr's application to his quantal theory of the hydrogen atom, his correspondence principle limit of large quantum numbers is not the only one. Planck earlier considered the limit of his quantal theory of blackbody radiation as his quantum of action $h \rightarrow 0$.[6] Heisenberg's correspondence principle[7] relates quantum mechanical matrix elements to Fourier components of the classical motion. Percival and Richards *(11)* derived a "strong coupling correspondence principle," generalizing Heisenberg's work. It has been successfully used by them and others in theories of collisions involving highly excited atoms as well as in theories of the microwave excitation and ionization processes that are the central focus of this chapter.

Applying the correspondence principle(s) to real situations can be tricky business. Pais *(12)* cautions that "It takes artistry to make practical use of the correspondence principle." Most of our experience with it throughout this century has involved integrable classical systems undergoing regular motions. We do not yet understand in any detailed way how solutions from the linear theory of quantum mechanics correspond with chaotic solutions of nonlinear classical dynamics. What we do know suggests that there is still plenty of room left for artistry.

Simple properties of the hydrogen atom

Table 1 lists useful numbers for hydrogen atoms based on classical theory or Bohr quantum theory. The third column shows how each property listed in the first column scales with n. The second and fourth columns show, respectively, numerical values for the ground state with principal quantum number $n = 1$ and for a highly excited state with $n = 100$.

Hartree's atomic units are based on setting the charge, e, and mass, m_e, of the electron and $\hbar$ all equal to one. The atomic units of length a_0, velocity v_1, frequency ν_1 (the angular frequency $\omega_1 = 2\pi\nu_1$), and electric field strength F_1 are those given in Table 1 for $n = 1$, with all numbers in the table being appropriate for the reduced mass of real hydrogen atoms, $\mu = 0.999455\, m_e$, set equal to one. The atomic unit of energy is $2R_H = 27.2$ eV.

Some of the numbers for an $n = 100$ atom are worthy of comment. Its Bohr radius is

---

6 The reader should be aware of semantic confusion in the literature. The term "semiclassical" as in "semiclassical limit" is often loosely used to refer to the limit $h \rightarrow 0$, which is the limit considered by Planck. (This is not to be confused with $h = 0$, which is *the* classical limit.) In terms of an action $S = \hbar j$, where j is a quantum number, Berry *(3)* is careful to emphasize the importance of the evolution time t. He calls the semiclassical limit $j \rightarrow \infty$, corresponding to $\hbar \rightarrow 0$ with S and t fixed. Because long-time quantal evolution is fundamentally different from long-time classically chaotic evolution, this limit does not commute with the long-time limit of $t \rightarrow \infty$ with j and S fixed. In many processes involving atoms and molecules, e.g., in their interactions with external fields, "semiclassical" as in "semiclassical theory" refers, instead, to a theoretical treatment in which the atoms or molecules are treated quantum mechanically, but the external fields are treated classically (see footnote 9).

7 D. Park *(82)* wrote that Bohr preferred the name "correspondence argument" until Heisenberg's matrix-mechanics quantum theory supplied strong theoretical support for the concept. Two recent articles *(83)* discuss applications of Bohr's and Planck's correspondence principles and include references to some earlier discussions. A. Norcliffe reviews "correspondence identities" for the special case of the Coulomb potential *(84)*; see also *(11)*.

Table 1. Useful numbers for hydrogen atoms based on classical theory or Bohr quantum theory.

Property	$n = 1$	Arbitrary n	$n = 100$
Binding energy, $\|E_n\|$	$1R_H = 13.6$ eV	R_H/n^2	1.36×10^{-3} eV
Bohr radius, a_n	$a_1 \equiv a_0 = 5.3 \times 10^{-11}$ m	$n^2 a_0$	0.53 μm
Electric field at a_n	$F_1 = 5.137 \times 10^{11}$ V/m	$n^{-4} F_1$	5137 V/m
rms electron velocity, v_n	$v_1 = 2.2 \times 10^6$ m/s	v_1/n	22 km/s
Radiative lifetime	$n = 2 \rightarrow 1: 1.6 \times 10^{-9}$ s	$\propto n^3$	$\approx$ milliseconds
Fund. classical period	$T_1 = 1.5 \times 10^{-16}$ s	$n^3 T_1$	0.15 ns
Fund. classical frequency, ν_n	$\nu_1 = 6.5761 \times 10^{15}$ Hz	$n^{-3} \nu_1$	6.5761 GHz
Avg. of $\Delta n = \pm 1$ freqs.	Cannot have $n < 1$	$\simeq n^{-3} \nu_1$	6.5774 GHz

comparable to the dimensions of microbes. Its *rms* electron velocity is comparable to that of artificial earth satellites, large but nonrelativistic ($v_{100}/c < 10^{-4}$). Its radiative lifetime is about one oscillation period of the musical tone concert A. Because this is a very long time compared to the less than microsecond time scale of the microwave ionization experiments discussed in this chapter, radiative decay could be safely ignored for them. The transition frequencies from $n_0 = 100$ to $n = 101$ or 99 are only about three times that in a microwave oven, and the Coulomb electric field holding it together is achieved when a nine volt radio battery charges a capacitor whose parallel plates are separated by about the thickness of an American dime.

The last two of these show why it is not a great challenge to produce in the laboratory a microwave electric field whose strength is comparable to that inside the atom, i.e., nonperturbative. Moreover, microwave technology is mature and generally available. Rather, the experimental challenge is shifted to creating, controlling, and monitoring the very fragile, highly excited atoms.

Tightly bound atoms in their ground states or low-lying excited states are easier to work with experimentally, but the numbers in the second column of Table 1 suggest where the difficulties lie for performing strong-field experiments with them. Only when nanosecond or shorter pulsed-laser beams are focused to spots measuring micrometers in diameter does one achieve the optical electric field strengths on the order of 10^{11} V/m that hold together tightly bound atoms. It is an experimental challenge to produce reproducible, well-characterized, focused pulses of laser light or to measure the spatio-temporal variation of the intensity for each pulse, but steady progress is still being made in these directions *(8)*.

Early microwave ionization experiments

In 1974, J.E. Bayfield and the present author had the fun and good luck to discover accidentally in an experiment designed for other purposes the ionization of highly excited hydrogen atoms by a 9.9 GHz microwave electric field *(13)*. Even though the hydrogen atoms with principal quantum numbers $n_0 \simeq 66$ were bound by only 3.1 meV, that was still huge compared to the miniscule 41 μeV of energy carried by each 9.9 GHz photon. To ionize an atom required at least the combined energy of about 80 photons. Additional data were obtained for frequencies near 1 GHz and 30 MHz, for which ionization required even many more photons to be absorbed. There was a dearth of theory available with which one could quantitatively interpret the experimental results, but the data did confirm a qualitative trend of a crude theory *(14)* for ionization in low-frequency oscillatory fields. As the frequency increased, the observed ionization threshold field strength decreased.

A second early experiment *(15)* reported the first experimental studies of the frequency dependence of the excitation and ionization probabilities over about a 20% range of frequencies within the X band of microwaves, 8–12.4 GHz. Broad, frequency-dependent structures were recorded for most of the n_0-values that were studied, 45–57. Again, there was a dearth of theory with which one could interpret the results quantitatively.

Enter dynamics

In 1978–79, papers appeared from three different groups, one in England and two in the Soviet Union, that began to offer the hope of a theoretical approach that might lead to quantitative intepretation of these early experimental results. Leopold and Percival *(16)* carried out a purely classical, numerical simulation of the atomic ionization process as a periodically driven Kepler system, producing the first calculated threshold fields that actually agreed with 1974 experimental data *(13)* within estimated errors. Delone, Zon, and Krainov *(17)* reasoned that excitation towards the ionization continuum took place along states highly perturbed and mixed by the field, leading them to give a new name, "diffusion ionization," to this process. That their theoretical estimate of the diffusion rate had the wrong n-dependence, however, was pointed out by Meerson, Oks, and Sasorov *(18)*, who also linked the onset of microwave ionization in the quantal atom to the onset of stochasticity (chaos) in a driven, nonlinear classical oscillator.

The first and last of these papers particularly emphasized the dynamical nature of the ionization process, reasoning that because of the large quantum numbers involved, one might apply classical dynamics to this quantal system. The papers by Leopold and Percival, including one with Jones [*(19)* and references therein], are especially worth noting. Although they were carried out for a different range of parameters from that covered in the 1977 experiments *(15)*, the classical calculations of *(19)* contained the same kind of frequency-dependent structure as had been observed experimentally. This was sufficient to draw this author's group back into microwave ionization experiments *(20)*. Moving the laboratory from one university to another one suspended experimental efforts by more than two years *(22)*, but a clear experimental goal was in sight — to confirm or refute the predictions of the classical, numerical calculations *(19)* that predicted local stability of the driven atom for scaled frequencies near one-half. This we found (see Fig. 1 and the section on resonances in Regime-III, page 89) and much, much more.

Experiments in Quantum Chaology

The Hamiltonian and scaled variables

The nonrelativistic, time-dependent Hamiltonian usually used to describe the microwave ionization of hydrogen is essentially identical to one that would be used for the 3D version of the periodically forced planet model described on page 76. It is

$$H(t) = \frac{p^2}{2} - \frac{1}{r} + zA(t)F\sin(\omega t + \varphi)\,, \tag{5}$$

where p^2 is the square of the electron's linear momentum; $\vec{r}$ is the displacement vector between the proton and the electron, with z its component along the axis of the linearly polarized electric field F, which oscillates at angular frequency ω; φ is a relative phase that, so far, all experiments have averaged; and $A(t)$ is the slowly varying envelope function that turns on, maintains, and turns off the pulse of oscillatory field used in the experiment. With operator substitutions, Eq. 3 is used to represent electric dipole coupling of the quantal hydrogen atom to the classical, oscillatory electric field; see footnotes 6 and 9.

As first emphasized by Leopold and Percival *(16)*, Eq. 5 has scaling properties. Ignoring the dependence on $A(t)$ and φ, if one measures the driving frequency in units of the

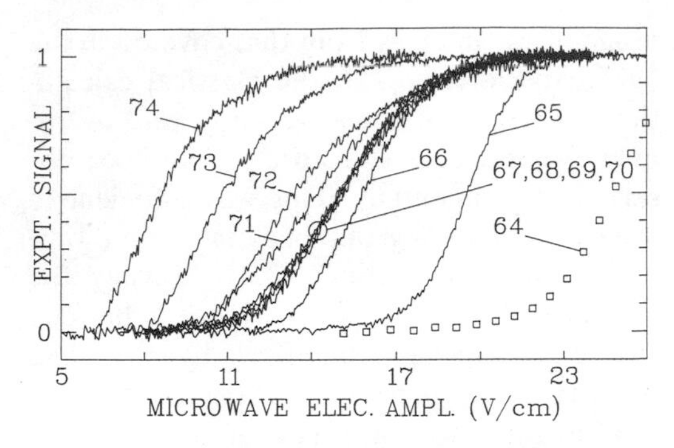

Fig. 1. 9.92 GHz "ionization" curves for hydrogen atoms in each of the principal quantum numbers n_0 between 64–74. The n-cutoff was $n_c \simeq 92$. Notice the clustering of some of the curves that occurs for n_0-values where $n_0^3\omega \simeq 1/2$. (For $n_0 = 69, n_0^3\omega = 0.496$.)

Kepler frequency for the initial orbit, which varies with initial principal action I_0 as I_0^{-3} (see Eq. 3) and if one measures the driving field strength in units of the mean Coulomb field strength for the initial orbit, which varies as I_0^{-4}, the classical dynamics depends on the scaled frequency $I_0^3\omega$ and scaled field strength $I_0^4 F$.

Quantizing to integral values of the principal action $I_0 \rightarrow n_0$, we have the important dynamical parameters $n_0^3\omega$ and $n_0^4 F$. When $n_0^3\omega = 1/1$, the driving frequency is equal to the classical Kepler frequency, or quantally the photon energy is very near the mean of the $\Delta n = \pm 1$ energy splittings. Our strategy for the experiments at Stony Brook has been motivated by a desire to investigate the dynamics over a very wide range of the scaled frequency.

For a single or small group of n_0-values, a large range is not easily covered through variation of ω because one needs new microwave sources and components every time the frequency changes by from 50% to a factor of two (see "Early microwave ionization experiments," page 80). Using band after band of microwaves would be an expensive and tedious way to cover a wide dynamic range in scaled frequency, so we decided to use a few fixed frequencies ω of resonant cavities and a wide variation of the value of n_0. Since 1985 our experiments at Stony Brook *(22–34)* with some frequencies between 7.58–36.02 GHz and n_0-values between 23–90 have covered a dynamic range in $n_0^3\omega$ of over 150. The principal disadvantage to fixing ω and varying n_0 is a stepwise variation of $n_0^3\omega$, but even a stepwise coverage of such a large dynamic range has led us to identify physical phenemonena in five different regimes of the scaled frequency, all below the one-photon ionization threshold, Regime-VI. Before describing them in "Regimes of Observed Behavior" (page 84), we will digress to describe briefly the experiments that found them.

The experimental method

Highly excited hydrogen atoms in a fast beam with typical energy 15 keV are prepared in a series of steps. Electron-transfer collisions of protons in a Xe gas-scattering cell produces fast neutral atoms with a tiny fraction in $n_0 = 7$ substates. Using a double-resonance excitation sequence *(35, 36)* employing two independent, continuous CO_2 lasers that cross *(29, 30, 32)* the fast atomic beam in two different regions of static electric field, atoms in one of the 28 parabolic substates of $n_0 = 7$ are excited into one of the 55 parabolic substates of $n_0 = 10$, most or all of the other 54 of which are also present in the beam. The second laser excites atoms from this one $n_0 = 10$ substate into a substate of a much higher lying manifold with n_0 in the range 23–90.

Although the atoms are prepared in individual parabolic substates so highly stretched out along the field axis that one may call them "quasi-one dimensional" (q1D), the atoms actually enter the microwave cavity with a different ensemble-averaged substate distribution. Because the atoms travel through a region of nominally zero (i.e., stray) fields on the way to the cavity, the quantization axis is not preserved.[8] With use of static field ionization as a diagnostic tool just before the cavity, we have found *(22, 29, 32)* the ensemble of atoms entering the cavity to be consistent with a uniform distribution of substates of the initially prepared n_0-value, at least when $n_0 \gtrsim 35$. This corresponds to a classical, microcanonical ensemble of initial Kepler orbits in all three spatial dimensions.

Thus far, all our experiments have used linear polarization of the microwave electric field, for which the selection rule $\Delta|m| = 0$ preserves the value of $|m|$. Careful theoretical modeling of the 3D experiments must take this all into account, but as covered briefly below, so far we have found that 1D theory furnishes surprisingly accurate threshold fields for the onset of ionization in the 3D experiments. The apparent reason for this is because near the ionization threshold the dynamics evolve toward states that are quite one dimensional, stretched out along the field axis. This has been found theoretically to be so for scaled frequencies up to the highest yet investigated experimentally, $n_0^3\omega = 2.8$, but some classical theoretical work *(44)* suggests that for significantly larger scaled frequencies the 3D threshold fields will rise appreciably above those in the 1D dynamics. This should be investigated.

The experiments are usually carried out by measuring at a given microwave frequency an experimental "ionization" or quench signal for each n_0-value in turn as a function of the microwave power dissipated in the cavity.[9] We have found it desirable to use the TM_{00q} mode of right-circular cylindrical cavities, with q = 2, 3, or 4. For the ideal mode, the longitudinal microwave electric field is independent of the axial coordinate inside the cavity. It drops quadratically as a J_0-Bessel function for small radial excursions from the axis; depending on how tightly the atomic beam was collimated in different experiments, this has varied between 7% and a few tenths of a percent.

With use of accelerator resonator design software, we are able to calculate the actual microwave field distribution inside and near a cavity, including the perturbing effects of the holes in its endcaps and of a metal collimator disk placed before the entrance endcap.

The pulse shape, i.e., the rise, middle, and fall in time of the microwave field amplitude given by A*(t)* in Eq. 5, is determined mainly by the geometry of the cavity and its endcap holes, the microwave frequency, and the beam velocity. In typical experiments, the field is maintained at nearly a constant amplitude during about 300–600 oscillations between a rise and fall of about 50–100 oscillations. For example, see *(26, 29, 32, 45, 47)*. The rise, middle, and fall can be independently changed by varying experimental parameters,

---

8 There is no reason why the quantization axis cannot be preserved and the atoms made to enter the microwave interaction region in the q1D substate, the approximate atomic analogue of the 1D, surface-state electron system (see footnote 4), but this requires maintaining a static electric field before and through the microwave interaction region *(37–39)*. Some aspects of the classical dynamics of the superimposed static and microwave fields are treated in *(40–42)*. For "ionization" or quenching experiments carried out with both fields present, comparisons with theory should require that the theory include the static field interaction in the initial Hamiltonian. This has not always been the case. For example, one may be suspicious of the comparison made in *(43)* between experimental data obtained with a superimposed static field and the results of a "localization theory" that did not explicitly include the static field. The static field has two main effects: (i) It lowers the (zero-microwave field) continuum from the free-atom continuum $E = 0$ to the (substate-dependent) static-field saddle energy; this may significantly reduce the number of photonic states between the initial state and the actual continuum relevant to the experimental conditions. (ii) The static-field–induced distortion of the potential seen by the electron changes the microwave interaction matrix elements, particularly for the states closest to the top of the static-field barrier; this certainly affects the dynamics evolving through them. The overall effect of the superimposed static field is not obvious and may depend significantly on details of the experimental situation.

9 The reader may find it amusing to see just how well the cavity experiments satisfy the conditions for application of semiclassical theory, in the second sense of "semiclassical" in footnote 6. The typical 9.9 GHz cavity used in our experiments has a (unloaded) quality factor $Q_0 \approx 10^4$. When it dissipates one watt of power, there are about 3×10^{16} 9.9 GHz photons in the cavity, and the microwave amplitude on its symmetry axis is about 5×10^2 V/cm. This is above the ionization threshold field for $n_0 \gtrsim 34$. The energy of about 3×10^2 9.9 GHz photons is needed to ionize the $n_0 = 34$ atom, about 14 orders-of-magnitude fewer photons than are stored in the cavity. In this case, one is certainly justified in treating the microwave field classically!

a distinct advantage of the microwave experimental method for studying atoms in strong fields. Our calibration procedure *(46)* also allows absolute determination of the microwave field inside the cavity typically to 5% or better; we have confirmed this experimentally *(45, 47)* in excitation experiments with helium Rydberg atoms. Contrast this with pulsed laser experiments *(8)*, where it is difficult to control and measure precisely the spatio-temporal variation of the focused laser optical field. Only with difficulty does one determine the peak intensity above 10^{12} W/cm^2 to a factor of two *(48)*. This corresponds to only a 40% determination of the peak optical electric field.

A microwave "ionization" experiment (the quotation marks are important!) records a signal registering $P_{\text{“}ion\text{”}}$, the sum of probabilities for *true* ionization and for excitation to final *n*-states above n_c^i, a *cutoff* *n*-value determined by static electric fields used in the apparatus for particle deflection and detection. A microwave quench experiment records a signal registering the probability P_{quench} for microwave-induced loss of the signal produced by static electric field ionization of the excited atom signal after the cavity; it includes true ionization plus excitation to above a cutoff *n*-value n_c^q.

Via comparisons of data that differ only in their *n*-cutoffs, one gains experimental information about the distribution of final bound states that remain after exposure to the microwave field pulse. It is reasonable to assume that when either *n*-cutoff is much higher than the initial n_0-value, a signal is dominated by true ionization; this is confirmed by data shown in Fig. 5a,b. When this is the case, we have observed that "ionization" and quench experiments give very nearly the same, but not exactly identical *(32)*, experimental threshold fields needed to produce the same $P_{\text{“}ion\text{”}}$ and P_{quench} *(24, 25, 32)*. Thus, when n_0 is far below the *n*-cutoff, we interpret quenching data in terms of "ionization." Conversely, as the *n*-cutoff and initial n_0-value move closer together, the "ionization" or quench signal is increasingly affected by microwave excitation to nearby, higher lying *n*-states, so that one must carefully specify all experimental conditions.

Figure 1 shows a composite of 9.92 GHz "ionization" curves and quench curves for each of the n_0-values between 64–74. The *n*-cutoff for each was near 85–90, high enough to make the curves relatively insensitive to its exact value. One should notice that below a certain driving field, each curve does not vary; there we normalize it to $P_{\text{“}ion\text{”}} = 0$. Similarly, we associate $P_{\text{“}ion\text{”}} = 1$ with the lack of variation of each curve above some larger F-value.

Each of the curves displayed in Fig. 1 is consistent with intuition: $P_{\text{“}ion\text{”}}$ increases monotonically with the driving field F. (This intuition is not always correct, however; see Regime-II, page 85, and Figs. 3a, b.) Noteworthy in Fig. 1 is the clustering of several of the curves, which physically means that even though the Coulomb binding field decreases as n_0^{-4}, the value of F needed to produce a given value of $P_{\text{“}ion\text{”}}$ does not vary over some ranges of initial n_0-values. We have learned to recognize such clustering as a clear sign of resonant behavior. The data in Fig. 1 are in Regime-III (see page 88), and one may associate this clustering behavior classically with the nonlinear trapping resonance ("island of stability") near a scaled frequency of 1/2. (See the penultimate sentence of the section entitled "Enter dynamics," page 81!)

From such experimental curves can be extracted field strengths F(X%) in V/cm, and scaled field strengths n_0^4 F (X%) at which, for each n_0-value, $P_{\text{“}ion\text{”}}$ first reaches X%. Plots of these (scaled) threshold fields vs. n_0 or $n_0^3\,\omega$ reveal much of the behavior of this intriguing physical system. Figures 2 and 4–6 in this chapter are examples of graphs of n_0^4 F vs. $n_0^3\,\omega$.

Regimes of Observed Behavior

Finally we return to the various regimes of behavior observed for the microwave "ionization" and excitation of excited hydrogen atoms.[10] Although most of the data discussed below have been published, some are not yet published in archival journals. Conclusions presented in this chapter, which are quite secure, are based on these data, supplemented in some cases by additional preliminary data that were only very recently obtained in our laboratory. The reader is urged to notice how quantal effects arise in each regime. In some regimes they systematically stabilize, in other regimes systematically destabilize, the quantal atom compared to its classical counterpart. In two regimes it goes both ways.

Regime-I: The tunneling regime

For near-static fields, our data for $n_0^3\omega \lesssim 0.07$ obtained with 9.9 GHz microwaves *(23–25,29–33)* are consistent with a principal ionization mechanism of quantal penetration of the slowly oscillating potential barrier. This systematically lowers ionization threshold fields below classical values. We interpret the following as evidence for tunneling: (i) The scaled field n_0^4F for any given, fixed value of $P_{\text{“ion”}}$ between, 5–95%, for example, is relatively insensitive to variation of $n_0^3\omega$ below about 0.07. (ii) These observed microwave-ionization–scaled threshold fields lie systematically below classically computed values by just the amount that one would expect from previous studies of static electric field ionization of individual parabolic quantal substates *(50)*.

One notices the onset of this behavior in the upper left-hand corner of Fig. 2, where experimental data for 3D atoms for $P_{\text{“ion”}}$ = 10% and 90%, respectively, are compared to classical, numerical Monte Carlo simulations performed by Rath and Richards *(51)* for a microcanonical distribution of initial classical orbits. An extrapolation of the classical $P_{\text{“ion”}}$ = 10% curve in Fig. 2 to a static field limit, $n_0^3\omega \to 0$, gives a scaled field of $n_0^4\,\text{F} \simeq 0.14$. Careful inspection of the experimental data in Fig. 2, reinforced by additional, preliminary data obtained for $n_0^3\omega$ as low as 0.018 *(33)*, shows the experimental 10% thresholds flattening off near $n_0^4\,\text{F} \simeq 0.125$, about 0.015 units lower. This 0.015 offset is close to by how much quantal tunneling lowers observed ionization fields for the most easily ionized parabolic substate of a given n_0-manifold below the ionization threshold fields for the corresponding classical case [compare *(50)* to results from *(52)*].

Regime-II: The low-frequency regime

For $0.05 \lesssim n_0^3\omega \lesssim 0.3$, some experimental "ionization" and quench curves have shown structures *(20, 23, 25–27, 29, 32)*, including non-monotonic bumps, steps, or changes in slope.[11] We have come to learn that these are

10 Experimental data for the ionization of $30 < n_0 < 70$ lithium and sodium Rydberg atoms by linearly polarized 15 GHz microwaves have also been obtained *(49)*. Without giving consideration to the difference in frequencies, the authors of *(49)* compared their 15 GHz field strengths for 50% ionization probability for "hydrogen-like" states of these alkali Rydberg atoms to the 9.9 GHz field strengths for 10% "ionization" probability reported in *(22)* for hydrogen atoms with n_0-values between 40–74. On a log-log plot of (unscaled) field strength vs. n_0^4 [see Fig. 3 of *(49)*] the trends of the data sets are similar. However, it seems questionable to compare threshold fields obtained for significantly different frequencies. Our own comparison *(32)* of the 50% threshold fields from *(49)* with the 50% threshold fields *(32)* from the full set of 9.92 GHz hydrogen data *(24, 25, 30, 31)* shows significant quantitative differences for all sodium threshold fields and for lithium threshold fields for $n_0 \gtrsim 48$.

11 We have also carried out experiments in Regime-II with hydrogen atoms in each n_0- level between 43–49 being driven simultaneously by microwave fields at two different frequencies, 7.58 and 11.89 GHz *(26, 27)*. Many, but not all, two-frequency quench curves exhibited structures similar to those observed in Regime-II with only one frequency. Quantal 1D calculations *(53)* have reproduced the observed two-frequency structures for the few cases that were modeled theoretically. Neglecting these structures in the two-frequency data, 50% "ionization" probability for each n_0-value was achieved for an approximately constant sum of the microwave amplitudes for the two fields *(26)*. This was also found in the 1D quantal calculations *(53)*.

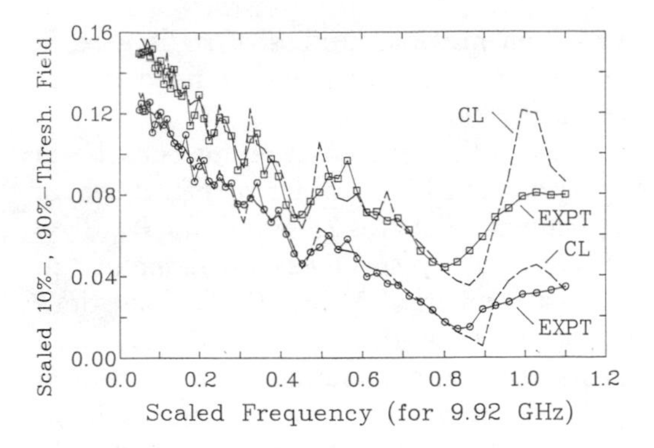

Fig. 2. Comparison of experimental data and classical calculations for 9.92 GHz "ionization" of hydrogen atoms in each of the principal quantum numbers n_0 between 32–90; the n-cutoff was $n_c \simeq 92$. (○): classically scaled experimental fields *(22,30,32)* for 10% "ionization" probability for 3D atoms. (□): experimental data for 90% "ionization" probability. Dashed lines: scaled fields for 10% and 90% "ionization" probability, respectively, obtained *(50)* from a classical, 3D Monte Carlo simulation of the experiment, with no adjustable parameters.

caused by quantal resonance phenomena at low-scaled frequencies. Here, unlike in Regime-I, quantal effects can make the real atom either less stable or more stable than its classical counterpart because of quantal excitation or de-excitation resonances, respectively. Figs. 3a,b show two of the former case; Figs. 3c,d show two of the latter case. To see this behavior most clearly, one should make like-with-like comparisons in Fig. 3; in each frame compare the 1D classical results (filled circles) with the 1D quantal curve (solid line), and compare the 3D classical results (open circles) with the 3D experimental results generated by the quantal analogue computer (other solid line). The curves in Fig. 3 are four of the many comparisons made in *(29)* between experimental "ionization" and quench curves in Regime-II, 1D and 3D classical calculations, and calculations based on a 1D quantal theory *(54)* that exploits the low scaled frequency via use of an efficient, adiabatic basis of quantal states.

Another quantal 1D theory *(55)*, which earlier explained these structures but considered only the quantal destabilization effect, based its numerical calculations on zero-field hydrogenic basis states and its physical interpretation on state-mixing effects near avoided crossings of Floquet states of the atom + field system *(56)*; see *(57)* for a review. [In the Soviet literature, the Floquet states are usually called "quasi-energy states" *(58)*.] Blümel and Smilansky *(55)* include instructive comparisons of their theoretical curves with five experimental "ionization" curves from our laboratory. One should be aware, however, that straightforward application of the Floquet method to this problem ignores an important time scale. Rigorously, the Floquet method requires the periodic-driving field amplitude to be constant for all time, whereas in real experiments, the amplitude varies in time according to the pulse shape, A*(t)*, in Eq. 5. The driving field period sets one time scale; the numbers of field oscillations over which the field rises, stays constant, or falls set others.

An interesting question is how slowly the field must turn on (or off) in order that the evolution be adiabatic. This will certainly depend on $n_0^3\omega$ and on the initial phase φ (to reiterate: over which all experiments, so far, have averaged) in Eq. 5. One reasonably expects that a field suddenly turned on with a relative phase of $\varphi = 0$, i.e., a suddenly switched sinewave field (but not a suddenly switched cosinewave field!), might become more and more adiabatic as the scaled frequency $n_0^3\omega$ gets very small compared to one. Classical *(59)* and quantal *(60–62)* calculations have supported this expectation.

All three quantal theories *(29, 54, 55, 62)* published to date for the low-frequency structures observed in the experiments with 3D hydrogen atoms are 1D theories. Whereas all correctly explain, now even predict, the n_0-values and field strengths F at which such structures appear for a given driving frequency, none reproduces the complete F-dependence of $P_{\text{"ion"}}$. Nor have experiments yet observed the "spiky" substructure predicted, in particular by *(55)*; each sharp spike in their calculated $P_{\text{"ion"}}$ vs. F curves correlates with an avoided crossing between Floquet potential curves. Because of the 7% spread of the microwave field amplitude F experienced by the atoms distributed across the finite cross

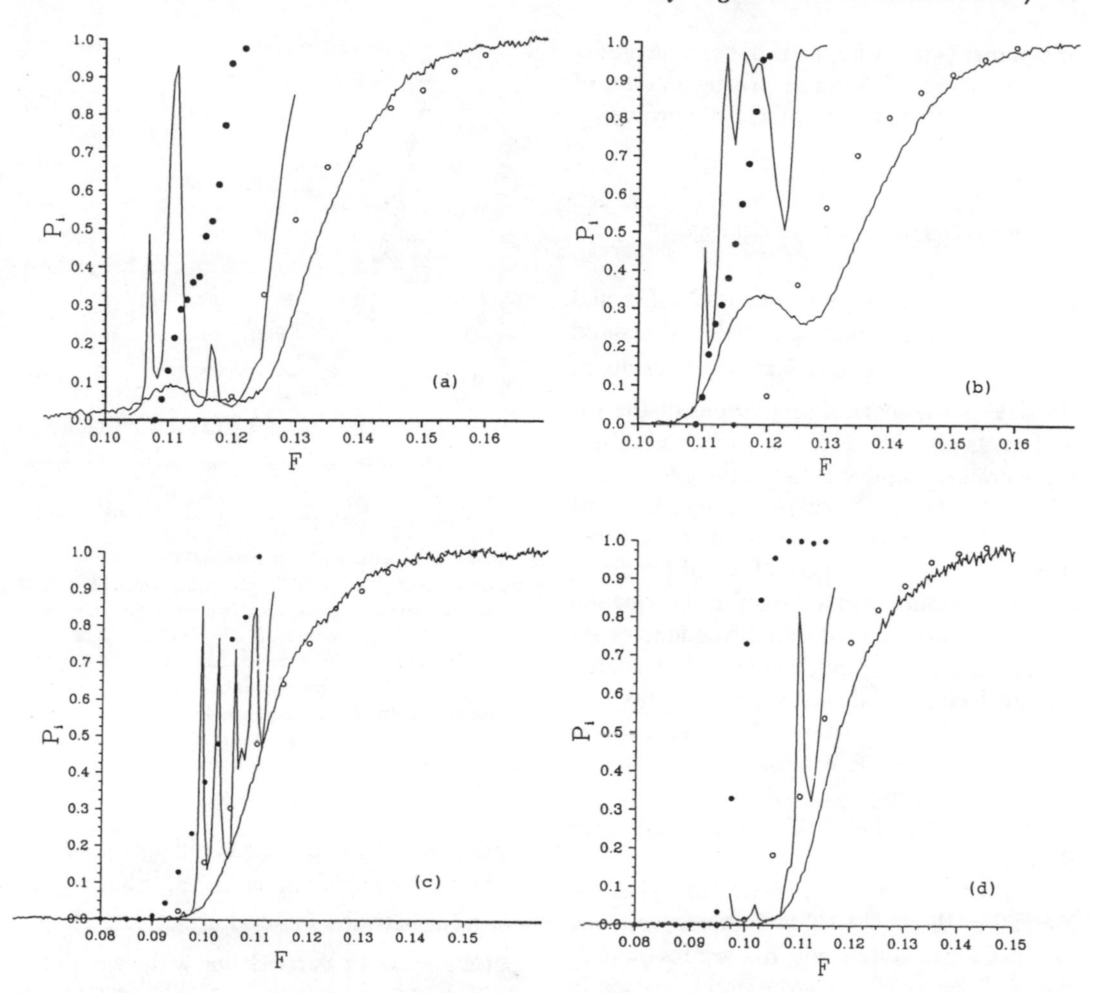

Fig. 3 Comparison *(29)* of experimental 9.92 GHz "ionization" curves (right-hand–most solid curve in each frame) for 3D atoms with quantal and classical calculations. The experimental n-cutoff in each case was higher than $n_c = 75$. (○): results of a 3D, classical Monte Carlo simulation. (●): results of a 1D, classical Monte Carlo simulation. Other solid curve: 1D, quantal calculation using the adiabatic-basis of states. Frames (**a**) and (**b**) show cases where quantal resonance effects destabilize the quantal atom compared to its classical counterpart, i.e., where quantal "ionization" begins at a lower field strength. Frames (**c**) and (**d**) show cases where quantal resonance effects stabilize the quantal atom compared to its classical counterpart, i.e., where quantal "ionization" begins at a higher field strength. Frame (a): $n_0 = 37$ and $n_0^3 \omega = 0.0764$. Frame (b): $n_0 = 38$ and $n_0^3 \omega = 0.0828$. Frame (c): $n_0 = 47$ and $n_0^3 \omega = 0.1567$. Frame (d): $n_0 = 48$ and $n_0^3 \omega = 0.1669$.

section of the atomic beam, however, "ionization" experiments reported so far were incapable of resolving individual, narrow ionization spikes.[12] In addition, one should not be surprised by the lack of perfect agreement between data from experiments with 3D atoms and the results of 1D theories. On the contrary, one ought to be impressed that the agreement is as good as it is. At least near the onset of ionization, the quantal analogue (see footnote 4) of the fictitious, 1D-model for a periodically forced solar system having only

12 We have observed the effects of individual, isolated avoided crossings of Floquet potential curves in other, very high field resolution, microwave-excitation experiments with helium Rydberg atoms *(45, 47)*. Along with the accompanying theory, they show that the amplitude-dependent lineshapes for the multiphoton transitions are sensitively dependent on the detailed pulse shape.

one planet (see "A fictitious but simple model system," page 76) has a surprising amount of the physics of the more general, 3D problem.

Regime-III: The semiclassical regime

Figure 4 compares published 9.92 GHz and 36.02 GHz 10%-ionization scaled threshold fields for $n_0^3\,\omega = 0.05–2.8$ with the results of classical 3D numerical simulations of the experiments *(28–30, 51, 63)*. That there is striking overall agreement between the experimental thresholds and the classical values for most scaled frequencies between about 0.1–1.2 demonstrates the utility of classical dynamics for understanding this manifestly quantal process in this range of scaled frequencies. At its low end, quantal resonances (see Regime-II) can locally spoil the agreement; this also happens in Regime-III (see Resonances, page 89, this section). Even lower in frequency, quantal tunneling systematically lowers the threshold fields below classical values (see Regime-I).

In and above Regime-III, the classical ionization threshold fields coincide with the onset of unstable (chaotic, but see footnote 3) classical trajectories, conveniently exposed in classical phase-space portraits *(5, 6, 21, 40, 42, 64–66)*. At lower scaled frequencies, in Regime-I and into Regime-II, computed ionizing orbits escape the nuclear Coulomb field so quickly that little seems chaotic (diffusive) about their motion *(85)*. The agreement between experimental and classically computed threshold fields in Regime-III, as shown in Figs. 2, 4, and 5 in this chapter, is now accepted to have established that a real, strongly driven quantal dynamical system is capable of mimicking the behavior of its nonlinear, classically chaotic counterpart, at least for some ranges of the parameters and for finite interaction times (see footnotes 3 and 6). This correspondence cannot be ignored.

An important feature of the classical dynamics is classical scaling (see "The Hamiltonian and scaled values," page 81). Therefore, one should notice the good agreement in Fig. 4 between two different sets of ex-

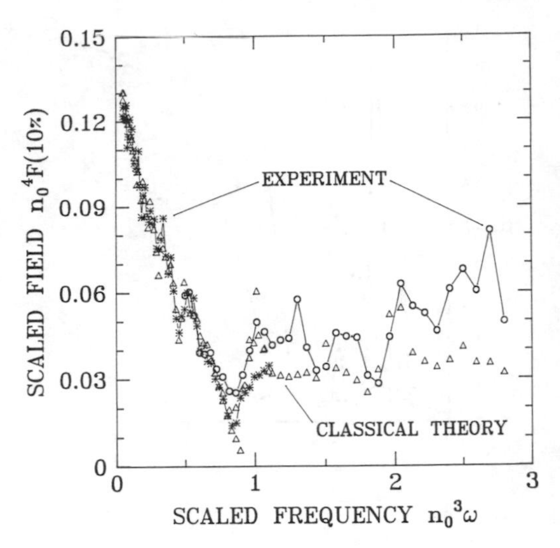

Fig. 4. A composite of experimental data for the scaled fields for 10% "ionization" probability for 3D hydrogen atoms compared to 3D, classical Monte Carlo simulations. (*): 9.92 GHz experiments *(22, 29–32)* with each principal quantum number n_0 in the range 32–90 and an n-cutoff $n_c \simeq 92$. (O): 36.02 GHz experiment *(28)* with each principal quantum number n_0 in the range 45–80 and an n- cutoff $n_c \simeq 89$. (Δ): 3D, classical Monte Carlo simulation *(29, 51)* of the experiments.

perimental 10% threshold field data [and with 3D classical calculations *(28, 51)* shown as Δ] for $n_0^3\,\omega$ between 0.5 and about 0.8. The experiments were carried out with two widely separated frequencies, $\omega/2\pi = 9.92$ GHz (*) and 36.02 GHz (O); this means that to get the same scaled frequency $n_0^3\,\omega$, the n_0-values differed by about 25 units and the "ionization" threshold field strengths in laboratory units (V/cm) differed by nearly a factor of six. Nevertheless, the scaled threshold fields n_0^4 F closely coincide for equivalent scaled frequencies $n_0^3\,\omega$.

With the agreement including the local maxima and minima in Fig. 4, it is an excellent example of classical scaling in this part of Regime-III. However, if one compares carefully the 9.92 and 36.02 GHz $P_{\text{"ion"}} = 10\%$ data in Fig. 4 for $n_0^3\,\omega$ between about 0.8–1.1, the agreement is not as good. This is probably an n-cutoff effect.

For both sets of data the n-cutoff was near 90, but for each given value of $n_0^3\,\omega$ the initial

n_0-values in the 9.92 GHz experiments were much closer to the n-cutoff than they were in the 36.02 GHz experiments. Therefore, it is not surprising that for $n_0^3 \omega$ between 0.8–1.1 in Fig. 4, the scaled threshold fields for 9.92 GHz are systematically lower than those for 36.02 GHz. This suggests that there were non-negligible amounts of microwave excitation probability to final bound states between about 80–90 for the $n_0^3 \omega = 0.8$–1.1 data taken at 9.92 GHz, compared to that for the 36.02 GHz experiments.

Resonances in Regime-III, and beyond. The distinct local maxima (minima) in n_0^4 F(10%) in Figs. 2, 4–6 mean locally more stable (unstable) behavior of the periodically driven hydrogen atom in the neighborhood of certain values of $n_0^3 \omega$. As we first showed in 1985 for Regime-III *(22)* and have recently continued to find in Regimes-IV, V (see the following two sections), this local behavior is a result of resonance phenomena. In Regime-III, they often correlate well with important nonlinear trapping resonances or islands in the classical phase space. These form around periodic orbits of the driven system when the scaled frequency is near low-order rational ratios p/q such as 1/1, 1/2, and 1/3 *(1)*.

In quantal language, the scaled frequency p/q coincides with being near a q-photon resonance for changing n_0 by $\pm p$-units, either up or down. However, Fig. 2 shows that the agreement between experimental and classical thresholds is less good on and very near these resonances; this suggests that quantal effects are important. The actual situation is quite subtle, because very recent theoretical work *(63, 66)* (see Regime-V below and "Scars," page 94) has linked quantal resonance effects – nonclassical stability – observed in our experiments to classical phase-space structures.

Regime-IV: The transition regime

For $1 \lesssim n_0^3 \omega \lesssim 2$, Figs. 4, 5 show the classical dynamics beginning to break down. For example, in Fig. 5A the experimental bumps around $n_0^3 \omega = 1/1$, 2/1 are partially reproduced by classical resonances there, but the experimental dip near 3/2 and rise just beyond anticorrelates with the classical local stability near 3/2. Notice that the large measured peak near 4/3 is entirely absent classically; we shall return to this below. In Regime-IV, as in Regime-II, quantal effects either lower experimental threshold fields below or raise them above classical values.

Regime-V: The high-frequency regime

When $n_0^3 \omega > 2$, observed threshold fields rise systematically – as much as a factor of two – above classical thresholds. Our 1988 paper *(28)* provided the first experimental confirmation of this previously predicted "quantal suppression of classical chaos" in the high-frequency regime *(67, 68, 70)* that we label Regime-V.

Experiments in Regime-V and localization theory. The 1984 prediction of an Italian/Soviet collaboration including Casati, Chirikov, and Shepelyansky *(68)* – subsequent papers included Guarneri and others – was based on a dynamical localization theory initially developed for the periodically kicked-rotor model problem. The quantal theory for the kicked-rotor was related earlier *(69)* to an Anderson-type localization theory for the effects of disorder on electronic wavefunctions in a spatially periodic lattice. The goal of the work reviewed in *(67, 70)* was an extension of this approach to the problem of the periodically driven hydrogen atom, but as they pointed out, in the condensed-matter problem the disorder is externally applied and random. In the kicked-rotor and microwave ionization problems, there is no random disorder applied from the outside: their dynamics are deterministic. Because the most recent "photonic localization" approach has been reviewed by its authors, the interested reader can consult *(67)* for details. Let us see next how well it confronts a real physics problem and then make a few comments.

The core of the photonic localization

theory *(67)* is formulae used to describe the quasi-stationary, time-averaged distribution of final n-states produced when the hydrogen atom is exposed to a sinusoidal driving field. At heart a theory for the bound-state distribution function, it blends classical diffusion theory and quantum mechanics. When the "localization length" of the distribution extends far enough for there to be appreciable transport in action to the continuum (or just above the n-cutoff n_C defined earlier), then "ionization" – recall the importance of the quotation marks – sets in. Paradoxically for a theory being used to understand the ionization process, the localization approach does not describe the dynamics in the continuum; however, we are also interested in how well it describes excitation above different experimental n-cutoffs n_C.

One should also appreciate that a localization theory describes only some average behavior of a system. Here it does not encompass the physics producing the resonances that are such an important feature of our data, nor does it include "pulse effects" of the rise and fall of the amplitude of the driving field, A(t), in Eq. 5. Neither has it (yet) treated the case of a static electric field superimposed with the microwave field (see footnote 8).

Useful comparisons require that the theory closely model important experimental details. Understanding and modeling what is actually measured as "ionization" are crucial. After being exposed to a pulse of microwave field, all atoms initially prepared in the experimental ensemble of substates of a level n_0 are not transported all the way to the zero-field continuum ("true" ionization); some remain in bound levels, higher or lower than n_0. The largest of the static electric fields in the experiment, usually one of those used intentionally for charged-particle deflection or detection purposes, determines the cutoff n-value n_c. Varying n_C gives some important experimental information *(28)* about the actual n-distribution of final bound states.

One way to correct their theoretical quantum delocalization border for the effect of a finite experimental n_C-value was presented in *(71)* and repeated in *(72)*. A simple approximation for those results is also given in *(73)*.

Figure 5 shows comparisons with two experimental data sets *(28)* for n_0^4 F (10%) vs. $n_0^3\,\omega$; 3D atoms having n_0 in the range 45–80 were exposed to about 300 oscillations of a 36.02 GHz field, with a slow turn-on and turn-off. Frame (a) shows data (O) obtained with n_C in the range 86–92, or a median value n_C = 89. Frame (b) shows data (O) obtained with a much higher n_C in the range 160–190, or a median value n_C = 175. Also shown as (×) connected by dashed lines are values obtained from careful, 3D classical numerical simulations *(28)* of each experiment, including the value of n_C and the experimental pulse shape, A(t), in Eq. 5. Each solid curve for $n_0^3\,\omega \geq 1$ is the result of using Eq. 3 of *(72)* to obtain the quantum delocalization border for 10% excitation above the cutoff n_C = 89, frame (a), or n_C = 175, frame (b). We next comment on the results of these comparisons.

For $n_0^3\,\omega > 2$, classical theory provides 10% "ionization" threshold fields that are systematically below the measured ones. This furnished *(28)* the first confirmation of the predicted "quantal suppression" effect *(68)* in the high-frequency regime. One should also note, however, that at least some of the undulations in the experimental data are mimicked by the classical resonances.

Although one does not expect an "ionization" border obtained from the photonic localization theory *(67)* to reproduce the undulations caused by these or other resonances, the delocalization border in frame (a) for the value n_C = 89 does roughly reproduce the mean behavior of the experimental data in Regime-V, $n_0^3\,\omega$ 2. However, the delocalization border in frame (b) for n_C = 175 significantly overestimates the 10% "ionization" threshold fields measured for this cutoff.

Is it valid to make both these comparisons? That is, were the data in Fig. 5 a,b for $n_0^3\,\omega \gtrsim 2$ collected at field strengths for which the localization theory should apply for both n-cutoffs according to a criterion presented by its authors [*(67)*, pages 1428 and

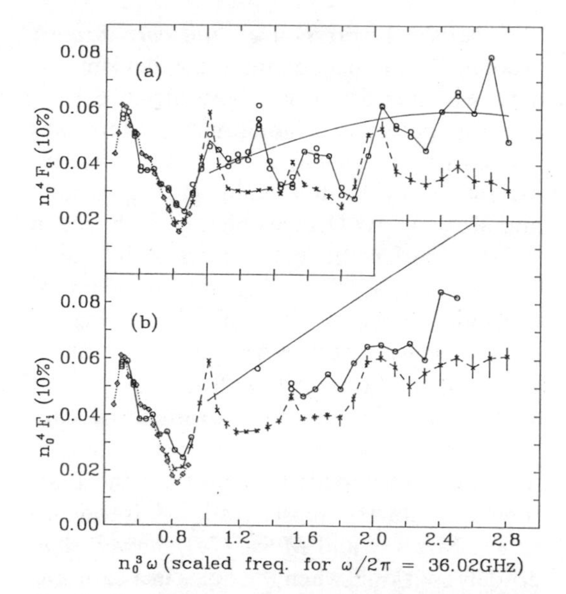

Fig. 5. A comparison of theoretical results with 36.02 GHz experimental data for the scaled fields for 10% quenching (**a**) and "ionization" (**b**) probability for 3D hydrogen atoms with principal quantum numbers n_0 in the range 45–80. (O): experimental data *(28)* with an n-cutoff $n_c^q \simeq 89$ in frame (a) and $n_c^i \simeq 175$ in frame (b). (×): 3D, classical Monte Carlo simulations *(28, 51)* of the 36.02 GHz experiments, including the experimental pulse shape and the effect of the different n-cutoffs for each set of experimental data. (◇): 3D, classical Monte Carlo simulations of earlier 9.92 GHz experiments with an n-cutoff $n_c \simeq 92$. Solid lines: the 1D quantal delocalization border *(67, 71)* for 10% excitation above $n_c^q = 89$ in frame (a) and above $n_c^i = 175$ in frame (b). We used Eqs. a1–a3 of *(72)* to evaluate these borders.

1429]? We find that the criterion $l_\varphi < N_I$ stated there — that the localization length l_φ, in number of photons, should be less than the number N_I of photon states to the (true) continuum, E = 0 — is indeed well satisfied for the experimental field strengths, so the comparison is valid. In our opinion, therefore, *(72)* incorrectly states that their localization theory does not apply to our data for $n_c = 175$. According to their own criterion, their localization theory may be applied equally as well to our 36.02 GHz data *(28)* for both n-cutoff values and to the 12–18 GHz data *(43)* from the Pittsburgh group.

How, then, does one interpret the result that the localization theory correctly describes the average behavior of the 10% threshold fields in Fig. 5a but fails to describe that for Fig. 5b? A clear implication is that the present version *(67, 71, 72)* of the photonic localization theory does not correctly describe transport to final, bound n-levels that are far from the initial level n_0 and close to the continuum. This suggests that the quasi-stationary, time-averaged distribution that decreases approximately exponentially over the bound states, Eq. 28 of *(67)*, which was "borrowed" [*(67)*, p. 1428] from the kicked-rotor model system, does not correctly describe the final n-distribution for the periodically driven hydrogen atom, at least for the experimental parameters considered here. (It may be useful to note that *(54d)* found that, when projected onto unperturbed, one-dimensional, hydrogen atom basis states, the eigenfunctions of the bound space projected dipole operator decayed not exponentially, but with and $n^{-7/3}$ power law in the "external-field-dominated regime.") The discrepancy shown in Fig. 5b suggests that the actual distribution of final high-lying photonic states that correspond to $n >> n_0$ might fall more slowly with increasing n than that of Eq. 28 of *(67)*.[13]

How can one reconcile this result with a central conclusion of *(43)*, which presented experimental 10% "ionization" data that ostensibly agreed with the localization theory corrected for the n-cutoff as described above. The data presented in *(43)* were obtained with several values of n_0 and n_c and for a number of frequencies between 12–18 GHz. We have used the prescription described in *(72)*; see also *(71)* to reevaluate the n_c-corrected, 10% ionization borders for each of the sets of data

13 Although the photonic localization theory does not correctly reproduce the (average) behavior for $n_0^3\omega \gtrsim 2$ of the high n_c-data in Fig. 5b, this behavior, as well as that for the low n_c-cutoff data in Fig. 5a, is reasonably well reproduced by theoretical results obtained with use of the "quantum Kepler map" [see *(72)*, Figs. 1a,b]. This suggests to this author that an approximation that was used to obtain the results of the photonic localization theory *(66)*, i.e., going from the dynamics of the quantum Kepler map to that of the quantized standard map, may contribute to the discrepancy noted above between the experimental data in Fig. 5b and the results of the n_c-corrected delocalization border.

presented in *(43)* for $n_0^3 \omega \geq 1.5$. We find that the n_c-corrected border evaluated in *(43)* and presented as the same solid line in their Figs. 2, 3 does not agree with any of the borders we calculate. In each case, the solid line presented in Figs. 2, 3 of *(43)* lies below the borders we calculate, and the discrepancy increases to as much as 20% as the difference $(n_c - n_0)$ increases.[14]

Notice that the discrepancy just noted between experimental data in *(43)* for $n_0^3 \omega \geq 1.5$ and the correctly evaluated n_c-corrected delocalization border *increases* as $(n_c - n_0)$ increases. This is consistent with the behavior discussed above for the results shown in Figs. 5a,b. We conclude, therefore, that the photonic localization theory described in *(67)*, using the prescription described in *(72)* [see also *(71)*] to correct for the effect of the n-cutoff, does not quantitatively reproduce experimental data for all conditions so far investigated. It appears that the distribution over final bound atomic states $n >> n_0$ may not be the one borrowed from the kicked-rotor model that is used in *(43, 67, 71, 72)*. These conclusions differ from those presented in *(43)*. Furthermore, more attention will need to be paid theoretically to ranges of parameters producing non-negligible excitation into the continuum, i.e., "true" ionization.

Although the localization theory does not agree quantitatively with all experimental data, one should recognize that its prediction of the quantal suppression effect, now verified by two different experiments *(28, 43)* was crucial in drawing both experimental and theoretical attention to the high-scaled frequencies here labeled Regime-V. Nevertheless, this is but one of six distinct regimes yet identified for this fascinating problem. Quantal stabilizing or destabilizing effects play important roles in the other regimes, via mechanisms quite different from "localization" in Regime-V, which (theoretically) is always stabilizing.

Quantal suppression of transport through "cantori." Another quantal mechanism for suppressing transport in a periodically driven, classical nonlinear Hamiltonian system was investigated by MacKay and Meiss *(74)*. Invariant "KAM" (after Kolmogorov, Arnold, and Moser) tori *(1)*, conveniently visualized in Poincaré surface-of-section plots of the classical phase space, are barriers to classical transport. Increasing the nonlinear forcing parameter causes each KAM torus in turn to break open, becoming a "cantorus." Over some range of the forcing parameter, a cantorus is an impediment to transport. The cantori that most impede transport are those associated with the most irrational frequency ratios. MacKay and Meiss *(74)* showed that, crudely speaking, when the holes in a cantorus are too small on a quantal scale, quantal transport through the cantorus is suppressed below its classical value.

Meiss *(75)* derived a threshold field border obtained from this cantorus mechanism for quantal suppression of classical transport and compared it with the low n-cutoff, 36.02 GHz microwave "ionization" data presented in *(28)* and in Figs. 5a, 6. The cantorus border *(75)* passed reasonably well through the middle of the experimental data for $n_0^3 \omega$ above about 2 up to the upper end of the data set at 2.8.

Following the prescription of *(74)* and *(75)*, Bayfield *et al.* calculated a cantorus border and compared it to their 12–18 GHz microwave "ionization" data *(43)*. The cantorus border roughly followed the trend of the experimental data for $n_0^3 \omega$ between 1–2, but it rose significantly less rapidly than the data for $n_0^3 \omega$ up to 2.6. Bayfield *et al.* concluded that the quantal suppression effect based on the cantorus mechanism did not have the correct frequency dependence.

Theoretical Regime-V behavior and "ATI." The quantal 1D calculations of Jensen *et al.* [see especially Fig. 2 of *(76)*] and Leopold and

14 To prevent possible confusion, we should point out that what we call n_0 and n_c here are called n and $\bar{n}$ in *(43)*. Moreover, each border we calculate lies above all of the corresponding measured 10% threshold data shown in Figs. 2, 3 of *(43)*.

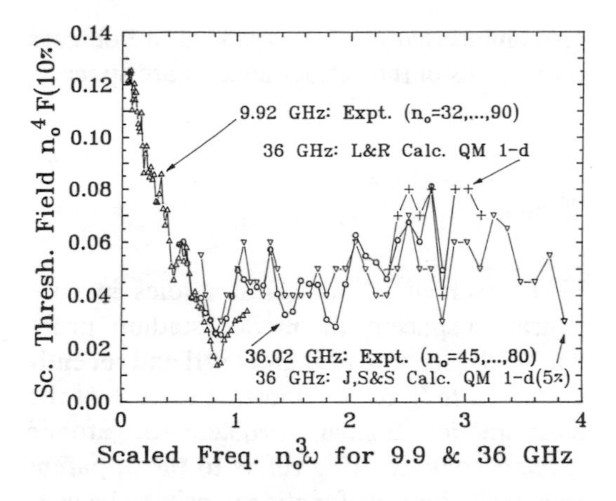

Fig. 6. A composite of experimental data for the scaled fields for 10% "ionization" probability for 3D hydrogen atoms compared to 1D quantal numerical calculations. (Δ): 9.92 GHz experimental data *(22, 30, 32)* with each principal quantum number n_0 in the range 32–90 and an n-cutoff $n_c \simeq 92$. (O): 36.02 GHz data *(28)* for each principal quantum number n_0 in the range 45–80 and an n-cutoff $n_c \simeq 89$. (∇): 1D quantal numerically calculated *(76)* 36.02 GHz fields for 5% excitation above an n-cutoff $n_c = 95$. (+): 1D quantal numerically calculated 36.02 GHz fields for 10% excitation above an n-cutoff $n_c = 86$, obtained with use of the compensated-energy representation *(77)*.

Richards [see especially Figs. 4–6 of *(77)*] both produced some numerical results in Fig. 6 in this chapter that agree with our experiments with 3D atoms. They have predicted final bound state distributions in Regime-V that are very sharply peaked (contrast ratio of more than four orders-of-magnitude) on final bound states separated by the microwave photon energy, even when the field strength is as much as a factor of two above the "quantal delocalization border"

$$n_0^4 \mathrm{F} = (n_0^3\,\omega)^{7/6}/(6.6 n_0)^{1/2}$$

that was derived for $n_c \rightarrow \infty$ with use of localization theory *(67, 70)*. (Using the notation of "Experiments in Regime-V and localization theory," pages 89–92), this border was derived *(67)* by finding the field strength at which $l_\varphi = \mathrm{N_I}$.) Leopold and Richards [see *(77)*, pp. 1492 ff.] have specifically challenged the claim made by Casati *et al.* that above this border "...one expects that the quantum electron will diffuse and ionize like the classical one." Although the numerically computed, time-averaged, quantal 1D bound-state distributions presented by Casati *et al.* have not shown such sharply peaked behavior above the quantal delocalization border [examples are Figs. 14a3, 14b3 of *(70)*], these authors commented [section 3.3 of *(70)*] that the series of peaks in the computed, final bound-state distributions was the analog in the discrete spectrum of the ATI peaks *(8)* [see also page 78 on "Multiphoton ionization"] seen in the continuum in ("long-pulse") laser experiments.

Notice that the calculated peaks presented in Fig. 2 of *(76)* and Figs. 4–6 of *(77)* are significantly sharper than those presented in *(70)*, particularly for cases above the quantum delocalization border. The differences seem to be related *(77)* to details in the respective calculations. Such peaks, which have not been produced from the localization theory, do seem to be related to the physics underlying the quasi-resonant–basis-state approximation *(76)* [see also *(77)*] and the quantum Kepler map *(67, 72)*; see footnote 13. The physical basis for the peaks at high scaled frequencies, Regime-V, is rather simple. When $n_0^3\,\omega >> 1$, the splitting between adjacent n-levels is much smaller than the photon energy. Neglect of those n-levels that are far from resonance (i.e., over which the photons "hop") is of little consequence. Moreover, notice that this line of reasoning joins the physics in Regime-V smoothly onto that in Regime-VI. When the frequency rises above the one-photon ionization limit, the simplest quantal approximation to the weak-field, long-pulse photoionization dynamics is a two-state approximation that couples only the initial bound state and the continuum-state, one-photon energy above it.

It is of more than passing interest, therefore, to do experiments using final bound-state analysis to investigate *(77)* whether prominent peaks exist in Regime-V for parameters above the quantum delocalization border. The only reported experimental study of the (q1D) bound distribution *(39)* was limited to Regime-III. Not unexpectedly, it confirmed the validity of both quantal and classical theory in this (semi-)classical regime and saw none of the prominent peaks predicted *(76, 77)* for Regime-V. Experiments in Regimes-IV, V are needed to examine the predicted changeover *(76, 77)* to ATI-type peaks *(8)* appearing in the bound final-state spectrum.

Regime-VI: the photoelectric effect

When photon energy rises above the threshold for one-photon ionization, $\hbar\omega > (2n_0^2)^{-1}$, all vestiges of classical behavior are gone, at least for weak field intensities.

Resonances

We shall begin to close this chapter by focusing on the importance of the resonances observed in our experiments in Regimes-III, IV, V, whose complete investigation motivates our current experiments. Figure 6 shows a composite of our 9.92 GHz *(22–25, 30–32)* and some 36.02 GHz *(28)* 10% threshold data along with the results of two independent quantal 1D-numerical calculations that modeled important experimental conditions, such as the pulse shape, A*(t)*, in Eq. 5. That of *(66, 76)* was a supercomputer calculation. That of *(77)* used a basis of states in the computationally efficient "compensated-energy representation," which is physically justifiable *(77)* for $n_0^3\omega \gtrsim 2.5$, i.e., in Regime-V; see page 93.

Notice that these quantal 1D numerical calculations approximately reproduce nearly all of the important resonant features of our 3D experiments, in particular, the large bump at $n_0^3\omega \simeq 1.3$. That this bump does not appear in the classical simulations of the experiment, see Figs. 4 and 5, means that the local stability, i.e., a locally higher "ionization" threshold field strength, is nonclassical. A recent nonperturbative quantal explanation *(63, 66)* for this and other resonances, however, has connected this quantal local stability to "scarred" wavefunctions that are tied to the tangle of unstable periodic orbits around classical resonances that can be visualized in Poincaré projections of the action-angle phase space.

"Scars"

First observed in numerical studies in configuration space of the model "stadium problem" *(78)* [see also *(79)* and *(80)*] and recently discussed *(81)* as an important feature of the diamagnetic Zeeman problem in atomic spectroscopy, scarring refers to the apparent propensity for wavefunction density to be concentrated along trajectories of classical unstable periodic orbits. Although they (densely) form a set of measure zero in chaotic classical dynamics, unstable periodic orbits seem to play an important role in quantization. This poorly understood phenomenon is now actively being investigated [*(79)* and references therein; *(66)*].

Jensen *et al.* *(66)* used this scarring idea to explain our observation of local stability, not only for $n_0^3\omega \simeq 1.3$ *(28, 63)*, but also for $n_0^3\omega \simeq 0.4, 0.57$ (see Fig. 2) and possibly others. If this explanation is correct, this phenomenon would emerge as a general feature of this type of dynamical system, here operative in at least Regimes-III, IV, and V. Furthermore, it continues to show the surprising robustness of 1D theory in accounting, even quantitatively, for experimental results obtained with 3D atoms.

There is a clear experimental test for the scarring hypothesis. Although the local stability is a quantal effect, it is linked to structures in the classical action-angle phase space. This suggests that the quantal local stability should scale classically: if the experiment is repeated with different frequencies and n_0-values, such that the scaled frequency $n_0^3\omega$ covers the same range — near 1.3, for example — the local stability should still appear and the scaled threshold field n_0^4 F should be the same. Preliminary experimental data *(34)* collected in our laboratory at Stony Brook with a 26.42 GHz field show local stability at $n_0 = 69$ that confirms the classical scaling of the quantal local stability at $n_0^3\omega \simeq 1.3$, previously observed with a 36.02 GHz at $n_0 = 62$. This is very strong support for the scarring hypothesis.

The scarring hypothesis also seems to explain another striking feature of our published data *(28)*, viz., the insensitivity of 10% threshold fields for a certain few n_0-values, which included $n_0 = 62$ at 36.02 GHz (see above) to variation in the experimental n-cutoff from $n_c^q \simeq 89$ to $n_c^i \simeq 175$. For example, compare the nearly equivalent experimental 10% threshold fields for $n_0^3\omega \simeq 1.3$ in Fig. 5A,B. No variation of the 10% threshold field with the n-cutoff means that there was no microwave excitation to bound states between the two cutoffs. At other n_0-values, large variation of n_0^4 F(10%) with the n-cutoff indicated that significant amounts of population were left in excited states between $n_c^q \simeq 89$ and $n_c^i \simeq 175$ after the pulse turned off.

The scarring hypothesis is just the latest of a number of developments demonstrating that, in this fascinating dynamical system – the excited hydrogen atom, driven by a strong oscillatory field – there is, indeed, plenty of room left for artistry in the investigation of the correspondence between its quantal and classical dynamics. We stop here on this hopeful note for the future.

Acknowledgments

This chapter is dedicated to the memory of my father, Edward Clifford Koch, who died suddenly and unexpectedly while I was working on it. I thank him and my mother Miriam for always fostering the curiosity that has been the lifeblood of this scientist, their son. Especially at this time I cherish the love and support of my wife Nancy, our daughter Amanda, and our son Nathan for living the joy of life looking forward.

Experimental data shown in this paper were coaxed out of our apparatus at Stony Brook by a number of co-workers: E.J. Galvez, K.A.H. van Leeuwen, L. Moorman, G. v. Oppen, B.E. Sauer. We thank the other graduate and undergraduate students and members of the technical staff at Stony Brook, especially Pete Davis in the Advanced Technology Lab (ATL), who contributed in important ways to the results presented here. Stimulating conversations aplenty with many theorists have been a wonderful part of this research work; it is a pleasure to acknowledge M. V. Berry, R. Blümel, H.-P. Breuer, G. Casati, B. Chirikov, K. Dietz, I. Guarneri, M. Holthaus, J. E. Howard, R. V. Jensen, J. G. Leopold, B. Meerson, D. Richards, D. L. Shepelyansky, U. Smilansky, and B. Sundaram. MVB, KD, RVJ, DR, and US get special thanks. They know why.

I appreciate primary financial support for this research from the Atomic, Molecular, and Plasma Physics Program of the U.S. National Science Foundation, with additional support from New York State, NATO (travel), and the U.S.-Israel Binational Science Foundation. I am also grateful to the Alexander von Humboldt-Stiftung for a Senior U.S. Scientist Award that is affording the opportunity to spend a memorable year in Garching-bei-München at the Max-Planck-Institut für Quantenoptik and the Sektion Physik der Universität München. My family and I are especially grateful for the warm hospitality of Professor Herbert Walther and his charming wife Margot.

References

1. Lichtenberg, A. J., and M.A. Lieberman, *Regular and Stochastic Motion*, (Springer-Verlag, New York, 1983).
2. Berry, M. V., *Proc. R. Soc. Lond. A* **413**, 183 (1987).
3. Berry, M., in *Chaos and Quantum Physics*, Les Houches Lectures, M.-J. Giannoni, A. Voros, J. Zinn-Justin, Eds. (Elsevier, Amsterdam, 1990).
4. Grimes, C. C., T. R. Brown, L. M. Burns, C. L. Zipfel, *Phys. Rev. B* **13**, 140 (1976).
5. Jensen, R. V., *Phys. Rev. Lett.* **49**, 1365 (1982).
6. Jensen, R. V., *Phys. Rev. A* **30**, 386 (1984).
7. Berry, M., *Physica Scripta* **40**, 335 (1989).
8. Bandrauk, A., Ed., *Atomic and Molecular Processes with Short Intense Laser Pulses*, NATO ASI Series B: Physics Vol. 171 (Plenum

Press, New York, 1988); Delone, N. B., and M. V. Fedorov, *Usp. Fiz. Nauk.* **158**, 215 (1989) [*Sov. Phys. Usp.* **32**, 500 (1989)]; see also many papers in *J. Opt. Soc. Am. B*, **7**, no. 4 (April 1990).
9. Reinhardt, W. P., and I. Dana, *Proc. Roy. Soc. Lond. A* **413**, 157 (1987).
10. Ramachandran, B., and K.G. Kay, *Phys. Rev. A* **41**, 1757 (1990).
11. Percival, I. C., and D. Richards, *Adv. Atom. Mol. Phys.* **11**, 1 (1975).
12. Pais, A., *Inward Bound* (Oxford University Press, New York, 1986), p. 247.
13. Bayfield, J. E., and P.M. Koch, *Phys. Rev. Lett.* **33**, 258 (1974).
14. Keldysh, L. V., *Zh. Eksp. Teor. Fiz.* **47**, 1945 (1964) [*Sov. Phys. –JETP* **20**, 1307 (1965)].
15. Bayfield, J. E., L. D. Gardner, P. M. Koch, *Phys. Rev. Lett.* **39**, 76 1977).
16. Leopold, J. G., and I. C. Percival, *Phys. Rev. Lett.* **41**, 944 (1978).
17. Delone, N. B., B. A. Zon, V. P. Krainov, *Zh. Eksp. Teor. Fiz.* **75**, 445 (1978) [*Sov. Phys. – JETP* **48**, 223 (1978)].
18. Meerson, B. I., E. A. Oks, P.V. Sasorov, *Pis'ma Zh. Eksp. Teor. Fiz.* **29**, 79 (1979) [*Sov. Phys. – JETP Lett.* **29**, 72 (1979)].
19. Jones, D. A., J. G. Leopold, I. C. Percival, *J. Phys. B* **13**, 31 (1980).
20. Koch, P. M., *J. Physique Colloq.* **43**, C2–187 (1982).
21. Leopold, J. G., and D. Richards, *J. Phys. B.* **18**, 3369 (1985).
22. van Leeuwen, K. A. H., *et al., Phys. Rev. Lett.* **55**, 2231 (1985).
23. Koch, P. M., in *Fundamental Aspects of Quantum Theory*, V. Gorini and A. Frigerio, Eds., (Plenum, New York, 1986).
24. Koch, P. M., K. A. H. van Leeuwen, O. Rath, D. Richards, R. V. Jensen, in *The Physics of Phase Space*, Lecture Notes in Physics, vol. 278 (Springer-Verlag, Berlin, 1987), pp. 106–13.
25. Koch, P. M., in *Electronic and Atomic Collisions*, H. B. Gilbody, W. R. Newell, F. H. Read, A. C. H. Smith, Eds., (Elsevier Science Publishers, Amsterdam, 1988), pp. 501–515.
26. Moorman, L., *et al.*,in *Atomic Spectra and Collisions in Intense Fields*, K. T. Taylor, M. H. Nayfeh, C. W. Clark, Eds. (Plenum, New York, 1988), pp. 343–357.
27. Moorman, L., *et al., Phys. Rev. Lett.* **61**, 771 (1988).
28. Galvez, E. J., B. E. Sauer, L. Moorman, P. M. Koch, D. Richards, *Phys. Rev. Lett.* **61**, 2011 (1988).
29. Richards, D., *et al., J. Phys. B* **22**, 1307 (1989).
30. Koch, P. M., L. Moorman, B. E. Sauer, E. J. Galvez, K. A. H. van Leeuwen, *Physica Scripta*, **T26**, 51 (1989).
31. Koch, P. M., L. Moorman, B. E. Sauer, E. J. Galvez, in *Classical Dynamics in Atomic and Molecular Collisions*, T. Grozdanov, P. Grujic, P. Kristic, Eds. (World Scientific, Singapore, 1989), pp. 348–67.
32. van Leeuwen, K. A. H., and P.M. Koch, "Ionization of Highly Excited Hydrogen Atoms by a Microwave Electric Field. I: Experiment,"to be published.
33. Sauer, B. E., L. Moorman, P.M. Koch, D. Richards, *Bull. Am. Phys. Soc.* **35**, 1183 (1990).
34. Sauer, B. E., L. Moorman, P. M. Koch, *Bull. Am. Phys. Soc.* **35**, 1183 (1990).
35. Koch, P. M., and D.R. Mariani, *J. Phys. B* **13**, L645 (1980).
36. Koch, P.M., in *Rydberg States of Atoms and Molecules*, R .F. Stebbings and F. B. Dunning, Eds. (Cambridge University Press, New York, 1983), pp. 473–512.
37. Bayfield, J. E., and L.A. Pinnaduwage, *Phys. Rev. Lett.* **54**, 313 (1985).
38. Bayfield, J. E., and D. W. Sokol, in *Atomic Spectra and Collisions in External Fields*, K. T. Taylor, M. H. Nayfeh, C.W. Clark. Eds. (Plenum, New York, 1988), pp. 315–326.
39. Bayfield, J. E., and D. W. Sokol, *Phys. Rev. Lett.* **61**, 2007 (1988).
40. Jensen, R. V., *Phys. Rev. Lett.* **54**, 2057 (1985).
41. Leopold, J. G., and D. Richards, *J. Phys. B.* **20**, 369 (1987).
42. Stevens, M. J., and B. Sundaram, *Phys. Rev. A* **36**, 417 (1987).
43. Bayfield, J. E., G. Casati, I. Guarneri, D. W. Sokol, *Phys. Rev. Lett.* **63**, 364 (1989).
44. Richards, D., in *Aspects of Electron-Molecule Scattering and Photoionization, New Haven, CT, 1989*, AIP Conf. Proc. No. 204, A. Herzenberg, Ed. (AIP, New York, 1990), pp. 55–64.
45. van de Water, W., *et al., Phys. Rev. A*, July 1990.
46. Sauer, B. E., K. A. H. van Leeuwen, A. Mortazawi-M, and P. M. Koch, "Precise Calibration of a Microwave Cavity with a Non-Ideal Waveguide System," to be published.
47. van de Water, W., *et al., Phys. Rev. Lett.* **63**, 762 (1989).
48. Petite, G., P. Agostini, F. Yergeau, *J. Opt. Soc. Am. B* **4**, 765 (1987).
49. Gallagher, T. F., C. R. Mahon, P. Pillet, P. Fu, J. B. Newman, *Phys. Rev. A* **39**, 4545 (1989).
50. Koch, P. M., and D.R. Mariani, *Phys. Rev. Lett.* **46**, 1275 (1981).
51. Rath, O., and D. Richards, "Classical Aspects of Microwave Ionisation of Atomic Hydrogen," to be published.
52. Banks, D., and J. G. Leopold, *J. Phys. B.* **11**, 37 (1978).
53. Blümel, R., G. Jaeckel, U. Smilansky, *Phys. Rev.*

A **39**, 450 (1989).
54. Richards, D., *J. Phys. B.* **20**, 2171 (1987).
55. Blümel, R., and U. Smilansky, (a)*Z. Phys. D* **6**, 83 (1987); (b) *Phys. Rev. Lett.* **58**, 2531 (1987); (c) *J. Opt. Soc. Am. B* **7**, 664 (1990); (d) Blümel, R., C. Hillermeier, U. Smilansky, *Z. Phys. D* **15**, 267 (1990).
56. Shirley, J. H., *Phys. Rev.* **138**, B979 (1965).
57. Chu, S.-I., *Adv. Chem. Phys.* **73**, 739 (1988).
58. Zeldovich, Ya. B., *Zh. Eksp. Teor. Fiz.* **51**, 1492 (1966) [*Sov. Phys. –JETP* **24**, 1006 (1967)].
59. Jensen, R. V., *Physica Scripta* **35**, 668 (1987).
60. Breuer, H. P., K. Dietz, H. Holthaus, *Z. Phys. D* **10**, 12 (1988).
61. Breuer, H. P., and M. Holthaus, *Z. Phys. D* **11**, 1 (1989).
62. Breuer, H. P., K. Dietz, H. Holthaus, *J. Phys. B* **22**, 3187 (1989).
63. Galvez, E. J., *et al.*, in *Proceedings of the Sixth International Rochester Conference on Quantum Optics*, L. Mandel, Ed. (Plenum Press, New York, 1990).
64. Leopold, J. G., and D. Richards, *J. Phys. B.* **19**, 1125 (1986).
65. Gontis, V., and B. Kaulakys, *J. Phys. B* **20**, 5051 (1987).
66. Jensen, R. V., M. M. Sanders, M. Saraceno, B. Sundaram, *Phys. Rev. Lett.* **63**, 2771 (1989).
67. Casati, G., I. Guarneri, D.L. Shepelyansky, *IEEE J. Quantum Electron.* **24**, 1240 (1988).
68. Casati, G., B. V. Chirikov, D. L. Shepelyansky, *Phys. Rev. Lett.* **53**, 2525 (1984).
69. Fishman, S., D. R. Grempel, R. E. Prange, *Phys. Rev. Lett.* **49**, 509 (1982).
70. Casati, G., B. V. Chirikov, D. L. Shepelyansky, I. Guarneri, *Phys. Rep.* **154**, 77 (1987).
71. Brivio, G. P., G. Casati, L. Perotti, I. Guarneri, *Physica D* **33**, 51 (1988).
72. Casati, G., I. Guarneri, D. L. Shepelyansky, *Physica A* **163**, 205 (1990).
73. Chirikov, B. V., in *Chaos and Quantum Physics*, Les Houches Lectures, M.-J. Giannoni, A. Voros, J. Zinn-Justin, Eds. (Elsevier, Amsterdam, 1990).
74. MacKay, R. S., and J. D. Meiss, *Phys. Rev. A* **37**, 4702 (1988).
75. Meiss, J. D., *Phys. Rev. Lett.* **62**, 1576 (1989).
76. Jensen, R. V., S. M. Susskind, M. M. Sanders, *Phys. Rev. Lett.* **62**, 1476 (1989).
77. Leopold, J. G., and D. Richards, *J. Phys. B* **22**, 1931 (1989).
78. Heller, E. J., *Phys. Rev. Lett.* **53**, 1515 (1984).
79. Eckhardt,B., G. Hose, E. Pollak, *Phys. Rev. A* **39**, 3776 (1989); Berry, M. V., *Proc. R. Soc. Lond. A* **423**, 219 (1989).
80. Radons, G., and R. E. Prange, *Phys. Rev. Lett.* **61**, 1691 (1988).
81. Wintgen, D., and A. Hönig, *Phys. Rev. Lett.* **63**, 1467 (1989).
82. Park, D., *Classical Dynamics and Its Quantum Anlogues*. Lecture Notes in Physics, vol. 110 (Springer-Verlag, Berlin, 1979), p. 209.
83. Crawford, F. S., *Am. J. Phys.* **57**, 621 (1989); Hassoun, G. Q., and D. H. Kobe, *ibid.*, p. 658.
84. Norcliffe, A., *Case Studies in Atomic Physics* **4**, 1 (1973).
85. Jensen, R., and D. Richards, private communications.
86. Einstein, A., *Verh. Deutsch. Phys. Ges. (Berlin)* **19**, 82 (1917); Brillouin, L., *J. Phys. Radium* **7**, 353 (1926); Keller, J. B., *Ann. Phys. (New York)* **4**, 180 (1958); see also Maslov, V. P., and M. V. Fedoriuk, *Semi-Classical Approximation in Quantum Mechanics* (Reidel, Boston, 1981).

9

Quantum Chaos

Roderick V. Jensen

Abstract

The realization that simple, nonlinear dynamical systems can exhibit deterministic behavior that is indistinguishable from a random process has given birth to a new field of interdisciplinary research called *nonlinear dynamics* or the study of *chaos*. However, quantum mechanics, which provides the fundamental description of mechanical systems, does not appear to be capable of exhibiting the "extreme sensitivity to initial conditions" that defines chaos in classical mechanics. Consequently, the phrase *quantum chaos* is used to refer to a broad class of interesting problems related to the quantum behavior of classically chaotic systems. For example, since the Bohr model of a highly excited hydrogen atom in a strong static magnetic field or strong microwave field can exhibit chaotic behavior, these physical systems have provided important paradigms for experimental and theoretical investigations of quantum chaos. In the past eight years a number of significant results have emerged from these studies which synthesize a variety of new ideas from classical nonlinear dynamics, atomic physics, chemical physics, and nuclear physics.

Introduction

In the last ten or fifteen years, the word chaos has emerged as a technical term to refer to the complex, irregular, and apparently random behavior of a wide variety of physical phenomena such as turbulent fluid flow, the irregular motion of charged particles in accelerators and fusion devices, and the wandering orbits of asteroids. In the past, these complex phenomena were often referred to as random or stochastic, which meant that researchers gave up all hope of providing a detailed microscopic description and restricted themselves to statistical descriptions alone. What distinguishes *chaos* from these older concepts is the recognition that many complex physical phenomena are actually described by *deterministic equations* such as the Navier-Stokes equations of fluid mechanics or Newton's equations of classical mechanics, and the important discovery that even very simple, deterministic equations of motion can exhibit exceedingly complex behavior that is indistinguishable from a random process. Consequently, a new term was required to describe the irregular behavior of these deterministic dynamical systems, which reflected the new-found hope for a deeper understanding of these various physical phenomena.

The study of chaos in classical systems such as turbulent fluids or strongly coupled nonlinear oscillators is now a fairly well-developed field, which has been described in great detail in a number of popular and technical books (and in other chapters in this volume) *(1–5)*. For example, one important application has been the work of Jack Wisdom at MIT on the chaotic motion of asteroids and the origin of meteorites *(6)*. By numerically integrating the classical Newtonian equations of motion for an asteroid perturbed in its orbit about the sun by the periodic gravitational pull of Jupiter, Wisdom found that some trajectories, originating in the asteroid belt between the orbits of Mars and Jupiter, could become chaotic. These chaotic orbits do not orbit the sun in perfect Kepler ellipses, but instead wobble, precess, and become highly elongated after hundreds of thousands of years until they cross the Earth's orbit and appear as meteors and meteorites. In turn, the depletion of these chaotic trajectories leaves gaps in the asteroid belt (the Kirkwood gaps) which have been observed for over 100 years *(6)*.

However, the many fascinating illustrations of chaos in classical systems are not the main focus of this brief review. Instead, I will discuss the more recent and more controversial studies of the quantum behavior of strongly coupled and strongly perturbed Hamiltonian systems, which are classically chaotic.

The discovery that simple nonlinear models of classical dynamical systems can exhibit chaos has naturally raised the question of whether this behavior persists in the quantum realm where the classical nonlinear equations of motion are replaced by the linear Schrödinger equation *(7)*. Consider, for example, Wisdom's classical problem of the periodically perturbed asteroid orbits, and imagine shrinking this system down to atomic dimensions so that the sun corresponds to the nucleus of a hydrogen atom, the asteroid to an electron, and the periodic perturbation to an externally applied microwave field. If the classical description of this physical system exhibits chaos, what does the quantum theory predict? How do the wavefunctions evolve? Does the quantum system show any signs of classical chaos? These are some of the questions and problems of "quantum chaos" *(7)*.

This is currently a lively area of research. Although there is general consensus on the key problems, the solutions remain a subject of debate. In contrast to the theory of classical chaos, there is not even a consensus on the definition of the term quantum chaos because the Schrödinger equation is a linear equation for the deterministic evolution of the quantum wavefunction, which is incapable of exhibiting the strong dynamical instability that defines chaos in nonlinear classical systems. Nevertheless, numerical studies of model problems and real experiments on atoms and molecules both reveal that quantum systems can exhibit behavior that resembles classical chaos for long periods of time. In addition, considerable research has been devoted to identifying the distinct signatures or symptoms of the quantum behavior of classically chaotic systems.

For example, experiments with highly excited hydrogen atoms in strong microwave fields have revealed a novel ionization mechanism that depends strongly on the intensity of the radiation but only weakly on the frequency *(8–10)*. This dependence is just the opposite of the quantum photoelectric effect, but the sharp onset of ionization, in many cases, is well described by the onset of chaos in the corrresponding classical system. However, in other cases, the ionization can be either enhanced or suppressed by quantum interference effects. Another simple example of a quantum system that is classically chaotic is a hydrogen atom in a strong static magnetic field. In this case, the quantum manifestations of classical chaos cannot be observed directly. However, the quantum energy levels and the transition matrix elements exhibit several distinct quantum "symptoms" of the underlying classical chaos *(11–14)*.

The Problem of Quantum Chaos

Guided by Bohr's correspondence principle it might be natural to conclude that quantum mechanics should agree with the predictions of classical mechanics for macroscopic systems. In addition, because chaos has played a fundamental role in improving our understanding of the microscopic foundations of classical statistical mechanics, one would hope that it would play a similar role in shoring up the foundations of quantum statistical mechanics. Unfortunately, quantum mechanics appears to be incapable of exhibiting the strong local instability that defines classical chaos as a mixing system with positive Kolmogorov-Sinai entropy.

One way of seeing this difficulty is to note that the Schrödinger equation is a linear equation for the wavefunction, and neither the wavefunction nor any observable quantities (determined by taking expectation values of self-adjoint operators) can exhibit the extreme sensitivity to initial conditions that defines classical chaos without first making some classical approximations (see, for example, chapter 10 by R. F. Fox). In fact if the Hamiltonian system is bounded, then the quantum mechanical energy spectrum is discrete, and the time evolution of all quantum mechanical quantities is doomed to quasiperiodic behavior.

Although the question of the existence of quantum chaos remains a controversial topic, nearly everyone agrees that the most important questions relate to the problem of how quantum systems behave when the corresponding classical Hamiltonian system exhibits chaotic behavior. For example, how does the wavefunction behave for strongly perturbed oscillators like an atom or a molecule in an intense electromagnetic field? What are the characteristic properties of the energy levels for a system of strongly coupled oscillators such as an atom in a strong magnetic field or a highly excited polyatomic molecule?

Symptoms of Quantum Chaos

Although the Schrödinger equation is linear, the essential nonintegrability of chaotic Hamiltonian systems carries over to the quantum domain. There are no known examples of chaotic classical systems for which the corresponding wave equations can be solved analytically. Consequently, theoretical searches for quantum chaos have relied heavily on numerical solutions. These detailed numerical studies by physical chemists and physicists of the dynamics of molecules and excitation and ionization of atoms in strong fields have led to the identification of several characteristic symptoms in the quantum wavefunctions and energy levels that reveal the manifestation of chaos in the corresponding classical systems.

One of the most studied characteristics of nonintegrable quantum systems that correspond to classically chaotic Hamiltonian systems is the appearance of "irregular" energy spectra. The energy levels in the hydrogen atom, which are described classically in terms of regular, elliptical Kepler orbits, form an orderly sequence, $E_n = -1/2n^2$ (a.u.), where, $n = 1,2,3,...$ is the principal quantum number. However, the energy levels of chaotic systems, like a highly excited hydrogen atom in a strong magnetic field, do not appear to have any simple order that can be quantified in terms of well-defined quantum numbers. This correspondence makes sense, since the quantum numbers that define the energy levels of integrable systems are associated with the classical constants of motion, like angular momentum, which are destroyed by the nonintegrable perturbation.

For example, Fig. 1 displays the calculated energy levels for a hydrogen atom in a static magnetic field which shows the irregular spectrum ("spaghetti") at high fields where the magnetic forces are comparable to the Coulomb binding.

This irregular spacing of the quantum energy levels can be conveniently characterized in terms of the statistics of the energy-level spacings. For example, Fig. 2 shows a histogram of the energy-level spacings for the hydrogen atom in a magnetic field that is strong enough to make most of the classical electron orbits chaotic. Remarkably, this distribution of energy level spacings is identical to that found for a much more complicated quantum system with irregular spectra – com-

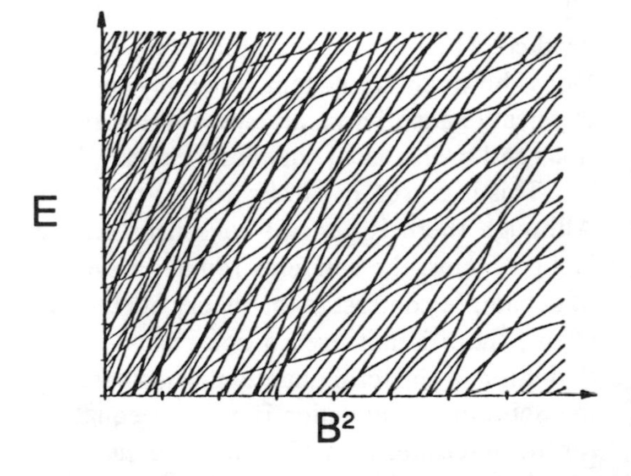

Fig. 1. The quantum mechanical energy levels for a very excited hydrogen atom in a strong magnetic field are highly irregular. This figure shows the numerically calculated energy levels as a function of the square of the magnetic field for a range of energies corresponding to quantum states with principal quantum numbers $n \approx 40 - 50$. Because the magnetic field breaks the natural spherical and Coulomb symmetries of the hydrogen atom, the energy levels and associated quantum states exhibit a "jumble" of multiple avoided crossings due to level repulsion, which is a common symptom of quantum systems that are classically chaotic *(29)*.

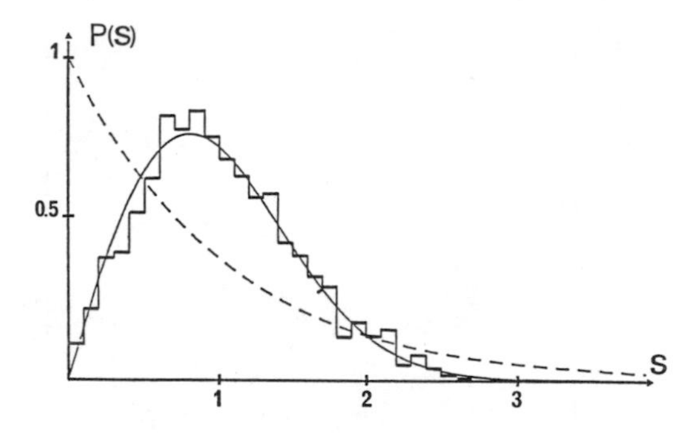

Fig. 2. The repulsion of the quantum mechanical energy levels displayed in Fig. 1 results in a distribution of energy level spacings, P(s), in which accidental degeneracies (s = 0) are extremely rare. This figure displays a histogram of the energy level spacings for 1295 levels like those in Fig. 1. This distribution compares very well with the Wigner-Dyson distribution (solid curve) that is predicted for the energy level spacing for random matrices. If the energy levels were uncorrelated random numbers, then they would be expected to have a Poisson distribution indicated by the dashed curve. [From *(14)* with permission.]

plex nuclei. Moreover, both distributions are well described by the predictions of random matrix theory, which simply replaces the nonintegrable (or unknown) quantum Hamiltonian with an ensemble of large matrices with random values for the matrix elements *(7, 13, 14)*. In particular, this distribution of energy level spacings should obey the Wigner-Dyson distribution $P(s) \sim s \exp(-s^2)$ displayed in Fig. 2. Although these random matrices cannot predict the location of specific energy levels, they do account for many of the statistical features relating to the fluctuations in the energy level spacings.

Despite the apparent statistical character of the quantum energy levels for classically chaotic systems, these level spacings are not completely random. If they were completely uncorrelated, then the spacings statistics would obey a Poisson distribution $P(s) \sim \exp(-s)$, which would predict a much higher probability of nearly degenerate energy levels. The absence of degeneracies in chaotic systems is easily understood because the interaction of all the quantum states induced by the nonintegrable perturbation causes a "repulsion" of nearby levels. In addition, the energy levels exhibit an important long-range correlation called "spectral rigidity," which means that fluctuations about the average level spacing are relatively small over a wide energy range. Recently, Berry has traced this spectral rigidity in the spectra of simple chaotic Hamiltonians, like the hydrogen atom in a strong static magnetic field, to the persistence of regular (but not necessarily stable) periodic orbits in the classical phase space *(15)*. Remarkably, these sets of measure zero classical orbits appear to have a dominant influence on the characteristics of the quantum energy levels and quantum states *(16–19)*.

Recent experimental studies of the energy levels of highly excited Rydberg atoms in strong magnetic fields by Karl Welge and collaborators at the University of Bielefeld *(11)* and by Dan Kleppner's group at MIT *(12)* appear to have confirmed many of these theoretical and numerical predictions. Unfortunately, the experiments can only resolve a limited range of energy levels, which makes the confirmation of statistical predictions difficult. However, the experimental observations of this symptom of quantum chaos are very suggestive. In addition, the experiments have provided very striking evidence for the important role of classical periodic orbits embedded in the chaotic sea of trajectories in determining gross features in the fluctuations in the irregular spectrum. In particular, there appears to be a one-to-one correspondence between regular oscillations in the spectrum and the periods of the shortest periodic orbits in the classical Hamiltonian system *(18, 19)*. Although the corresponding classical dynamics of these simple systems is fully chaotic, the quantum mechanics appears to cling to these last remnants of regularity.

Another more direct symptom of quantum chaos is simply to look for quantum behavior that resembles the predictions of classical chaos. In the cases of atoms or molecules in strong electromagnetic fields where classical chaos predicts ionization or dissociation, this symptom is unambiguous. The patient dies. However, quantum systems appear to be capable only of mimicking classical chaotic behavior for finite times determined by the density of quantum states (or the size of the quantum numbers). In the case of as few as 50 interacting particles, this "break-time" may exceed the age of the Universe *(20)*; however, for small quantum systems like those described by the simple models of Hamiltonian chaos, this time scale, where the Bohr correspondence principle for chaotic systems breaks down, may be accessible to experimental measurements.

Microwave Ionization of Highly Excited Hydrogen Atoms

In particular, there has been a considerable experimental effort by Jim Bayfield's group at Pittsburgh *(9)* and by Peter Koch's group at Stony Brook *(10)* to see whether the quantum manifestations of classical chaos and the modifications by quantum interference effects both could be observed experimentally in a real quantum system consisting of a hydrogen atom prepared in a highly excited state, which is then exposed to intense microwave fields. Since the experiments can be performed with atoms prepared in states with principal quantum numbers as high as $n = 100$, one could hope that the dynamics of this electron with a 0.5 micron Bohr radius would be well described by classical mechanics.

In the presence of an intense oscillating field, this classical nonlinear oscillator is expected to exhibit a transition to global chaos *(21)*. For example, Fig. 3 shows a Poincaré section of the classical action-angle phase space for a one-dimensional model of a hydrogen atom in an oscillating field for parameters that correspond closely to those of the experiments. For small values of the classical action I, which corresponds to low quantum numbers by the Bohr-Somerfeld quantization rule, the perturbing field is much weaker than the Coulomb binding fields, and the orbits lie on smooth curves that are bounded by invariant KAM tori *(21)*. However, for larger values of I, the relative size of the perturbation increases, and the orbits become chaotic, filling large regions of phase space and wandering to arbitrarily large values of the action and ionizing. Since these chaotic orbits ionize, the classical theory predicts an ionization mechanism

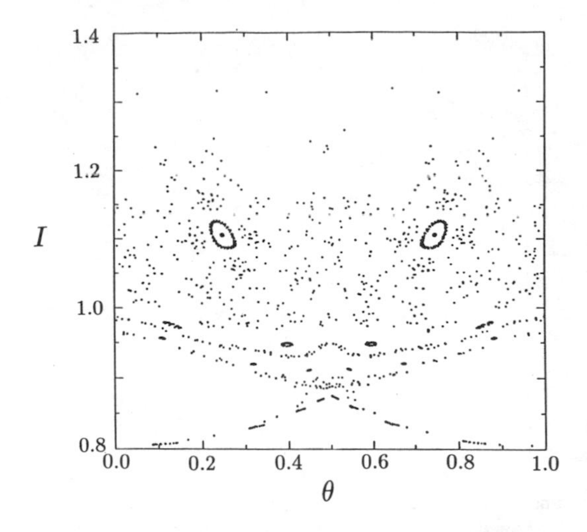

Fig. 3. This Poincaré section of the classical dynamics of a one-dimensional hydrogen atom in a strong oscillating electric field was generated by plotting the value of classical action, I, and angle, θ, once every period of the perturbation with strength $I^4F = 0.03$ and frequency $I^3\Omega = 1.5$. In the absence of the perturbations, the action (which corresponds to principal quantum number n by the Bohr-Sommerfeld quantization rule) is a constant of motion. In this case, different initial conditions (corresponding to different quantum states of the hydrogen atom) would trace out horizontal lines in the phase space. Since the Coulomb binding field decreases as $1/I^4$ (or $1/n^4$), the relative strength of the perturbation increases with I. For a fixed value of the perturbing field F, the classical dynamics is regular for small values of I with a prominent nonlinear resonance below $I = 1.0$. A prominent pair of islands also appears near $I = 1.1$, but it is surrounded by a chaotic "sea." Since the chaotic orbits can wander to arbitrarily high values of the action, they ultimately lead to ionization of the atom.

that depends strongly on the intensity of the radiation and only weakly on the frequency, which is just the opposite of the dependence of the traditional photoelectric effect.

In fact, this chaotic ionization mechanism was first experimentally observed in 1974 by Jim Bayfield and Peter Koch, whose pioneering experiments revealed a sharp onset of ionization in atoms prepared in the $n = 66$ state when a 10 GHz microwave field exceeded a critical threshold *(8)*. Subsequently, the agreement of the predictions of classical chaos with the quantum measurements has been confirmed for a wide range of parameters *(10)* corresponding to principal quantum numbers from $n = 32$ to 90. Figure 4 compares the measured thresholds for the the onset of ionization with the theoretical predictions for the onset of classical chaos in a one-dimensional model of the experiments by Koch *et al.* *(10)*.

Moreover, detailed numerical studies of the solution of the Schrödinger equation for this one-dimensional model have revealed that the quantum mechanism that mimics the onset of classical chaos is an abrupt "delocalization" of the evolving wavepacket when the perturbation exceeds a critical threshold *(22–24)*. However, the quantum calculations also found that, in a parameter range just beyond that studied in the experiments, the threshold fields for quantum delocalization would become larger than the classical predictions for the onset of chaotic ionization *(22)*. Very recently, experiments in this new regime *(25–26)* have been performed, and the experimental evidence supports the theoretical prediction for quantum suppression of classical chaos, although the detailed mechanisms remain a topic of controversy *(27)*.

For example, Fig. 5 shows a comparison of the recent experimental results of Koch *et al.* in this high-frequency regime *(25)* with the results of detailed one-dimensional (1D) numerical simulations of the quantum dynamics *(27)*. Both the experimental measurements and the numerical calculations show a clear increase in the ionization thresholds above the classical values with increasing scaled frequency as predicted by Casati *et al.* *(22)*. However, both the experiment and the calculation continue to exhibit large variations that appear to reflect properties of classical dynamics *(28)*. Understanding the quantum mechanism responsible for this suppression of the chaotic ionization and the origin of the large fluctuations in the ionization threshold continue to be active areas of theoretical research.

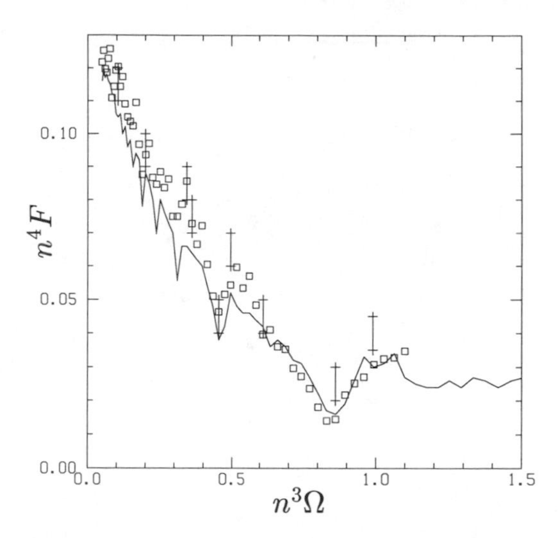

Fig. 4. A comparison of the threshold field strengths for the onset of microwave ionization predicted by the classical theory for the onset of chaos (solid curve) with the results of experimental measurements on real hydrogen atoms with n = 32 to 90 (open squares) and with estimates from the numerical solution of the corresponding Schrödinger equation (crosses). The threshold field strengths are conveniently plotted in terms of the scaled variable, $n^4F = I^4F$, which is the ratio of the perturbing field, F, to the Coulomb binding field, $1/n^4$, versus the scaled frequency, $n^3\Omega = {}^{13}\Omega$, which is the ratio of the microwave frequency, Ω, to the Kepler orbital frequency, $1/n^3$. The prominent features near rational values of the scaled frequency, $n^3\Omega$ = 1, 1/2, 1/3, 1/4, which appear in both the classical and quantum calculations as well as the experimental measurements, are associated with the presence of nonlinear resonances in the classical phase space.

Conclusions

At present, the principal contribution of these studies of quantum chaos has been the demonstration that atomic and molecular physics of strongly perturbed and strongly coupled systems can be very different from that predicted by the traditional perturbative methods of quantum mechanics. New theoretical tools and new intuition based on recent advances in classical nonlinear dynamics are required to explain a variety of new physical phenomena.

These theoretical and experimental studies are helping to open up a new interdisciplinary frontier of modern physics and chemistry, where microscopic systems meet the macroscopic world, where the deterministic descriptions cross over to statistical theories, where the classical theory is chaotic, and where the quantum theory is complicated. In addition to rewriting parts of textbooks in atomic and molecular physics, the results of these studies promise many new theoretical and experimental advances in the future.

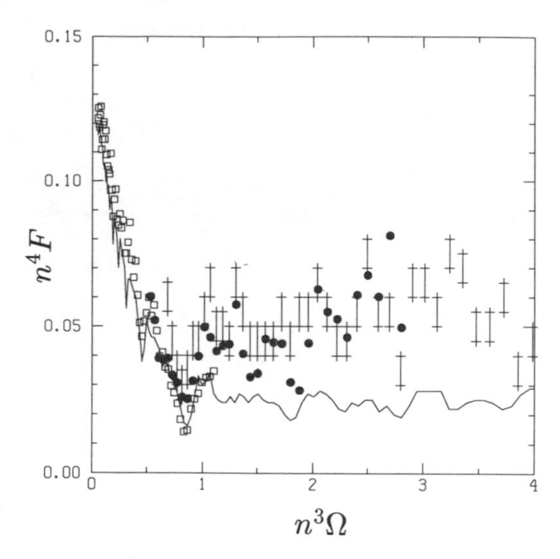

Fig. 5. The classical predictions for the threshold fields (solid curve) are extended to higher scaled frequencies where they are compared with the recent experimental measurments for the ionization of highly excited hydrogen atoms with $n = 50$ to 80 in a 36 GHz field *(25)* (solid circles) and with detailed quantum 1D simulations of the experiment (crosses). In contrast with the low-frequency, 10-GHz results (open squares), which show excellent agreement with the classical results, the new high-frequency measurements show a clear suppression of the classical chaotic ionization *(22)*. Although the 1D numerical simulations of the experiment are not perfect, they do provide very good overall agreement with both the magnitudes of the threshold fields as well as the strong fluctuations as a function of scaled frequency in this new high-frequency regime *(27, 28)*.

Acknowledgments

The author thanks his experimental and theoretical colleagues P. Koch, J. Bayfield, D. Kleppner, K. Welge, M. Berry, J.-C. Gay, and J. Ford for many stimulating discussions of the problems of quantum chaos.

References

1. Gleick, J., *Chaos: Making of a New Science* (Viking, New York, 1987).
2. Cvitanovic, P., Ed., *Universality in Chaos* (Adam Hilger, Bristol, U.K., 1984). This volume contains a collection of the seminal articles by M. Feigenbaum, E. Lorenz, R. M. May, and D. Ruelle as well as an excellent review by R. H. G. Helleman.
3. MacKay, R. S., and J. D. Meiss, Eds., *Hamiltonian Dynamical Systems* (Adam Hilger, Bristol, U.K., 1987).
4. Crutchfield, J. P., J. D. Farmer, N. H. Packard, R. S. Shaw, *Sci. Am.* **255**, 46 (1986).
5. Jensen, R. V., *Am. Sci.* **75**, 166 (1987).
6. Wisdom, J., *Astron. J.* **87**, 577 (1982); ______, *Nature* **315**, 731 (1985); ______, *Phys. Today* **38**, 17 (1985).
7. Casati, G., Ed., *Chaotic Behavior in Quantum Systems* (Plenum, New York, 1985).
8. Bayfield, J. E., and P. M. Koch, *Phys. Rev. Lett.* **33**, 258 (1974).
9. Bayfield, J. E., and L. A. Pinnaduage, *Phys. Rev. Lett.* **54**, 313 (1985); Bayfield, J., and D. W. Sokol, *Phys. Rev. Lett.* **61**, 2007 (1988).
10. van Leeuwen, K. A. H., *et al., Phys. Rev. Lett.* **55**, 2231 (1985).
11. Main, J., G. Wiebush, A. Holle, K. H. Welge, *Phys. Rev. Lett.* **57**, 2789 (1986).
12. Welch, G. R., M. M. Kash, C. Iu, L. Hsu, D. Kleppner, *Phys. Rev. Lett.* **62**, 1975 (1989).
13. Wintgen, D., and H. Friedrich, *Phys. Rev. Lett.* **57**, 571 (1986).
14. Delande, D., and J. C. Gay, *Phys. Rev. Lett.* **57**, 2006 (1986).
15. Berry, M. V., *Proc. R. Soc.* **A400**, 229 (1985).
16. Gutzwiller, M., *J. Math. Phys.* **12**, 343 (1971).
17. Bogomolny, E. B., *Physica D* **31**, 169 (1988).
18. Du, M. L., and J. B. Delos, *Phys. Rev. A* **38**, 1896 (1988).
19. Wintgen, D., *Phys. Rev. Lett.* **61**, 1803 (1987).
20. Peres, A., *Phys. Rev. Lett.* **49**, 1118 (1982).
21. Jensen, R. V., *Phys. Rev. A* **30**, 386 (1984).
22. Casati, G., B. V. Chirikov, D. L. Shepelyansky, *Phys. Rep.* **154**, 77 (1987); Casati, G., I. Guarneri, D. L. Shepelyansky, *IEEE Quantum Electron.* **24**, 1420 (1988).
23. Blümel, R., and U. Smilansky, *Z. Phys. D* **6**, 83 (1987).
24. Jensen, R. V., in *Atomic Physics 10,* H. Narumi and I. Shimamura, Eds. (North Holland, New York, 1987), p. 319; Jensen, R. V., S. M. Susskind, M. M. Sanders, *Phys. Rep.,* submitted for publication.
25. Galvez, E. J., B. E. Sauer, L. Moorman, P. M. Koch, D. Richards, *Phys. Rev. Lett.* **61**, 2011 (1988).
26. Bayfield, J. E., G. Casati, I. Guarneri, D. W. Sokol, *Phys. Rev. Lett.* **62**, 1476 (1989).
27. Jensen, R. V., S. M. Susskind, M. M. Sanders, *Phys. Rev. Lett.* **62**, 1476 (1989); Jensen, R. V., J. G. Leopold, D. Richards, *J. Phys. B* **21**, L527 (1988); Richards, D., J. G. Leopold, R. V. Jensen, *J. Phys. B* **22**, 417 (1989).
28. Jensen, R. V., M. Saraceno, B. Sundaram, M. M. Sanders, *Phys. Rev. Lett.* **63**, 2771 (1989).
29. Delande, D., Ph.D. thesis, Université Pierre et Marie Curie, Paris (1988).

10

Quantum Chaos in Two-Level Quantum Systems

Ronald F. Fox

Abstract

A system of N two-level spins inside a resonant laser cavity interacts with a radiation field tuned to the level spacing. The time evolution of the population of spins in the two states is followed. This variable exhibits chaotic sensitivity to initial conditions for which a positive Lyapunov exponent is computed. It is shown that the dynamics can be transformed into that for a periodically perturbed planar pendulum operating near its separatrix motion. The periodic perturbations, which are responsible for the chaos in this quantum system, are caused by virtual quantum transitions that have no classical analogue.

Introduction

The work discussed in this chapter was performed in collaboration with my recent graduate student, John C. Eidson. It resulted in his Ph.D. thesis *(1)* and in three journal publications *(2–4)*. Our interest in quantum chaos was originally stimulated by my colleague, Joseph Ford, who expressed serious concerns regarding the possibility of chaos in a quantum system. We looked for a simple system to study in order to put Ford's concerns to the test. By chance, we came across a very provocative paper by Milonni *et al.* *(5)* that discussed quantum chaos in a two-level quantum system inside a laser cavity, where it interacted with a mode of laser radiation. Their approach was based on the behavior of power spectra that, as it turned out, had very eye-catching peculiarities. In their paper, there was a reference to the earlier work of Belobrov *et al.* *(6)* that brought to this quantum problem the utilization of a concept that is central in discussions of classical mechanical chaos, the Lyapunov exponent. Ultimately, our studies showed that these two antecedent studies were slightly different, and our analysis turned out to be nearly identical with the model studied by Belobrov *et al.* Therefore, we call our model system the BZT model.

In classical mechanical settings, the concept of chaos is now very well established. Nearly every researcher agrees that the characterization of chaos in such settings is achieved by study of phase-space trajectories. When these trajectories show extreme sensitivity to initial conditions, the system is said to be chaotic. The initial condition sensitivity may be quantified by the Lyapunov exponent, which measures the rate of exponential separation of initially close trajectories *(7)*. In classical mechanics, phase-space trajectories are determined by the solutions to autonomous, coupled, first-order, ordinary differential equations. These equations are just Newton's equations, or equivalently, Hamilton's equations of motion.

This straightforward approach to chaos in classical mechanical settings appears to run into difficulty in quantum mechanical settings because there is no phase space in quantum mechanics. Heisenberg's uncertainty principle tells us that coordinates and their conjugate momenta cannot be known simultaneously with the arbitrary precision of classical mechanical trajectories. While this is true, our approach is based upon the idea that it is meaningful in quantum mechanical settings to study the trajectories of the expectation values of the quantum variables in an expectation-value phase space.

Heisenberg's uncertainty principle tells us that these expectation-value trajectories have nonvanishing variances associated with them, but it does not prevent us from following all of the expectation values. For such quantum mechanical trajectories, a Lyapunov exponent may be determined in precisely the same manner used in classical mechanical settings *(1, 3, 6)*. In contrast with the classical mechanical case in which the trajectories are determined by autonomous coupled equations, in the quantum mechanical case the trajectories of the expectation values are determined, in accord with Ehrenfest's theorem, by a separate solution of Schrödinger's equation (or equivalently of Heisenberg's equations) from which the expectation values are computed. This is a nonautonomous solution, which goes over into the autonomous classical description only if Planck's constant, $\hbar$, becomes vanishingly small.

The equations governing chaotic behavior are not integrable. Consequently, we had to use computer-generated numerical simulations to perform our study. Initially, many of our numerical studies were performed as experiments to give us insight into the model. This approach paid off, and eventually we found powerful analytical methods that justified and explained our numerical discoveries.

The BZT Model

Our model consists of a laser cavity tuned for a single mode of electromagnetic radiation with frequency Ω. Inside the cavity are N two-level quantum systems (Fig. 1). These two-level systems might be spin 1/2 particles (for which there would also be an external, constant, uniform magnetic field along a direction

Fig. 1. The laser cavity is the space between the mirror planes. The wiggly arrows denote the electromagnetic field in the cavity. The small arrows denote the two-level systems.

perpendicular to the axis of the laser cavity, which has the effect of splitting the degeneracy of the two levels of the spin 1/2 particles), or they might correspond to atoms in high Rydberg states such that two adjacent Rydberg states serve as the two levels. In either case, the energy splitting of the two levels is given by $E_2 - E_1 = \hbar\Omega_o$. In this model, the N two-level systems do not interact by collisions or by any other direct means. For a cavity one meter long with a cross section of one centimeter squared, N could be 10^{12} and the two-level systems would constitute an extremely dilute gas!

In the earlier work *(1, 5, 6)*, only a single two-level system inside the cavity was contemplated. However, as will be explained below, such a system does not permit the treatment used here, whereas for sufficiently large N, our treatment is valid. It is also important to emphasize that this model consists of three equally essential ingredients: an electromagnetic field, two-level systems, and the laser cavity. The laser cavity creates the feedback effects of the state of the two-level systems on the electromagnetic field, which ultimately causes chaos.

The Hamiltonian that describes our version of the BZT model *(3)* is given by

$$H = 1/2\,\hbar\Omega_o S_z + \hbar\Omega(a^+a + 1/2) + \hbar\Gamma S_x(a + a^+) \qquad (1)$$

in which

$$S_z = \sum_{j=1}^{N} \sigma_{zj} \text{ and } S_x = \sum_{j=1}^{N} \sigma_{xj}, \qquad (2)$$

where σ_{zj} and σ_{xj} are Pauli matrices for the jth two-level system, a^+ and a are creation and annihilation operators for the photons that constitute the quantized electromagnetic field, and Γ is the coupling constant for the interaction between the two-level systems and the photons. In the Schrödinger picture, these operators are constants and the wave function evolves in time. In the Heisenberg picture, these operators evolve in time. While in the Schrödinger picture, the Pauli operators and the photon operators commute with each other for all times, in the Heisenberg picture they become combined in a complicated way by the time evolution, such that at all times subsequent to the initial time they no longer commute with each other. Consequently, in the Heisenberg picture, we should formally make the interaction term in Eq. 1 manifestly hermitean by the replacement

$$S_x(a+a^+) \rightarrow 1/2\, S_x(a+a^+) + 1/2\,(a+a^+)\,S_x. \qquad (3)$$

We will avoid this cumbersome technicality below, although we have it in mind throughout.

The Heisenberg operator equations may be written in the form

$$\dot{S}_x = -\,\Omega_o S_y \qquad (4)$$

$$\dot{S}_y = \Omega_o S_x - 2\Gamma\,(a + a^+)\,S_z \qquad (5)$$

$$\dot{S}_z = 2\Gamma\,(a + a^+)\,S_y \qquad (6)$$

$$\dot{a} + \dot{a}^+ = -\,\Omega i\,(a - a^+) \qquad (7)$$

$$i(\dot{a} - \dot{a}^+) = \Omega(a + a^+) + 2\Gamma S_x \qquad (8)$$

in which the over dots denote the time derivative. (In Eq. 5 and 6 we have in mind the notion expressed in Eq. 3.) These operator equations cannot be solved in closed form; they are not integrable. However, by an ad hoc modification of the interaction term, one obtains the Jaynes-Cummings model *(8)*, which is equivalent to the rotating wave approximation (RWA)

$$S_x(a+a^+) \rightarrow S_-a^+ + S_+a, \qquad (9)$$

where

$$S_\pm = 1/2(S_x \pm iS_y). \qquad (10)$$

With this slight modification, the corresponding Heisenberg operator equations become integrable in closed form. I emphasize this feature of the analysis because it means we can think of the BZT model as being "very close" to an integrable (and hence nonchaotic) model. The complete model somehow behaves like a perturbation of the integrable model. For the rest of our analysis, we also impose the resonance condition, $\Omega_o = \Omega$.

Numerical Analysis of the BZT Model

Rather than study how operators evolve according to Heisenberg's equations, we want to use the perspective of expectation value phase space on this problem. In the Heisenberg picture, the expectation values of the operators at time *t* are determined by the state function at the initial time and the evolved operators at time *t*. We introduce the symbol Ex( ) for expectation value and some rescaling of variables by N to define *x, y, z*, Q, P, A, and B:

$$Nx = Ex(S_x) \qquad (11)$$

$$Ny = Ex(S_y) \qquad (12)$$

$$Nz = Ex(S_z) \qquad (13)$$

$$Q = Ex(a + a^+) \qquad (14)$$

$$P = Ex(i(a - a^+)) \qquad (15)$$

$$\Gamma_N = \sqrt{N}\Gamma,\ A = Q/\sqrt{N},\ \text{and}\ B = P/\sqrt{N}. \qquad (16)$$

Taking expectation values of Eqs. 4–8 yields the ordinary, coupled, first-order differential equations

$$\dot{x} = -\,\Omega y, \qquad (17)$$

$$\dot{y} = \Omega x - 2\Gamma_N Az, \qquad (18)$$

$$\dot{z} = 2\Gamma_N Ay, \qquad (19)$$

$$\dot{A} = -\,\Omega B, \qquad (20)$$

$$\dot{B} = \Omega A + 2\Gamma_N x. \qquad (21)$$

Now, we have done one manipulation above that is not rigorous but only approximate. In Eqs. 18 and 19, we have replaced the expectation value of the products of two operators by the products of their expectation values. This factorization of expectation values is generally invalid, but in this case, it is correct up to an error of order 1/N for sufficiently large N. In this context, such a factorization has been extensively studied *(9–12)* and, in particular, application to both the Jaynes-Cummings model and the BZT model has been investigated *(13)*. For an N of order 10^{12}, as contemplated in the previous section, we have a very dilute gas and extremely good factorization. The feedback effect of the cavity on the electromagnetic field shows itself in the *x*-dependence of Eq. 21.

The set of Eqs. 17–21 is an autonomous set of equations. While this is not the situation expected generally for the expectation-value, phase-space approach, it is the situation in the special case of the BZT Model because of factorization of expectation values. These five variables are not independent. There are two conserved quantities which may be expressed by

$$x^2 + y^2 + z^2 = C_1 \tag{22}$$

$$\Omega z/2 + \Omega(A^2 + B^2)/4 + \Gamma_N Ax = C_2 \tag{23}$$

where C_1 and C_2 are constants. The first conserved quantity expresses the conservation of total probability for the states of the two-level systems. This is an absolute conservation law for the original, fully quantal description implicit in the Hamiltonian Eq. 1. Factorization of expectation values has not altered this conservation law. The second conserved quantity expresses the conservation of total energy for the two-level system and the electromagnetic field together. It too is an absolute conservation law for the fully quantal description implicit in the Hamiltonian Eq. 1, and it too has not been affected by factorization. These are good signs regarding our treatment. Five variables and two conserved quantities imply three independent variables. This is the minimum number of variables necessary for chaos according to the Poincaré-Bendixson theorem *(14)*. Thus, our model is the minimal model that might exhibit chaos.

The numerical integration of Eqs. 17–21 was done on a SUN 2/120 computer in double precision (14 decimal places). IMSL library routines for solving differential equations and for computing fast Fourier transforms (FFT) were employed. [See *(2)* for more details regarding the results illustrated below.]

In Fig. 2, the time dependence of *z(t)* is shown for the initial condition in which all N two-level systems are in the upper energy

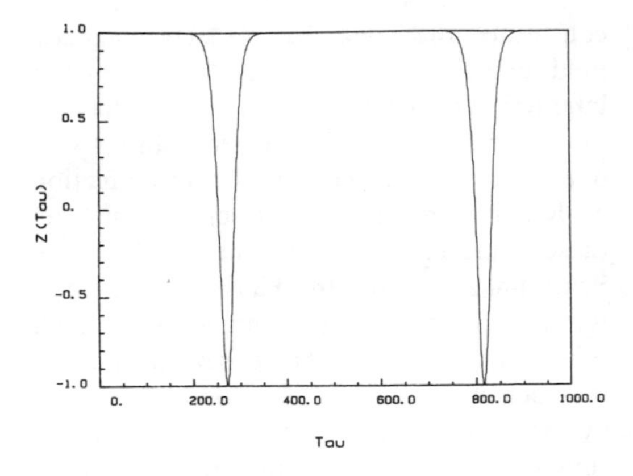

Fig. 2. *z* is along the ordinate, and scaled time (tau) is along the abscissa. N = 0.05.

level. The rapid dips from +1 to −1 repeat over and over again with a regular minimum to minimum period of about 550 tau units. This corresponds to a fundamental frequency of 0.0018 Hz. Following Milonni *et al.* *(5)*, we computed the power spectrum for *z(t)* using a FFT, and we show a semi-log plot of the power spectrum in Figs. 3 and 4. In Fig. 3, we reproduce the eye-catching spectrum originally published by Milonni *et al.* *(5)*. Notice that the ordinate is $\log_{10}$. The power spectrum covers nearly 10 decades of power. In Fig. 4, we show a highly refined version of the spectrum obtained from a longer time series. The scale of the abscissa shown is one Hertz. Therefore, 0.0018 Hz is way over to the left against the ordinate. The features at 0.32 Hz

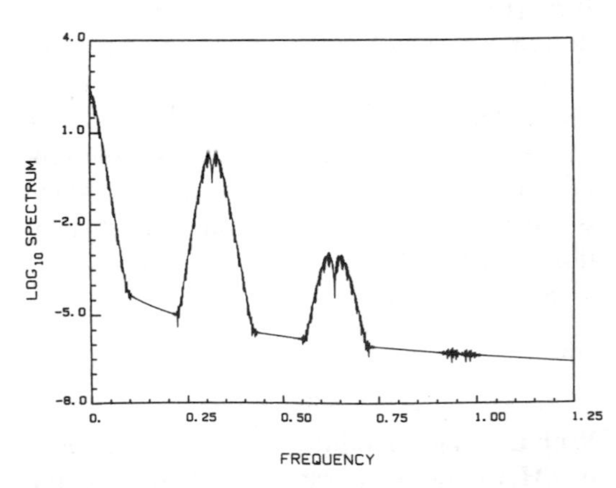

Fig. 3. The ordinate is log 10 of the FFT spectrum, and the abscissa is the frequency (in Hertz for our scaled time units).

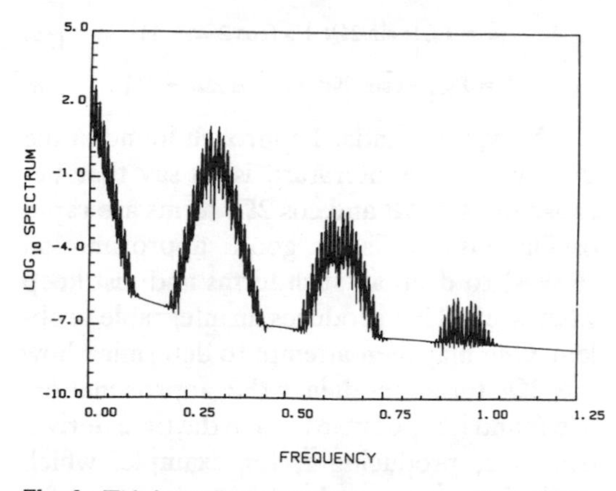

Fig. 4. This is a refinement of the power spectrum in Fig. 3 created from five times as many time points.

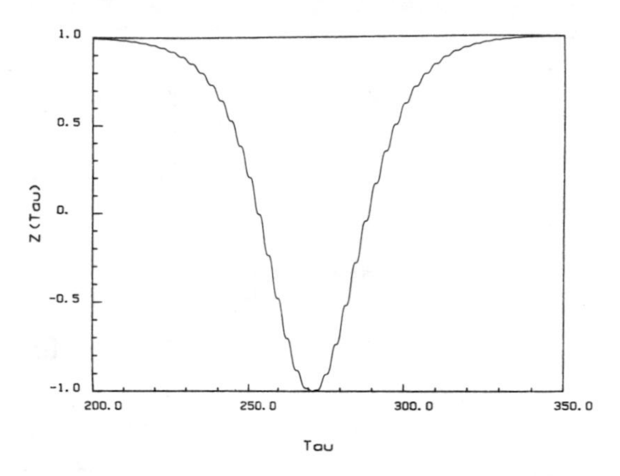

Fig. 6. This is the first dip in Fig. 2 with the time axis expanded by eight-fold.

and at two and three times 0.32 Hz appear to be a series of harmonics of a frequency approximately 178 times bigger than the fundamental at 0.0018 Hz. We were very much intrigued by these features but at first had no idea what caused them.

In Fig. 5 we show the left-hand initial segment of Fig. 4 with the time axis along the abscissa expanded by about eight-fold. Now the fundamental at 0.0018 Hz is very clearly visible as the very first peak at the left-hand edge, and a beautiful series of perfect harmonics of this fundamental is seen, falling off by ten decades in power. These harmonics reflect the intrinsic nonlinearity in our model. However, the features in Fig. 4 at 0.32 Hz and its harmonics are still a mystery. In Fig. 6, we show a single dip from Fig. 2, with the time axis (abscissa) expanded by about eight-fold. What we now see is that the dip carries a more rapid modulation. When we measure the period of this rapid modulation, we find that it corresponds precisely with the frequency 0.32 Hz. Thus, we have discovered the origin of the additional features in the power spectrum in Fig. 4, but we still need to identify their underlying cause.

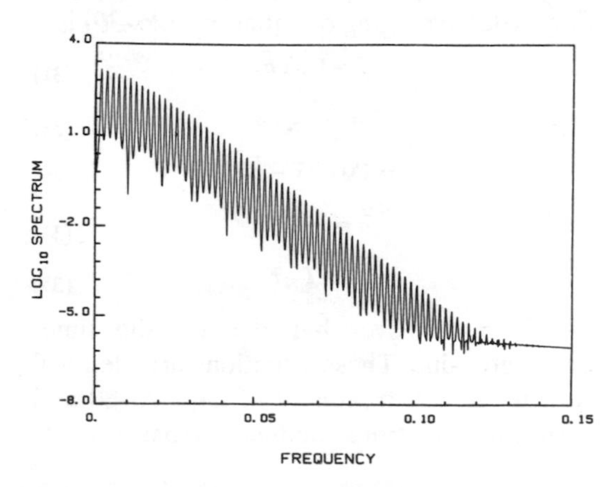

Fig. 5. This is the left-hand initial segment of Fig. 4 with the frequency axis expanded by eight-fold.

Virtual Quantum Processes

The physical effect that produces the rapid modulations is a very basic quantum phenomenon called a virtual transition. Figure 7 shows four intrinsic transitions that are created by the interaction portion of the Hamiltonian Eq. 1. Two of these transitions are energy conserving and are familiar in classical physics, but the other two, the virtual transitions, are purely quantum mechanical. In classical physics, the nonlinear coupling of two oscillators with frequencies Ω and Ω_o will produce output containing frequencies $\Omega \pm \Omega_o$. This is very close to what is happening here, where $\Omega - \Omega_o$ corresponds to the energy conserving transitions, and $\Omega + \Omega_o$ corresponds to the virtual transitions. At resonance, the virtual transitions have a frequency of 2Ω in radians per unit time. When converted to the scaled time units of the figures, $\Omega \rightarrow 1$, and conversion to Hertz yields $2\Omega/2\pi = 1/\pi = 0.318$ Hz, which is precisely where we see the unusual feature

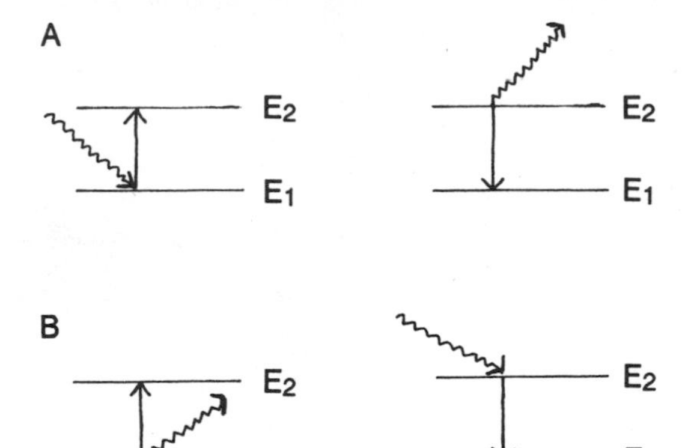

Fig. 7. Basic quantum processes: (A) energy conserving (RWA); (B) virtual, energy nonconserving. The wiggly arrows represent photons that are either absorbed or emitted, as indicated.

(first harmonic) in the power spectrum.

The Jaynes-Cummings interaction (Eq. 9) is designed so that it removes the virtual transitions and only retains the energy-conserving transitions. This is the same effect caused by the RWA. Since this approximation to our model also eliminates chaos (the RWA model is integrable), we may conclude that the virtual transitions are the cause of chaos in the BZT model. However, we still have not yet shown the chaos in the model. In anticipation of demonstrable chaos, let us use our current understanding to represent the dynamics of our model in a more illuminating fashion.

Equations 17–21 show that x and y are coupled to each other like a pair of harmonic oscillator variables, as are A and B. This suggests the transformations[1]

$$\begin{bmatrix} r \\ s \end{bmatrix} = \begin{bmatrix} \cos \Omega t & \sin \Omega t \\ -\sin \Omega t & \cos \Omega t \end{bmatrix} \begin{bmatrix} x \\ y \end{bmatrix} \tag{24}$$

$$\begin{bmatrix} a \\ b \end{bmatrix} = \begin{bmatrix} \cos \Omega t & \sin \Omega t \\ -\sin \Omega t & \cos \Omega t \end{bmatrix} \begin{bmatrix} \mathrm{A} \\ \mathrm{B} \end{bmatrix} \tag{25}$$

The BZT equations (17–21) now become

$$\dot{r} = -\Gamma_N z \left\{ a \sin 2\Omega t + b (\cos 2\Omega t - 1) \right\} \tag{26}$$

$$\dot{s} = -\Gamma_N z \left\{ -b \sin 2\Omega t + a (\cos 2\Omega t + 1) \right\} \tag{27}$$

$$\dot{z} = \Gamma_N \left\{ (ar - bs) \sin 2\Omega t + (as + br) \cos 2\Omega t + (as - br) \right\} \tag{28}$$

$$\dot{a} = \Gamma_N \left\{ r \sin 2\Omega t + s (\cos 2\Omega t - 1) \right\} \tag{29}$$

$$\dot{b} = \Gamma_N \left\{ -s \sin 2\Omega t + r (\cos 2\Omega t + 1) \right\} . \tag{30}$$

Now, the standard approach found in the quantum optics literature is to say that because the sin $2\Omega t$ and cos $2\Omega t$ terms are rapid oscillations, it is a good approximation (RWA) to drop all such terms and just keep what is left. That produces an integrable problem. One may then attempt to determine how the $2\Omega t$ terms modulate the approximation. We found it is possible to take the time derivative of $\dot{z}$, producing $\ddot{z}$, for example, which depends on $\dot{a}, b, \dot{r}$, and $\dot{s}$. After substituting for these first derivatives from Eqs. 26–30, we find terms in z that contain $(\sin 2\Omega t)^2$ and $(\cos 2\Omega t)^2$. Now, while sin $2\Omega t$ is on the average equal to zero, its square, on the average, yields 1/2. Since there are many different ways to get higher time-derivative equations, we were able to generate all sorts of averaged equations that were inequivalent. In short, averaging over rapid oscillations and generating higher-order derivative equations are noncommuting operations on Eqs. 26–30. This left us in a quandary until we discovered *(4)* a systematic averaging procedure for rapid oscillations that removes this ambiguity.

Hale's Averaging Theorem and Eberly's Transformation

Hale's averaging theorem *(15, 16)* is designed for problems just like ours. The result *(4)* of first-order averaging of equations *(26–30)* is

$$\dot{\bar{r}} = \Gamma_N \bar{z} \bar{b} \tag{31}$$

$$\dot{\bar{s}} = -\Gamma_N \bar{z} \bar{a} \tag{32}$$

$$\dot{\bar{z}} = \Gamma_N (\bar{a}\bar{s} - \bar{b}\bar{r}) \tag{33}$$

$$\dot{\bar{a}} = -\Gamma_N \bar{s} \tag{34}$$

$$\dot{\bar{b}} = \Gamma_N \bar{r} \tag{35}$$

in which the over bar denotes the time-averaged value. These equations are identical with the RWA treatment of Eqs. 26–30 and with the equations deduced from the in-

1 In *(4)*, we used a different notation for these transformations in which *(x,y)*→*(a,b)* and (A,B)→*(r,s)*. This seems perverse, given the naturalness of choosing (A,B)→*(a,b)* instead, so we have changed notation for this chapter.

tegrable Jaynes-Cummings model (without involving expectation value factorization).

Inspection of Eqs. 31–35 verifies that there are now three conserved quantities

$$\bar{r}^2 + \bar{s}^2 + \bar{z}^2 = C_1 \tag{36}$$

$$\bar{a}^2 + \bar{b}^2 + 2\bar{z} = C_2 \tag{37}$$

$$\bar{a}\bar{r} + \bar{b}\bar{s} = P . \tag{38}$$

This reduces the number of independent variables from three to two, which means there cannot be chaos in this approximate description. However, we are now able to render this approximation in an especially revealing form.

Because $\bar{z}$ varies between +1 and −1, Eberly *(17)* long ago suggested the transformation

$$\bar{z} = \cos \Phi . \tag{39}$$

From Eqs. 31–35, we first deduce the equivalent equation

$$\ddot{\bar{z}} = -\Gamma_N^2 (C_1 + C_2 \bar{z} - 3\bar{z}^2) , \tag{40}$$

and from Eq. 39 we finally get

$$\ddot{\Phi} = \Gamma_N^2 \sin \Phi + \Gamma_N^2 \frac{P^2 \cos \Phi}{\sin^3 \Phi} , \tag{41}$$

which is the equation for a spherical pendulum. For this pendulum, Φ is 0 for the vertical up position, and P is the azimuthal angular momentum. Equation 40 can also be recognized *(4)* as the equation for Jacobi elliptic functions, which are the solutions for the pendulum dynamics.

For the initial conditions used to obtain the figures shown in this presentation, the constants C_1 and P are 1 and 0, respectively. P = 0 implies that the pendulum is actually a planar pendulum (i.e., it has no azimuthal motion). Moreover, our initial conditions further imply that this planar pendulum is operating very close to its separatrix motion, which is especially clear in Fig. 2 where the following pattern is apparent. The pendulum starts off in the up position, where it stays for awhile, and then it rapidly shoots down and up again, where it again hangs out in the up position for awhile before repeating this behavior over and over again. This is characteristic, near-separatrix pendulum motion.

Hale's theorem allows us to obtain the systematic conditions *(4)* to this pendulum dynamics caused by the rapid modulations that were eliminated by first-order averaging. These corrections are obtained by second-order averaging *(4, 18)*. The result is a periodically modulated, near-separatrix pendulum dynamics governed by the equation

$$\ddot{z} = -\Gamma_N^2 (C_1 + C_2 z - 3z^2) + 2\Gamma_N \{ f(t) \cos 2\Omega t + g(t) \sin 2\Omega t \} \tag{42}$$

in which $f(t)$ and $g(t)$ are determined separately from the second-order averaged equations *(4)*, which are autonomous. Now, P is no longer conserved, so that we are back to the three independent variable situation for which chaos is a possibility. Moreover, a periodically modulated, near-separatrix pendulum motion is known *(19, 20)* to be generic for chaos. This means that the virtual quantum transitions produce a periodic modulation of an integrable system in just the right manner to make it chaotic.

To confirm this analysis, which tells us that our original BZT model is really very well described by a periodically perturbed pendulum, we look at $z(t)$ and its power spectrum as determined from Eq. 42. Because Eq. 42 is a generic example of a nonintegrable system, we must again use numerical simulation to obtain its behavior.[2] But by now, the reader may well exclaim, "Wait a minute, I don't see anything chaotic about these figures!"

The chaos in these figures cannot be seen because the coupling constant, Γ_N, is too small (0.05 in the numerical simulations). The periodically perturbed pendulum is chaotic for any value of Γ_N, but when Γ_N is small, the chaos shows up as an extremely narrow band around the separatrix motion. For $\Gamma_N = 0.05$, it is too small to see in $z(t)$ or in its power spectrum. Nevertheless, we can see it in Eqs. 17–21 by computing the Lyapunov exponent for the trajectories in the expectation value phase space. For $\Gamma_N = 0.05$, we obtained a

2 The figures corresponding to this numerical simulation are found in *(2)*. They are omitted here because they reproduce Figs. 2–6 extremely well.

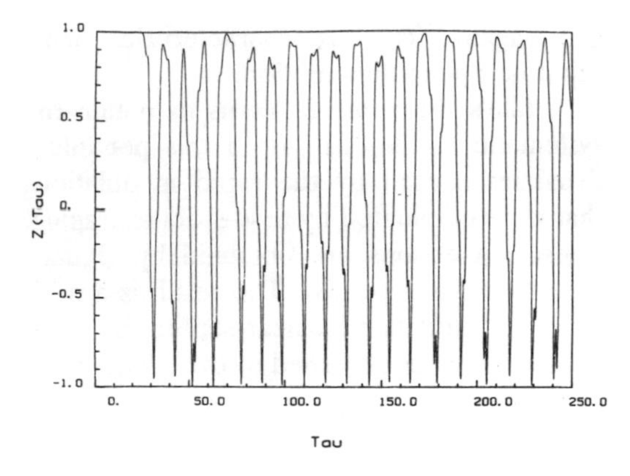

Fig. 8. z is along the ordinate, and scaled time (tau) is along the abscissa. $\Gamma_N = 0.5$.

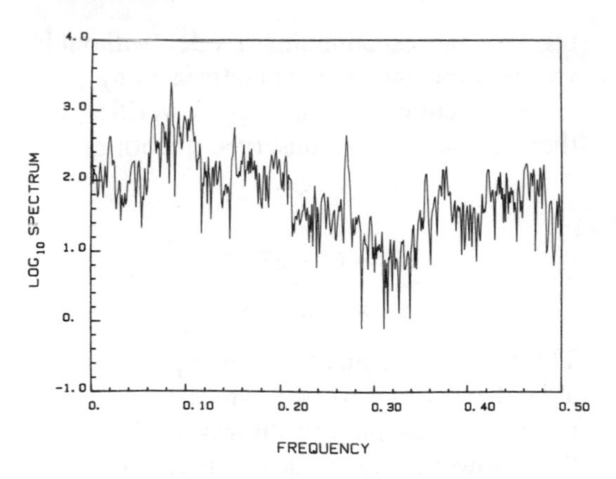

Fig. 10. Eight-fold frequency axis (abscissa) expansion of left-hand initial segment of Fig. 9.

Lyapunov exponent of 0.03 which, while small, is rigorously nonzero. To make the chaos more manifest requires increasing Γ_N, which we did *(2)*. Using $\Gamma_N = 0.5$ produces a dramatic change in the figures and increases the Lyapunov exponent to 0.196.

In Fig. 8, the time dependence of $z(t)$ is shown for $\Gamma_N = 0.5$. The dips are much more frequent (notice the time scale differences between Figs. 2 and 8) and no longer go all the way to -1 or back up to $+1$. This figure is identical with one originally published by Milonni *et al.* *(5)* and is not exhibiting numerical artifacts. Its corresponding power spectrum, shown in Fig. 9, looks manifestly noisy. However, power spectra by themselves cannot definitively determine chaos, a point we discussed in detail in *(3)*. When the power

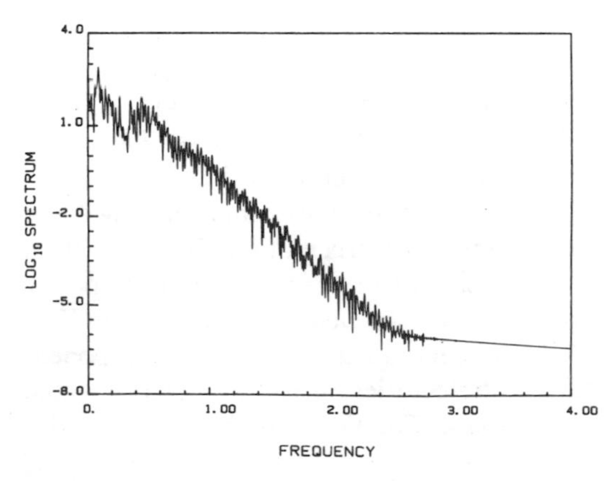

Fig. 9. Semi-log plot of the power spectrum for Fig. 8.

spectrum of Fig. 9 is viewed with the frequency axis (abscissa) expanded by eight-fold, we get Fig. 10, which continues to look noisy. This is characteristic of a truly noisy spectrum. There is a noticeable peak at about 0.27 Hz. In Fig. 8, the dips occur with an irregular period, but one can measure an averaged period, and this turns out to correspond precisely with the power spectrum peak at about 0.27 Hz. Overall, however, we have evidence for manifest chaos when $\Gamma_N = 0.5$ because the Lyapunov exponent is nearly 0.2 and the power spectrum is noisy.

This large value of Γ_N does not permit us to use Eq. 42 as a model for these observations because Eq. 42 only incorporates the first correction to the pure pendulum dynamics, which is valid for small Γ_N only. For large Γ_N, we would need many more correction terms, and the averaging procedure is no longer an effective approach to the analytic justification of Figs. 8–10. It occurred to us that the $\Gamma_N = 0.5$ case is essentially a pendulum dynamics perturbed by noise rather than by a periodic modulation. To test this idea, we simply ran Eq. 40 with the same initial conditions as for Figs. 8–10 and reduced our integration routine from double precision to single precision (seven decimal digits) so that numerical round-off errors would act like a noisy perturbation. The corresponding power spectrum is shown in Fig. 11, and it clearly captures the essence of Fig. 9.

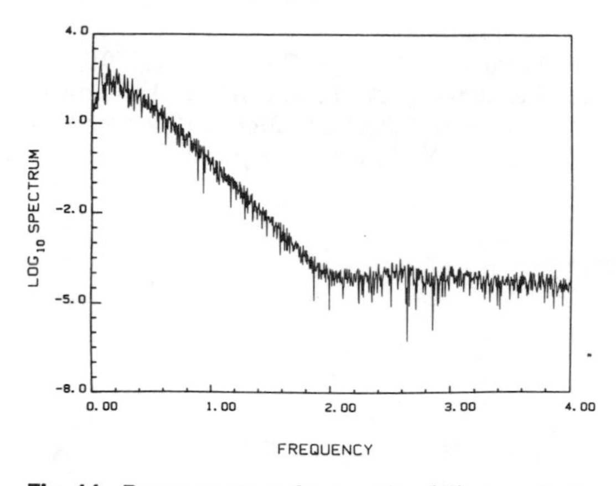

Fig. 11. Power spectrum for equation (40) numerically integrated in single precision.

Concluding Remarks

We have demonstrated that our quantum mechanical model behaves just like periodically modulated, near-separatrix, pendulum dynamics for weak coupling (Γ_N = 0.05). We have done this both by numerical simulation and by analytical treatment with Hale's averaging theorem. The periodic modulations were seen to arise from virtual quantum transitions. When these virtual processes are omitted (RWA or Jaynes-Cummings model) the quantum problem is exactly integrable (and nonchaotic). The presence of the cavity enables these virtual transitions to cause chaos.

Chaos in our model depends on the initial conditions. Choosing initial conditions so that we have near-separatrix motion of a planar pendulum leads to chaos. But it is also possible to have initial conditions that are very different and in the RWA correspond to $P \neq 0$. When $P \neq 0$, near-separatrix motion is prevented by the P-dependent term in Eq. 41. Consequently, even the periodically modulated pendulum dynamics are kept away from the separatrix for Γ_N = 0.05. We studied this case by starting the two-level systems in identical eigenstates of x instead of z. The result was a dynamics for which the computed Lyapunov exponent asymptotically approached zero (in practice, the Lyapunov computation can never give zero in a finite time period, but it does show an asymptotically monotone-decreasing tendency, and for our studies, it was well below 0.03).

Obviously, these remarks mean that the existence of chaos in a quantum system is not solely determined by the Hamiltonian (or its energy eigen-spectrum or its eigenfunctions) but also crucially depends on initial conditions, just as in classical mechanical settings.

It is hoped that this study will enlarge the perspective of other researchers regarding the question of the existence of quantum chaos and its definition.

Acknowledgment

This work was supported by National Science Foundation (NSF) Grant No. PHY-86-03729.

References

1. Eidson, J.C., Ph.D. Dissertation, Georgia Institute of Technology (1987).
2. Fox, R. F., and J. C. Eidson, *Phys. Rev.* **A34**, 482 (1986).
3. Eidson, J. C., and R. F. Fox, *Phys. Rev.* **A34**, 3288 (1986).
4. Fox, R. F., and J. C. Eidson, *Phys. Rev.* **A36**, 4321 (1987).
5. Milonni, P. W., J. R. Ackerhalt, H. W. Galbraith, *Phys. Rev. Lett.* **50**, 966 (1983).
6. Belobrov, P. I., G. M. Zaslavskii, G. Kh. Tartakovskii, *Zh. Eksp. Theor. Fiz.* **71**, 1799 (1976) [*Sov. Phys. JETP* **44**, 945 (1977)].
7. Lichtenberg, A. J., and M. A. Lieberman, *Regular and Stochastic Motion* (Springer-Verlag, New York, 1983), Chapter 5.
8. Jaynes, T., and F. W. Cummings, *Proc. IEEE* **51**, 89 (1963).
9. Bonifacio, R., and G. Preparata, *Phys. Rev.* **A2**, 336 (1970).
10. Arecchi, F. T., E. Courtens, R. Gilmore, H. Thomas, *Phys. Rev.* **A6**, 2211 (1972).
11. Glauber, R. J., and F. Haake, *Phys. Rev.* **A13**, 357 (1976).
12. Heidmann, A., J. M. Raimond, S. Reynaud, N. Zagury, *Opt. Commun.* **54**, 189 (1985); *Phys. Rev. Lett.* **54**, 326 (1985).
13. Yaffe, L., *Rev. Mod. Phys.* **54**, 407 (1982).
14. Lichtenberg, A. J., and M. A. Lieberman, *Regular and Stochastic Motion* (Springer-Verlag, New York, 1983), Chapter 7.
15. Hale, J. K., *Ordinary Differential Equations* (Wiley, New York, 1969), Chapter 5.

16. Guckenheimer, J., and P. Holmes, *Nonlinear Oscillations, Dynamical Systems, and Bifurcations of Vector Fields* (Springer-Verlag, New York, 1983).
17. Eberly, J. H., *Phys. Lett.* **26A**, 499 (1968).
18. Holmes, C., and P. Holmes, *J. Sound Vib.* **78**, 161 (1981).
19. Chirikov, B. V., *Phys. Rep.* **52**, 265 (1979).
20. Lichtenberg, A. J., and M. A. Lieberman, *Regular and Stochastic Motion* (Springer-Verlag, New York, 1983), Chapter 2.

11

Chaos and the Business Cycle

Chera L. Sayers

Abstract

This chapter addresses empirical applications of chaos in the context of economic business cycles. The overall goal is to detect structures in seemingly random data series and then incorporate these structures into increasingly accurate model specifications. Chaos, determinism, and related subjects are defined. Methods of searching for evidence of determinism are discussed and critiqued, given the lack of formal distribution theory for measures of the correlation dimension and sensitive dependence upon initial conditions. A method of statistical inference based upon the correlation integral is presented, which utilizes the null hypothesis of independent and identically distributed data series.

Basic linear time series models are presented for three economic time series: U.S. monthly pig iron production, January 1877 to January 1937; U.S. quarterly real gross national product, 1952:I to 1988:III; and the U.S. monthly unemployment rate, January 1949 to December 1986. The putatively white residuals from the linear models are then tested for evidence of nonrandomness, using the methods presented. While a linear model appears adequate for the real gross national product series, linear models are inadequate for the pig iron production and the unemployment rate series.

Introduction

Economic business cycles are difficult to explain using standard economic theory. Although one can offer observations as to the characteristics of an average business cycle, such as in Zarnowitz and his references *(1)*, in practice, exceptions can always be found. For example, the typical post World War II business cycle is four years in duration, with three years for expansion and one year for contraction. An exception to this observation is the current expansion, which is now into its eighth year. While business cycles vary in duration and amplitude, a common observation is that expansions are long and slow, and contractions are sharp and quick. Thus, the pattern of business cycle data such as gross national product, unemployment, and investment appears asymmetric over time. Examples of work addressing business cycle asymmetry are Blatt *(2)*, Falk *(3)*, Neftci *(4)*, Neftci and McNevin *(5)*, and Sichel *(6)*. [For overviews of economic business cycle literature, see Lucas *(7, 8)* and Manuelli and Sargent *(9)*.]

In spite of evidence supporting asymmetry, model specifications for macroeconomic models have tended toward log linear specifications driven by symmetric, exogenous shocks. Due to the fact that such linear models generate symmetric output, many researchers, including Ashley and Patterson *(10)*, Blatt *(2)*, and Brock and Sayers *(11)*, have criticized the linear specification approach.

In an effort to better understand business cycles, several researchers have postulated deterministic models that are characterized by endogenous instability. [Readers interested in the theoretical aspects of chaotic models and business cycle applications should see Benhabib and Day *(12, 13)*, Boldrin *(14)*, Brock *(15)*, Grandmont *(16)*, and Rosser *(17)*.] These authors discuss models capable of generating behavior that appears random by standard time-series methods.

The purpose of this chapter is to examine empirical applications of chaos in the context of the subject of business cycles. Could it be that the business cycle is determined by some underlying deterministic law of motion? Could it also be that the law of motion can be uncovered, allowing the possibility for perfect modeling and, therefore, potentially perfect short-run prediction capability? These questions prompted Brock and Sayers *(11)* to test for evidence of deterministic chaos in business cycle data. [For overviews of chaos in the natural sciences and its application to economics, see Baumol and Benhabib *(18)*, Brock *(15)*, and Lorenz *(21)*.]

Relation to Economics and Time Series

Intuitively, a process is characterized by deterministic chaos if it is generated by a completely deterministic system, yet appears random by standard time-series methods. For example, consider the tent or triangle map (see Fig. 1),

$$X_{t+1} = F(X_t),$$

where $F(X_t) = 2X_t$, provided that X is between 0 and 1/2; however, if X is an element of [1/2, 1], then the functional form

$$F(X_t) = 2(1-X_t)$$

is used, and X_0 is given. Bunow and Weiss *(22)* show that data generated by the tent map are indistinguishable from pseudorandom numbers, using standard linear time-series methods such as time-series plots, autocorrelation functions, and spectral power density functions *(23)*.

The tent map is characterized by low-dimensional deterministic chaos. Imagine being handed a sequence of data points from the tent map without being told their origin and assigned the task of uncovering the original structure or model. How would one proceed? Linear time-series methods indicate that the tent map is random. However, given that the tent map is characterized by low-dimensional deterministic chaos and has a dimension equal to one, the structure can be uncovered by plotting X_t versus X_{t+1} in two space. The tent shape will become obvious.

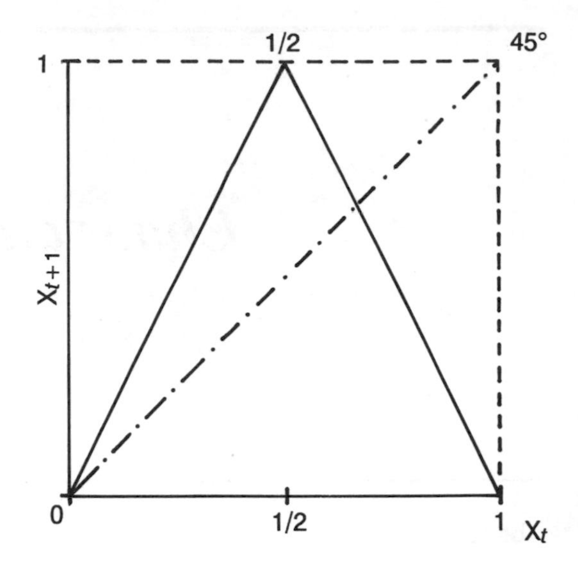

Fig. 1. The tent map. Realizations of this deterministic map appear random by standard linear time-series methods.

The tent map was chosen as an instructive example because of its simplicity. While it is unlikely that a structure as incomplex as the tent map will determine business cycles, it is likely that economic processes are characterized by more complex (i.e., higher dimensional) processes. Consider plotting a dynamical process in four or five space, and you will conclude that dynamic analysis becomes increasingly difficult in higher spaces.

Now, imagine an economist being handed another sequence of observations without knowing their origin and being assigned the task of uncovering the original structure. The usual procedure in building a model is to base the hypothesized model on economic theory and previous information concerning correlations between variables, estimate that model by using econometric methods, and then analyze the performance of the original model and make structural changes if necessary. These concepts are discussed in Lucas [see *(7)*, pp. 104–130 and 215–239].

Prior to the 1970s, macroeconometric models were evaluated by their empirical accuracy. Keynesian-type, fixed-price models prevailed, and the role of money was not greatly emphasized. During the 1970s, a large transition occurred in the field of macroeconomics as the large econometric models that fit well throughout the 1960s failed to capture the high inflation and high unemployment situation in the economy. Models built upon the classical ideas of flexible prices and models incorporating the role of expectations formed by rational agents were prevalent. These models emphasized monetary policy and its role in economic fluctuations. Lucas gives an overview of the policy debates during this decade [see *(7)*, pp. 248–261].

In spite of the rather eclectic nature of macroeconometric models presently utilized, prediction accuracy has not greatly improved. Thus, due to the lackluster performance of economic models with respect to predicting turning points over the business cycle and, worse yet, their inability to estimate when the current expansion will end, perhaps it is time to adopt a new way of examining data and to develop potentially revolutionary model-building techniques based upon the ideas of deterministic chaos and the methods of nonlinear dynamics.

Detecting Chaos in Economic Data

The general problem can be stated as such: given an observed time series, $\{a_t\}, t = 1, \ldots,$ T, what inferences can be made about the underlying system or model generating this sequence of observations? Then adopt the following definition: $\{a_t\}$ has a deterministic explanation if there exists a system $(h, \mathrm{F}, \mathrm{X}_0)$ such that $a_t = h(\mathrm{X}_t)$ for all t, $\mathrm{X}_t = \mathrm{F}(\mathrm{X}_{t-1})$, X_0 given. Here $\{a_t\}$ denotes the observed time series, which is known to the researcher at t. The observation function, denoted by h, maps R^n to R^1, is unknown to the researcher and may scramble the signal from $\{\mathrm{X}\}$ as it is mapped onto $\{a\}$. F denotes the unknown deterministic law of motion, where F maps R^n to R^n, and X_0 is the unknown initial condition. The existence of a deterministic explanation implies the potential for perfect model building. Determinism implies the absence of random components that are unforecastable. Thus, given the correct deterministic system specification and information up to time t, one can forecast the value of the deterministic system at time $t+1$.

The series $\{a_t\}$ is chaotic if it satisfies the deterministic definition and if nearby trajectories locally diverge from one another exponentially. The local divergence property, or sensitive dependence on initial conditions, is formalized by the condition that the largest Lyapunov exponent, L, be positive. Following Brock *(19)*, we define

$$\mathrm{L} = \lim \ln [\| \mathrm{D}\, \mathrm{F}^t(x) \cdot v \|]/t,$$

where the limit is taken as $t \rightarrow \infty$, D denotes derivative, ln is natural log, v denotes directional vectors, and $\mathrm{F}^t(x)$ denotes t applications of the map F to x.

The signs of the Lyapunov exponents provide qualitative information on the expansion or contraction of nearby trajectories in phase space. Positive Lyapunov exponents are characteristic of chaos, zero exponents characterize a marginally stable orbit, and negative exponents characterize a periodic orbit. Positive exponents imply that nearby trajectories diverge from one another exponentially and are characteristic of local instability.

The issue of stability is an important one in economics, as many of the above citations discuss. Should one model economic fluctuations by utilizing an endogenously unstable system, or should one adopt a system that is stable in the absence of disturbances and in which stochastic shocks drive the system?

Researchers such as Alogoskoufis and Stengos *(24)*, Barnett and Chen *(25)*, Brock and Sayers *(11)*, Eckmann *et al.* *(26)*, and Frank *et al.* *(27)* have estimated various measures of local instability and have found some evidence of sensitive dependence upon initial conditions in economic data.

Much controversy surrounds the use of Lyapunov exponent algorithms in an attempt to adduce evidence of sensitive dependence upon initial conditions. Three common approaches to measuring instability are the Wolf *et al.* *(28)*, Eckmann and Ruelle *(29)*, and Kurths and Herzel *(30)* algorithms. The issues of Lyapunov exponents and related measures have drawbacks in that no distribution theory exists for the estimated measures. Given that a truly deterministic system has no stochastic components, standard errors are not theoretically obtainable. However, given that experimental data consist of noisy data sets, methods of repeated simulation may be utilized in the future to obtain approximate distributions for finite and noisy data sets. The algorithms appear to be quite sensitive to the presence of noise and seem to display an upward bias in the presence of noise for finite data sets. [See Abraham *et al.* *(31)* for a recent discussion of these topics.]

At a minimum, there are three requirements needed to substantiate the claim for deterministic chaos. First, one needs evidence of low dimension. Second, one needs evidence of sensitivity to initial conditions by demonstrating positive Lyapunov exponents. Third, the researcher must be able to reconstruct the underlying dynamical system generating the sequence of observations. It is not difficult in economics to generate results that appear to satisfy the first two conditions; however, great caution must be exercised by researchers in this regard.

In practice, one should expect noise to infect the observed series, both by measurement noise through h, and noise in the dynamics of the state vector x. [See Brock *(19)* and Brock and Dechert *(32)* for a discussion.] For our purposes, we require a high signal-to-noise ratio; that is, the variance of the noise is small relative to the variation in the data.

R^n gives the space or dimension in which our observations exist. The definition of dimension is consistent with previously learned concepts in that a point has dimension zero, a line has dimension one, a plane has dimension two, and a cube has dimension three. Now, imagine an infinite sequence, $\{r_t\}, t = 1, \ldots \infty$, of uniform random numbers on the [0,1] interval. Plot these random numbers against each other, $\{r_t, r_{t+1}\}$, in two space, and they should fill the square. Now plot these numbers against each other, $\{r_t, r_{t+1}, r_{t+2}\}$, in three space. They should fill the cube. Continue with spaces of higher dimension, and one can see that in the limit, random numbers are infinite dimensional.

Thus, one can state a major difference between a chaotic process and a truly random process — they will both appear random to the naked eye and to standard linear time-series methods, yet the truly random process will have high dimension, whereas the chaotic process will have low dimension. Since the truly random process will contain no structure useful for model building and forecasting, we will be interested in testing the null hypothesis of whether or not a seemingly random process exhibits a low-dimensional structure.

Many economic time series are characterized by near-unit root processes, as documented by Nelson and Plosser *(33)*. Consider an AR(1) near-unit root process, $x_{t+1} = bx_t + u_{t+1}$, where $\{u_t\}$ is independent and identically distributed, b is near one, and the standard deviation of the stationary distribution of $\{x\}$ is one. If the distribution of $\{u\}$ is $N(0, \sigma^2)$, then $\{x\}$ is distributed $N(0, \sigma^2/(1-b^2))$. Given that $\{x\}$ has been scaled so that $[\sigma^2/(1-b^2)] = 1$, this implies that $\sigma^2 = (1-b^2)$, or, if b is near one, $\sigma 2$, the variance of innovations is near zero. Thus, realizations of $\{x\}$, i.e., $x_{t+1}, x_{t+2}, \ldots$, will lie close to a line and the estimated dimension will be close to one.

In order to avoid this problem and to test for determinism in economic time series, the residual diagnostic for deterministic chaos *(11, 19)* will be utilized. If a time series $\{x_t\}$, $t = 1, \ldots, T$ has a deterministic explanation, and if one fits a smooth time-series model with a finite number of leads and lags to $\{x_t\}$, then the residuals $\{u_t\}$ of that model must have the same dimension as $\{x_t\}$. For example, let $x_t = h(z_t)$ where $z_t = F(z_{t-1})$. Then postulate an autoregressive process of order two,

$$(\text{AR}(2)): x_t = \alpha x_{t-1} + \beta x_{t-2} + u_t,$$

where α and β are the estimated coefficients obtained using any time-series software package, and u_t is the residual. Equivalently,

$$u_t = x_t - \alpha x_{t-1} - \beta x_{t-2}.$$

Substituting for x yields

$$u_t = h(z_t) - \alpha h(z_{t-1}) - \beta h(z_{t-2})$$

or

$$u_t = h\,(F^2(z_{t-2}) - \alpha\, h(F(z_{t-2})) - \beta h(z_{t-2}).$$

This implies that u_t = M (z_{t-2}), and the dimension of $\{u_t\}$ should be equivalent to the dimension of $\{x_t\}$, which is a function of $\{z_t\}$.

In order to utilize the residual test, we pre-whiten business cycle data by fitting autoregressive moving average models ARMA(p,q), where the orders p and q are chosen according to standard Box and Jenkins techniques *(23)*. The autoregressive nature of the model implies that there exists serial correlation in the dependent variable, whereas the moving average nature implies the existence of dependencies in the innovations. Our goal is to filter obvious linear correlations in the data and then scrutinize the residuals of our ARMA models in order to test for additional structure.

We adopt the notion of correlation dimension following Grassberger and Procaccia *(34)*. We apply the Grassberger-Procaccia (GP) algorithm, forming m-histories, X_t^m, from the data set $\{a_t\}, t = 1, \ldots, T$, where m is the embedding dimension. For example, for $m = 1$,

$$X_t^1 = (a_1), (a_2), (a_3), \ldots, (a_T).$$

Likewise, for $m = 2$,

$$X_t^2 = (a_1, a_2), (a_2, a_3), \ldots, (a_{T-1}, a_T.$$

Then form the correlation integral,

$$C(m,e) = \{\#(i,j) \mid \| X_i^m - X_j^m \| < e\text{, and } i \neq j\}/N,$$

where m is the embedding dimension and $N = M^2 - M$. $M = T-(m-1)$ is the number of m-histories that can be formed from a series of length T. The tolerance distance, e, is chosen by the researcher. Notice that the correlation integral can be interpreted as the probability that, out of all possible choices, a given number of distances are within the tolerance distance.

Grassberger and Procaccia show that for small e, the correlation integral, $C(m,e)$, grows as a power, $k \cdot e^n$, where k is a constant and n is the dimension estimate. Taking natural logs of the power law and dividing by the natural log of the tolerance distance, one obtains the relationship $\ln C(m,e) / \ln(e) = \ln(k) / \ln(e) + n$. Then, if one assumes that $\ln(k) / \ln(e)$ is small, one estimate of dimension is $n_1 = \ln C(m,e) / \ln(e)$. For large e, all m-histories will be comparable, the value of the correlation integral will be one, and the dimension estimate will be zero. For e less than the smallest possible distance between m-histories, the value of the correlation integral will be zero, and the dimension estimate becomes infinite. Thus, one obtains a GP plot resembling a waterfall when plotting $\ln C(m,e)$ versus $\ln(e)$.

In practice, one searches for a zone of stability on the GP plots within which the power law fits well. Another estimate of dimension is then $n_2 = \{\ln C(m,e_i) - \ln C(m,e_{i-1})\}/\{\ln(e_i) - \ln(e_{i-1})\}$. Measurement of the slope within this zone of stability provides one estimate of the dimension of the data set. This method is subjective and prone to great error. In addition, the use of simple regression to obtain the slope estimate results in biased estimates. In particular, the dimension estimate and its variance depend upon the choice of embedding dimension and the number of observations in the data series. The reader is referred to Ramsey and Yuan *(35)* and Ramsey *et al. (36)* for further discussion of this matter.

Given that no distribution theory exists for the correlation dimension estimate, great caution must be exercised when using this method, especially when utilizing small, noisy data sets such as those common in economics. In an attempt to provide a standard of comparison, Brock and Sayers *(11)* utilized data diagnostics called the "wing" and the "shuffle." The wing diagnostic compares GP plots for the series under examination with a comparable series of pseudorandom numbers of the same length, mean, and variance. If the series under investigation is truly random, then the GP plots should lie close together and the dimension of the series should be large. A wing-shaped plot indicates inherent structure in the data series under examination.

The shuffle diagnostic compares GP plots for the series under examination with a GP plot for the shuffled series. The intuition behind this approach is that if the series under examination contains useful structure, then shuffling that series randomly should break up that structure, and the shuffled series should have a steeper GP plot than the original series.

Brock and Sayers *(11)* found that the hypothesis of determinism was rejected for all United States business cycle data under investigation. These series were quarterly real gross national product (1972$), 1947:I to 1985:I; quarterly real gross private domestic investment (1972$), 1947:I to 1985:I; quarterly employment, 1950:I to 1983:IV; quarterly unemployment rate, 1949:I to 1982:IV; monthly industrial production, February 1949 to February 1983; and monthly pig iron production, January 1877 to January 1937.

The dimension estimates were rough, given the very small data sets involved, but, in general, the dimensions of the time-stationary series ranged between two and four, whereas the dimensions of the pre-whitened series jumped to between seven and ten. Hence, the residual test for determinism failed for all series. In addition, our estimates of Lyapunov exponents were generally positive. Again, the strength of our testing was limited by very small data sets. The Lyapunov exponent algorithm used *(28)* could not easily differentiate between autoregressive processes and pseudorandom data in that both types of stochastic processes can generate spurious positive Lyapunov exponent estimates. Kurths and Herzel *(30)* found similar results using different algorithms to measure endogenous instability. Thus, the hypothesis of deterministic chaos was rejected for all data sets under scrutiny.

For purposes of this chapter, we examined three data sets: U.S. monthly pig iron production, January 1877 to January 1937, seasonally unadjusted, as published in Macaulay *(37)*; U.S. quarterly real gross national product (GNP), 1952:I to 1988:III, 1982 dollars, seasonally adjusted at annual rates; and the U.S. monthly unemployment rate, January 1949 to December 1986, seasonally adjusted. The data source for the GNP and unemployment data was the U.S. Department of Commerce, Bureau of Economic Analysis, *Business Conditions Digest*.

We first rendered our data series time stationary by either detrending or taking the log first differences. There exists a great debate in economics as to the preferred method. [For information on this controversy, see Singleton *(38)*.] We choose to report only the difference-stationary results here, although the time-stationary results are qualitatively similar. Beginning with the pig iron series, we fit an ARMA(2,1) model to the log first differences,

$$p_t: p_t = 0.90 p_{t\text{-}1} - 0.40 p_{t-2}$$

$$+ e_t - 0.28 e_{t-1},$$

$$t_1 = 8.01, t_2 = -7.04, t_3 = -2.33,$$

$$\mathrm{N} = 718, \mathrm{R}^2 = 31\%.$$

Here, t_i denotes the t-statistic associated with the ith coefficient, N denotes the number of observations, and R^2 represents the coefficient of determiniation. We found no evidence of seasonalities in the pig iron series. We fit an AR(2) with drift to the log first differences of real GNP,

$$x_t: x_t = 0.0074 + 0.30 x_{t-1} + 0.085 x_{t-2} + v_t,$$

$$t_1 = 5.60, t_2 = 3.64, t_3 = 1.02, \mathrm{N} = 144, \mathrm{R}^2 = 12\%.$$

We fit an AR(4) model with three seasonal yearly components to the log first differences of the unemployment rate, u_t. It is interesting to note that even though this series was seasonally adjusted, evidence of significant seasonalities were found. The model is

$$u_t = 0.11\, u_{t-1} + 1.19\, u_{t-2} + 0.08\, u_{t-3}$$

$$+ 0.12\, u_{t-4} - 0.26\, u_{t-12} - 0.22\, u_{t-24}$$

$$- 0.11\, u_{t-36} + n_t, t_1 = 2.14, t_2 = 3.78,$$

$$t_3 = 1.65, t_4 = 2.33, t_5 = -5.23,$$

$$t_6 = -4.67, t_7 = -2.43, \mathrm{N} = 415, \mathrm{R}^2 = 20\%.$$

Figure 2 gives a phase portrait of the log first differences of real GNP, and Fig. 3 gives a phase portrait of the residuals of the AR(2) with drift model. Note that both figures appear quite random, although the observations in Fig. 2 are concentrated disproportionately in the northeast quadrant. This concentration of observations is consistent with a positive drift parameter. Figure 4 gives a phase portrait of the log first differences of pig iron production, and Fig. 5 gives a phase portrait of

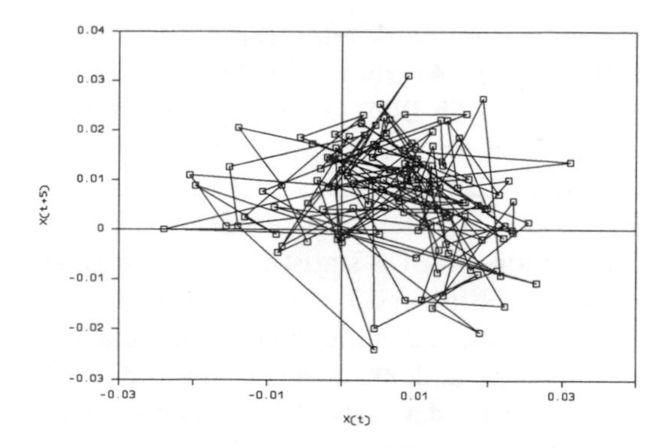

Fig. 2. Phase portrait of real GNP, log first differences, 1952:1 to 1988:III. Number of observations = 146.

the ARMA(2,1) residuals. Note again that both figures appear quite random, although Fig. 5 appears more random.

Figure 6 demonstrates the shuffle diagnostic for the pig iron ARMA(2,1) residuals and the shuffled residuals. The wing shape indicates the presence of structure in the residuals. Approximate dimension estimates for the pig iron series are 1 to 2 for the ARMA(2,1) residuals and 2 to 5 for the shuffled residuals. Note that great caution must be exercised in interpreting these estimates, due to small sample sizes and noisy data sets. Point estimates are reported here so that the reader may gain a sense of the complexity of the data series utilized. The error bars associated with these estimates are probably very large. [See Ramsey *et al. (36)* for further discussion.] The only conclusion that can be made is that the

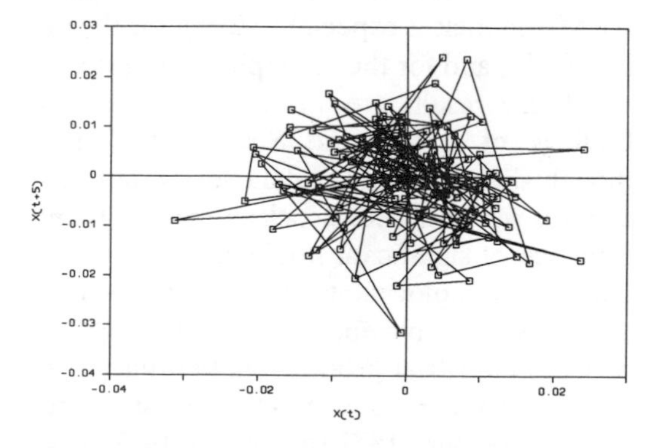

Fig. 3. Phase portrait of real GNP residuals, ARMA (2,0) with drift, 1952:I to 1988:III. Number of observations = 144.

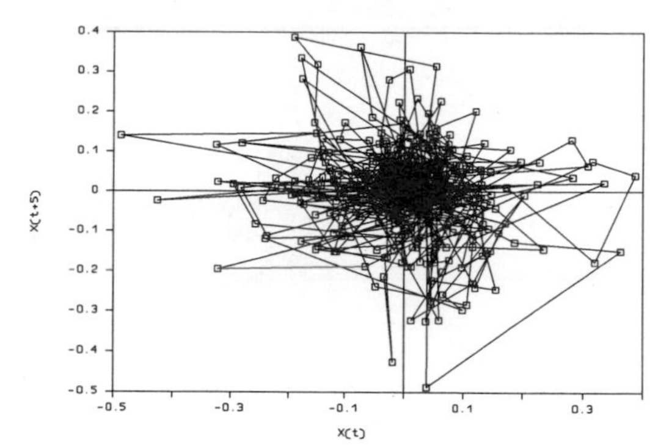

Fig. 4. Phase portrait of pig iron production, log first differences, January 1877 to January 1937. Number of observations = 720.

estimated dimensions of the pre-whitened data sets all jump substantially, leading to the rejection of the null hypothesis of determinism.

Detection of Nonlinear Structure in Time-Series Data

Although the hypothesis of determinism was rejected for business-cycle data, the methods used in searching for evidence of determinism can be mined into useful tests for general nonlinear structure in time-series data. In this section, we adopt the new null hypothesis that the series under investigation are consistent with independent and identically distributed (*i.i.d.*) random variables. This is done in order to test the residuals of time series-models for evidence of remaining structures that potentially may be identified and incorporated into more accurate model specification.

Brock, Dechert, and Scheinkman *(39)* have devised a test based upon the correlation integral under the null hypothesis of independent and identically distributed data series. Using the definition of the correlation integral defined above, define the Brock, Dechert, Scheinkman (BDS) statistic as follows:

$$\mathrm{BDS}\,(m, e)$$

$$= \mathrm{N}^{1/2}\,\{\mathrm{C}_m(\mathrm{e}) - [\mathrm{C}_1(e)]^m\}\,/\,b_m,$$

$m > 1$, where b_m is the standard deviation of

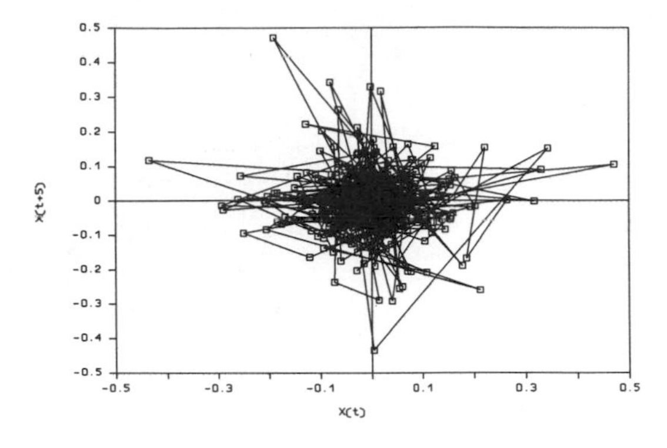

Fig. 5. Phase portrait of pig iron production residuals, ARMA (2,1), January 1877 to January 1937. Number of observations = 718.

the BDS statistic, the formulation of which varies with m, the embedding dimension. Under the null hypothesis of *i.i.d.*, for a given level of e, the quantity $\{C_m(e) - [C_1(e)]^m\}$ should equal zero. Brock, Dechert, and Scheinkman show that under the null hypothesis of *i.i.d.*, the BDS statistic follows an asymptotic normal distribution, with mean zero and variance one, as N goes to ∞. Thus, large values of the BDS statistic indicate evidence against the hypothesis of randomness and in favor of nonlinear structure, thereby serving as a signal of additional predictive power inherent in the putatively white data series under examination.

While the BDS statistic is a univariate one, Baek *(40)* and Baek and Brock *(41)* contain important work concerning multivariate formulations. As shown by Hsieh and LeBaron *(42)*, the BDS test has good power in comparison with a number of alternatives. The finite sample results in Hsieh and LeBaron *(43)* are used to determine the critical regions for the BDS statistic, given our small data sets. Hsieh and LeBaron *(43)* show that the BDS test has good power for data sets between 250 and 500 observations and excellent power for data sets of 500 or more observations, when using a value of e between 0.5 and 1.5 times the standard deviation divided by the spread of the data.

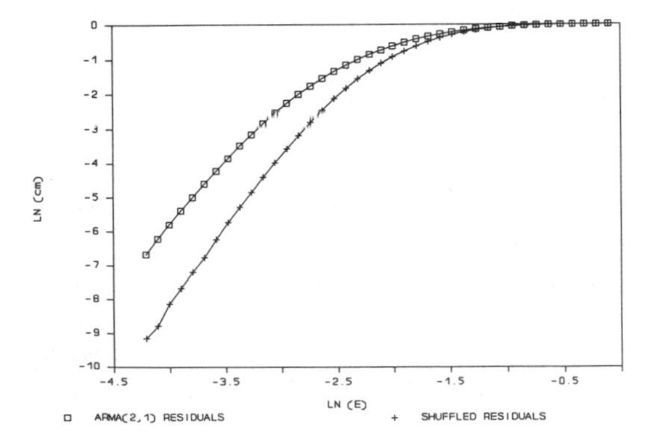

Fig. 6. Grassberger-Procaccia plot of residuals of ARMA (2,1) model versus shuffled ARMA (2,1) residuals. The wing shape indicates the presence of additional structure in the putatively white residuals.

We calculated BDS statistics for the log first differences of our data as well as the residuals of the time-series models. These statistics, along with statistics for the shuffled series, are listed in Table 1. Note that the BDS statistics for the log first difference data are large, leading to rejection of the *i.i.d.* null hypothesis. This comes as no surprise, as all the data series examined contain obvious linear structure. The BDS statistics for the GNP series are more borderline; however, the sample size is quite small.

The BDS statistics in Table 1 indicate that the pre-whitened real GNP data contain little evidence of remaining structure. The pre-whitened unemployment residuals show evidence of structure, above that of simple linear structure. The pre-whitened pig iron series contains the strongest evidence of nonlinear structure. Note that the BDS statistics for the shuffled series are all close to zero, indicating that the shuffled series appear random. The case for nonlinear structure in the residuals of ARMA models is especially strong for the pig iron series and for the unemployment rate.

These results are all consistent with the findings of Brock and Sayers *(11)*, even though slightly different data series and procedures were utilized. Brock and Sayers found evidence of significant nonlinear structure in quarterly employment, 1950:I to 1983:IV; quarterly unemployment, 1949:I to 1982:IV; monthly industrial production, February 1949 to February 1983; and monthly pig iron production, January 1877 to January 1937. They found little evidence for the presence of non-

Table 1. BDS statistics.

	REAL GROSS NATIONAL PRODUCT, 1952:I TO 1988:III					
	Log first differences		AR(2) with drift residuals		Shuffled residuals	
m	C_m	BDS	C_m	BDS	C_m	BDS
2	3118	1.94	2868	0.12	3020	0.51
3	1770	2.56	1574	1.18	1607	0.25
4	1019	2.79	864	1.38	837	–0.11
5	606	3.19	499	1.95	420	–0.55
	PIG IRON PRODUCTION, JANUARY 1877 TO JANUARY 1937					
	Log first differences		ARMA (2,1) residuals		Shuffled residuals	
m	C_m	BDS	C_m	BDS	C_m	BDS
2	125,422	16.92	124,106	13.93	108,542	–0.25
3	100,019	19.48	97,048	16.32	69,731	–0.57
4	81,851	21.41	78,309	18.15	44,787	–0.61
5	68,344	23.70	64,605	20.01	28,956	–0.48
	UNEMPLOYMENT RATE, JANUARY 1949 TO DECEMBER 1986					
	Log first differences		AR(2) with seasonalities residuals		Shuffled residuals	
m	C_m	BDS	C_m	BDS	C_m	BDS
2	36,474	4.61	27,810	2.24	26,926	–0.38
3	23,687	6.85	16,805	3.87	15,029	–0.27
4	16,253	8.82	10,484	4.91	8,376	–0.25
5	11,345	10.17	6,478	5.07	4,634	–0.33

m = embedding dimension; C_m = number of comparable pairs; BDS = $N^{1/2}\{C_m(e) - [C_1(e)]^m\}/b_m$; e = standard deviation/spread of the data.

linear components in detrended, pre-whitened real GNP.

Conclusion

Examples of work dealing with the correlation dimension and such related concepts as the BDS statistic as applied to economic data can be found in Alogoskoufis and Stengos *(24)*, Barnett and Chen *(25)*, Brock *(15)*, Diebold and Nason *(44)*, Frank and Stengos *(45–47)*, Frank *et al.* *(27)*, Frank *et al.* *(48, 49)*, Hiemstra *(50)*, Hsieh *(51)*, LeBaron *(52–54)*, Papell and Sayers *(55)*, Sayers *(56, 57)*, and Scheinkman and LeBaron *(58, 59)*. While there is little compelling evidence supporting the hypothesis of deterministic chaos in the current literature, there appears to be abundant evidence to support the hypothesis of nonlinear dynamical components in many economic time series. This conclusion is reached in many of the above citations as well as Ramsey *et al.* *(36)*; see Barnett and Hinich *(60)* for another viewpoint. Gallant *(61)* contains parametric approaches to nonlinear economics and statistics. Hinich and Patterson *(62)* apply bispectral tests in order to search for the presence of nonlinearities.

The evidence for significant nonlinear structure in many economic data series is strong. Researchers should be aware of the nonlinear components of the data when postulating model specifications. The BDS statistic can indicate the presence of structure in residuals of fitted time-series models, although it is unable to identify the form of the structure. If researchers are successful in identifying the correct structural form, model specification and prognostication accuracy may be greatly improved. Existing literature on specification tests such as Brock and Dechert *(63)* and Thursby *(64)* presents valid hope for researchers.

Acknowledgments

Much of the content of this chapter arises from previous research on nonlinear science with William A. Brock to whom the author's intellectual debt is great. I am very grateful to Mikel Ham for dedicated computational assistance. W. D. Dechert provided a copy of the BDS software. I am grateful to an anonymous referee for helpful comments.

References

1. Zarnowitz, V., *J. Econ. Lit.* **23,** 523 (1985).
2. Blatt, J. M., *Ox. Econ. Pap.* **32,** 467 (1980).
3. Falk, B., *J. Polit. Econ.* **94,** 1096 (1986).
4. Neftci, S. N., *J. Polit. Econ.* **92,** 307 (1984).
5. Neftci, S. N., and B. McNevin, "Some Evidence on the Non-Linearity of Economic Time Series: 1845-1981"(The Graduate School and University Center of The City University of New York, 1986).
6. Sichel, D. E., *J. Polit. Econ.* **97,** 1255 (1989).
7. Lucas, R. E. Jr., *Studies in Business-Cycle Theory* (MIT Press, Cambridge, MA, 1981).
8. Lucas, R. E. Jr., *Models of Business Cycles* (Basil Blackwell, Oxford, U. K., 1987).
9. Manuelli, R., and T. J. Sargent, *J. Monetary Econ.* **22,** 523 (1988).
10. Ashley, R. A., and D. M. Patterson, *Int. Econ. Rev.* **30** no. 3 (1989).
11. Brock, W. A., and C. L. Sayers, *J. Monetary Econ.* **22,** 71 (1988).
12. Benhabib, J., and R. Day, *Rev. Econ. Stud.* **48,** 459 (1981).
13. Benhabib, J., and R. H. Day, *J. Econ. Dyn. Control* **4,** 37 (1982).
14. Boldrin, M., in *The Economy as an Evolving Complex System. SFI Studies in the Sciences of Complexity*, P. Anderson, K. Arrow, and D. Pines, Eds. (Addison-Wesley, Reading, MA, 1988), pp. 49–75.
15. Brock, W. A., in *The Economy as an Evolving Complex System. SFI Studies in the Sciences of Complexity*, P. Anderson, K. Arrow, D. Pines, Eds. (Addison-Wesley, Reading, MA, 1988), pp. 77–97.
16. Grandmont, J.-M., *Econometrica* **53,** 995 (1985).
17. Rosser, J. B. Jr., "Chaos Theory and the New Keynesian Economics" (Department of Economics, Manchester School of Economic and Social Studies, James Madison University, Harrisonburg, VA, 1989).
18. Baumol, W. J., and J. Benhabib, *J. Econ. Perspectives* **3,** 77 (1988).
19. Brock, W. A., *J. Econ. Theory* **40,** 168 (1986).
20. Brock, W. A., in *Acting Under Uncertainty: Multidisciplinary Conceptions*, G. M. von Furstenberg, Ed. (Kluwer Academic Publishers, Boston, 1989).
21. Lorenz, H.-W., *Nonlinear Dynamical Economics and Chaotic Motion. Lecture Notes in Economics and Mathematical Systems, No. 334* (Springer-Verlag, Berlin, 1989).
22. Bunow, B., and G. H. Weiss, *Math. Biosci.* **47,** 221 (1979).
23. Box, G., and A. Jenkins, *Time Series Analysis: Forecasting and Control* (Holden-Day, Oakland, CA, 1976).
24. Alogoskoufis, G., and T. Stengos, "Testing for Nonlinear Dynamics in Historical Unemployment Series" (Department of Economics, Birbick College and Department of Economics, University of Guelph, Canada, 1990).
25. Barnett, W., and P. Chen, in *Dynamic Econometric Modeling. Proceedings of the Third International Symposium on Economic Theory and Econometrics*, W. Barnett, E. Berndt, H. White, Eds. (Cambridge University Press, Cambridge, U. K., 1988).
26. Eckmann, J.-P., S. O. Kamphorst, D. Ruelle, J. Scheinkman, in *The Economy as an Evolving Complex System, SFI Studies in the Sciences of Complexity,* P. Anderson, K. Arrow, D. Pines, Eds. (Addison-Wesley, Reading, MA), pp. 301–304.
27. Frank, M., R. Gencay, T. Stengos, *European Econ. Rev.* **32,** 1569 (1988).
28. Wolf, A., J. B. Swift, H. L. Swinney, J. A. Vastano, *Physica* **16D,** 285 (1985).
29. Eckmann, J.-P., and D. Ruelle, *Rev. Mod. Phys.* **57,** 617 (1985).
30. Kurths, J., and H. Herzel, *Physica* **25D,** 165 (1987).
31. Abraham, N.B., A. M. Albano, N. B. Tufillaro, in *Measures of Complexity and Chaos*, N. B.

Abraham, A. M. Albano, A. Passamante, P. Rapp, Eds. (Plenum Press, New York, 1990), pp. 1–28.
32. Brock, W. A., and W. D. Dechert, in *Dynamic Econometric Modeling, Proceedings of the Third International Symposium on Economic Theory and Econometrics,* W. Barnett, E. Berndt, H. White, Eds. (Cambridge University Press, Cambridge, U. K., 1988).
33. Nelson, C. R., and C. I. Plosser, *J. Monetary Econ.* **10,** 139 (1982).
34. Grassberger, P., and I. Procaccia, *Physica* **9D,** 189 (1983).
35. Ramsey, J. B., and H.-J. Yuan, *Physics Letters A* **134,** 287 (1989).
36. Ramsey, J. B., C. L. Sayers, P. Rothman, *Int. Econ. Rev.*, in press.
37. Macaulay, F. R., *The Movements of Interest Rates, Bond Yields, and Stock Prices in the United States Since 1856* (National Bureau of Economic Research, New York, 1938).
38. Singleton, K. J., *J. Monetary Econ.* **21,** 361 (1989).
39. Brock, W. A., W. D. Dechert, J. A. Scheinkman, *SSRI Working Paper No. 8702* (Department of Economics, University of Wisconsin-Madison, 1987).
40. Baek, E. G., Ph.D. dissertation, University of Wisconsin-Madison, 1988.
41. Baek, E. G., and W. A. Brock, *SSRI Working Paper No. 8816* (Department of Economics, University of Wisconsin-Madison, 1988).
42. Hsieh, D. A., and B. LeBaron, "Finite Sample Properties Under Alternative Hypotheses" (Fuqua School of Business, Duke University, Durham, NC, and Department of Economics, University of Wisconsin-Madison, 1988).
43. Hsieh, D. A., and B. LeBaron, "Finite Sample Properties of the BDS Statistic I: Distribution Under the Null Hypothesis" (Fuqua School of Business, Duke University, Durham, NC, and Department of Economics, University of Wisconsin-Madison, 1988).
44. Diebold, F. X., and J. M. Nason, *J. Int. Econ.,* in press.
45. Frank, M. Z., and T. Stengos, *J. Econ. Surveys* **2,** 103 (1988).
46. Frank, M. Z., and T. Stengos, *J. Monetary Econ.* **22,** 423 (1988).
47. Frank, M. Z., and T. Stengos, *Rev. Econ. Stud.*, in press.
48. Frank, M., C. Sayers, T. Stengos, "Evidence Concerning Nonlinear Structure in Provincial Unemployment Rates" (Faculty of Commerce, University of British Columbia, Vancouver, Canada; and Department of Economics, University of Houston; and Department of Economics, University of Guelph, Canada, 1989).
49. Frank, M., C. Sayers, T. Stengos, "Seasonality and Nonlinear Structure in the United States Unemployment: Some Disaggregated Evidence" (Faculty of Commerce, University of British Columbia, Vancouver, Canada; Department of Economics, University of Houston; and Department of Economics, University of Guelph, Canada, 1990).
50. Hiemstra, C., "Applications of New Tests For Nonlinear Dependence to the Efficient Market and Rational Expectations Hypothesis" (Board of Governors of the Federal Reserve, Washington, DC, 1989).
51. Hsieh, D. A., *J. Bus.* **62,** 339 (1989).
52. LeBaron, B., "Nonlinear Puzzle in Stock Returns" (Department of Economics, University of Wisconsin-Madison, 1987).
53. LeBaron, B., "Stock Return Nonlinearities: Comparing Tests and Finding Structures" (Department of Economics, University of Wisconsin-Madison, 1988).
54. LeBaron, B., "The Changing Structure of Stock Returns" (Department of Economics, University of Wisconsin-Madison, 1988).
55. Papell, D. H., and C. L. Sayers, "Nonlinear Dynamics and Exchange Rate Frequency" (Department of Economics, University of Houston, 1989).
56. Sayers, C. L., "Diagnostic Tests for Nonlinearity in Time Series Data: An Application to the Work Stoppages Series" (Department of Economics, University of Houston, 1987).
57. Sayers, C. L., "Work Stoppages: Exploring the Nonlinear Dynamics" (Department of Economics, University of Houston, 1988).
58. Scheinkman, J., and B. LeBaron, *J. Bus.* **62,** 311 (1986).
59. Scheinkman, J., B. LeBaron, in *Economic Complexity: Chaos, Sunspots, Bubbles and Nonlinearity*, W. Barnett, J. Geweke, K. Shell, Eds. (Cambridge University Press, Cambridge, U. K., 1989), pp. 213–227.
60. Barnett, W. A., and M. J. Hinich, in *Evolutionary Dynamics and Nonlinear Economics*, P. Chen and R. Day, Eds. (Oxford University Press, U. K., 1989).
61. Gallant, A. R., *Nonlinear Statistical Models* (John Wiley, New York, 1987).
62. Hinich, M., and D. Patterson, *J. Bus. Econ. Stat.* **3,** 69 (1985).
63. Brock, W. A., and W. D. Dechert, "A General Class of Specification Tests: The Scalar Case" (Department of Economics, University of Wisconsin-Madison, and Department of Economics, University of Houston, 1988).
64. Thursby, J. G., *In. Econ. Rev.* **30,** 217 (1989).

12

Dynamic Competitive Equilibria and Chaos

Michele Boldrin

Abstract

In this chapter, we survey some recent contributions to dynamic economic theory that have advanced our understanding of the sources of chaotic dynamics in economic models.

Our attention is restricted to the class of so-called optimal growth models, which are a reduced-form representation of an economy with infinitely lived agents and complete markets and/or perfect foresight.

For this class of models, an asymptotic stability theorem — the turnpike theorem — holds under certain circumstances. We survey recent work that has proved that these circumstances are very special and that, in general, any kind of dynamics may be the optimal outcome in growth models. We also report the results of other recent research in which a fully worked-out theory of optimal economic chaos has been proposed and illustrated with meaningful economic examples.

Introduction

The idea that market mechanisms are *inherently dynamically unstable* has not played a great role in studies of aggregate fluctuations over the past quarter century. Instead, the dominant strategy, both in equilibrium business cycle theory and in econometric modeling of aggregate fluctuations, has been to assume model specifications for which equilibrium is determinate and intrinsically stable, so that in the absence of continuing exogenous shocks, the economy would tend toward a steady state growth path. The existence of a stationary pattern of fluctuations is then attributed to the existence of exogenous shocks of one kind or another — most often either technology or taste shocks or stochastic shifts in government policies.

Recent work, however, has seen a revival of interest in the hypothesis that aggregate fluctuations might represent an endogenous phenomenon that would persist even in the absence of stochastic "shocks" to the economy. Even without giving credence to the extreme (and surely implausible) view that macroeconomic fluctuations are purely deterministic, the possibility that exogenous shocks might play a relatively minor role in generating the size of aggregate fluctuations that we observe must be judged of no small importance.

The endogenous cycle hypothesis is not new. Indeed, the earliest formal models of business cycles were largely of this type, including, most notably, the business cycle models proposed by Sir John Hicks, Nicholas Kaldor, and Richard Goodwin. In all these models, the stationary growth path for the economy is unstable, but deviations from it are eventually contained by "floors and ceilings" such as shortages of productive factors on the upside or technological limits to the amount by which investment can be made negative on the downside.

By the late 1950s, however, this way of attempting to model aggregate fluctuations had largely fallen out of favor, the dominant approach having become instead the Slutsky-Frisch-Tinbergen methodology of exogenous stochastic "impulses" that are transformed into a characteristic pattern of oscillations through the filtering properties of the economy's "propagation mechanism."

A major reason for the decline from favor of the endogenous cycle hypothesis concerns the inadequate behavioral foundations of the early models of this kind. The stability results obtained for many simple equilibrium models based upon optimizing behavior with perfect foresight – in particular, the celebrated "turnpike theorems" for optimal growth models doubtless led many economists to suppose that the endogenous cycle models not only lacked explicit foundations in terms of optimizing behavior but depended upon behavioral assumptions that were necessarily inconsistent with optimization. This latter issue will be the focus of this chapter.

We begin by recognizing that, at least in the one-sector, neoclassical growth model, it is true that cyclic or chaotic policy functions cannot be optimal. Next, we argue that this is indeed the only relevant framework in which such a critique holds: we prove this claim by showing that any kind of policy function could be optimal in a model of the same type but with two sectors of production. This is also true, obviously, for the general n-sector model. The global stability property of the one-sector model should therefore be considered as a very special implication of the highly restrictive technological hypotheses upon which it is built. Finally, we summarize recent results on the presence of cycles and chaos in a parameterized version of the two-sector model.

The One-Sector Model

Day *(1)* considers a one-sector, neoclassical growth model in which the dynamics of capital accumulation has the form

$$k_{t+1} = \frac{s(k_t) \cdot f(k_t)}{1+\lambda} = h(k_t)\,, \qquad (1)$$

where s is the saving function, f the production function, and $\lambda > 0$ is the exogenous population's growth rate. This is a discrete-time version of the famous Solow's growth model. In the discrete-time form (Eq. 1), Solow's assumption of a constant, exogenous saving rate and of a neoclassical, concave production function give rise to a map $h(k_t)$, which is monotonically increasing and has one and only one interior steady state $k^* = h(k^*)$. Therefore, not even damped oscillations are possible in this case.

The trouble with Solow's model is that it is not an optimizing one, i.e., the aggregate saving function is not explicitly derived from considerations of intertemporal efficiency. One is therefore free to pick "reasonable" shapes for $s(k_t)$ [and $f(k_t)$, obviously] in order to prove his claim. A typical Solow-like pair would be a constant saving ratio σ and a Cobb-Douglas form for f, Eq. 1 then becomes

$$k_{t+1} = \frac{\sigma \mathrm{B} k_t^{\beta}}{1+\lambda}\,, \qquad (2)$$

which is monotonic and therefore stable. The first modification Day suggests is to the production function. By introducing a "pollution effect," one obtains

$$k_{t+1} = \frac{\sigma \mathrm{B} k_t^{\beta} (m - k_t)^{\gamma}}{1+\lambda}\,, \qquad (3)$$

which is unimodal and has period three for certain ranges of parameter values. Returning to the Cobb-Douglas form and allowing instead for a variable saving rate, $s(k) = a(1-b/r)k/y$, Day obtains

$$k_{t+1} = \left[\frac{a}{1+\lambda}\right] k_t \left[1 - \frac{b}{\beta \mathrm{B}} k_t^{1-\beta}\right], \qquad (4)$$

using the fact that the rate of interest must be $r = \beta y/k$. This equation also displays chaos for feasible parameter values.

Day's examples (as well as many others) show that extremely simple behavioral hypotheses and model structures can produce very complicated dynamics. However, one may question whether the sort of behavior assumed is, in fact, consistent with optimization within the assumed environment. For example, the assumption of a constant saving ratio was often used in the early "descriptive" growth models and can indeed be derived from intertemporal utility maximization under certain hypotheses (for example, logarithmic utility function), but it becomes especially implausible when a production function of the type embodied in Eq. 3 is proposed. Why should a maximizing agent ever save up to the point at which marginal returns to capital are negative if he can obtain the same output level with much less capital stock? This implies that (given the assumptions on the technology) a "policy function" of the type shown by Eq. 3 would never occur in an "optimal growth model" of the Cass *(2)* type.

Although it is less obvious, Day's case of a variable saving ratio and a monotonic production function (i.e., Eq. 4) is equally inconsistent with intertemporal utility maximization. This was pointed out (in a general form) in Dechert *(3)*. Dechert argues as follows: let $y_t = f(k_t)$ be total output at time t, as a function of the existing stock of capital. The consumer-producer chooses how to split it between consumption and future capital in order to maximize $\sum_{t=0}^{\infty} u(c_t)\,\delta^t$, where u is a concave utility function, δ is a time-discount factor, $\delta \in (0,1)$, and k_0 is given as an initial condition. It turns out that, even if the production function is not concave, the optimal program $\{k_0, k_1, k_2, \ldots\}$ can be expressed by a policy function $k_{t+1} = \tau(k_t)$, which is monotonic. The dynamical system induced in this way cannot, therefore, produce cycles or chaos. The economic prediction is that such a society will asymptotically converge to some stationary position. The latter is unique when f is concave.

The Multisector Model

Here is the general model of intertemporal competitive equilibrium-optimal growth, of which the one-sector model of the preceeding section is just a special case [see Bewley *(4)* or Boldrin and Montrucchio *(5)*, Chapt. 1, for a more detailed illustration].

In every period $t = 0,1,2,\ldots$, the representative agent derives satisfaction from a "consumption" vector $c_t \in \Re^m_+$, according to a utility function $u(c_t)$, which is taken increasing, concave, and smooth as needed. The state of the world is fully described by a vector $x_t \in \Re^n_+$ of stocks and by a feasible set $F \subset \Re^n_+ \times \Re^m_+ \times \Re^n_+$ composed of all the triples of today's stocks, today's consumptions, and tomorrow's stocks that are technologically compatible, i.e., a point in F has the form (x_t, c_t, x_{t+1}). Now define

$$V(x,y) = \max_c u(c) \text{ such that } (x,c,y) \in F, \quad (5)$$

and let $D \subset \Re^{2n}_+$ be the projection of F along the c's coordinates. Then V, which is the short-run or instantaneous return function, will give the maximum utility achievable at time t if the state is x and we have chosen to go into state y by tomorrow. It should be easy to see that to maximize the discounted sum

$$\sum_{t=0}^{\infty} u(c_t)\,\delta^t \text{ such that } (x_t, c_t, x_{t+1}) \in F$$

is equivalent to

$$\text{Max} \sum_{t=0}^{\infty} V(x_t, x_{t+1})\,\delta^t \text{ such that } (x_t, x_{t+1}) \in D.$$

The parameter δ indicates the rate at which future utilities are discounted from today's standpoint (impatience): it takes values in [0,1).

The following assumptions on V and D may be derived from more basic hypotheses on u and F:

(A.1)
$V: D \to \Re$ is strictly concave and smooth (if needed). $V(x, y)$ is increasing in x and decreasing in y.

(A.2)
$D \subset X \times X \subset \Re^{2n}_+$ is convex, compact, and with nonempty interior. X is also convex, compact, and with nonempty interior.

The initial state x_0 is given.

The optimization problem we are facing can be equivalently described as one of dynamic programming:

$$W(x) = \max\{V(x,y) + \delta W(y), \text{ such that } (x,y) \in D\}. \quad (6)$$

The latter is the Bellman equation, and $W(x)$ is the value function for such a problem. A solution to Eq. 6 will be a map $\tau_\delta: X \to X$

describing the optimal sequence of states $\{x_0, x_1, x_2, \dots\}$ as a dynamical system $x_{t+1} = \tau_\delta(x_t)$ on X. The time evolution described by τ_δ contains all the relevant information about the dynamic behavior of our model economy. In particular, the price vectors p_t of the stocks x_t that realize the optimal program as a competitive equilibrium over time follow a dynamic process that (when the solution $\{x_t\}$ is interior to X) is homeomorphic to the one for the stocks. In other words, $p_{t+1} = \theta(p_t)$ with $\theta = \delta DW \cdot \tau_\delta \cdot (DW\delta)^{-1}$, where D is the derivative operator.

The question that concerns us is, what are the predictions of the theory about the asymptotic behavior of the dynamical system τ_δ? Where should a stationary economy converge under competitive equilibrium and perfect foresight? A first, remarkable answer is given by the following [see McKenzie *(6, 7)* for details and attributions].

Turnpike Theorem (Discrete Time)

Under Assumptions (A.1) and (A.2) with smoothness of V, there exists a value $\bar{\delta} < 1$ of the discount factor such that for all the δ's in the interval $[\bar{\delta}, 1)$, the function τ_δ that solves Eq. 6 has a unique globally attractive fixed point $x^* = \tau_\delta(x^*)$.

But the turnpike property is not the end of the story. Many more complicated dynamic patterns can be originated by τ_δ for fixed V and D as δ moves below $\bar{\delta}$. We will use the two-sector model as an illustrative device. For a detailed analysis of the more general case, see Boldrin and Montrucchio *(5, 8, 9)*.

A Two-Sector Chaotic Economy

Here is a brief description of a two-sector economy. For more details, see Boldrin *(10)* and Boldrin and Deneckere *(11)*.

Two goods (capital goods and consumption goods) and only one representative agent exist. The triples $\{w_t, r_t, q_t\}$, $t = 0,1,2,\dots$ denote the labor wage rate, the gross capital rental, and the price of capital in every period t. They are expressed in units of consumption goods which have the price fixed at one in all periods. We assume perfect foresight.

In each period, the consumer is endowed with one unit of labor time, which he supplies inelastically at the current wage rate, and with an amount k_t of capital stock, which is left over from previous consumption-saving decisions and that he supplies inelastically to the productive sectors. His budget constraint is then

$$c_t + q_t\left[k_{t+1} - \mu k_t\right] = r_t k_t + w_t, \qquad (7)$$

where $(1-\mu)$ is the capital depreciation rate. Given the initial capital stock k_0, the problem of the consumer amounts to picking up sequences of consumption $\{c_t\}$ and gross saving $\{k_{t+1} - \mu k_t\}$ to maximize the present value of his lifetime consumption stream under the period-by-period budget constraint (Eq. 7). Formally we write

$$\text{Max} \sum_{t=0}^{\infty} c_t \delta^t \text{ such that Eq. 7 is satisfied for}$$

$$\text{all } t = 0, 1, 2, \dots \text{ for a given } k_0. \qquad \text{(PC)}$$

Goods are produced by two industries. We summarize this with two production functions

$$y^1 = F^1(k^1, l^1), y^2 = F^2(k^2, l^2), \qquad (8)$$

where the superscript "1" denotes the consumption sector, and "2" denotes the capital goods sector; $k^i, l^i, i = 1,2$ are the quantities of capital and labor used as inputs in either one of the two industries. We assume

(T1)
$F^i, i = 1,2$ is an increasing and concave function from $[0, \infty) \times [0, \infty)$ into $[0, \infty)$. F^i is strictly concave in each separate factor. Also $F^2(0,x) = 0$ for all x in R_+.

(T2)
There exists a $\bar{k} \in (0,\infty)$ such that $F^2(k,1) < (1-\mu)k$ for all $k > \bar{k}$ and $F^2(k,1) > (1-\mu)k$ for all $k < \bar{k}$.

Firms take the price sequence $\{w_t, r_t, q_t\}$ as given. Their optimal decision problems reduce to the choice of factors-demand sequences $\{k_t^i, l_t^i\}$, which maximize the present discounted value of the stream of future profits. Therefore, the consumption good sector solves

(PF[1])
Max $y_t^1 - r_t k_t^1 - w_t l_t^1$,

such that $y_t^1 \le F^1(k_t^1, l_t^1)$, all t,

and the capital good solves

(PF[2])
Max $q_t y_t^2 - r_t k_t^2 - w_t l_t^2$,

such that $y_t^2 \le F^2(k_t^2, l_t^2)$, all t.

Given this description of agents' behavior, it is natural to define a competitive equilibrium (with perfect foresight) in the following way.

Definition 1. An intertemporal competitive equilibrium (ICE) is given by price sequences $\{w_t, r_t, q_t\}$ and quantity sequences $\{y_t^1, y_t^2, k_t^1, k_t^2, l_t^1, l_t^2, c_t, k_t\}$ such that

(a)
$\{c_t\}$ and $\{k_t\}$ solve (PC) given $\{w_t, r_t, q_t\}$;

(b)
y_t^1, k_t^1 and l_t^1 solve (PF[1]) given $\{w_t, r_t\}$, all $t = 0, 1, 2, \dots$;

(c)
y_t^2, k_t^2 and l_t^2 solve (PF[2]) given $\{w_t, r_t, q_t\}$, all $t = 0, 1, 2, \dots$;

(d)
$c_t = y_t^1, y_t^2 = k_{t+1} - \mu k_t, k_t = k_t^1 + k_t^2, 1 = l_t^1 + l_t^2$, all $t = 0, 1, 2, \dots$;

The existence of such ICE can be proved by standard arguments. Moreover, we can write down an infinite-horizon maximization problem whose solutions are ICE for our two-sector economy.

Proposition 1. Consider the economy described by (PC), (PF[1]), and (PF[2]) under assumptions (U1) and (T1)–(T2). Consider a set of quantity sequences satisfying Definition 1. Then they also solve the following problem:

(P1)

$$\text{Max} \sum_{t=0}^{\infty} c_t \delta^t \text{ such that } c_t \le T(k_t, y_t^2), k_{t+1} = \mu k_t + y_t^2,$$

where

(T)
$$T(k_t, y_t^2) = \text{Max } y_t^1 = F^1(k_t^1, l_t^1)$$

such that $y_t^2 \le F^2(k_t^2, l_t^2), k_t \ge k_t^1 + k_t^2$,

$1 \ge l_t^1 + l_t^2,\ l_t^i \ge 0, k_t^i \ge 0,\ i = 1, 2.$

The reciprocal is also true.

Example.

$$F^1(k^1, l^1) = \left[ak_1^\rho + (1-a)l_1^\rho\right]^{1/\rho},$$

$$a \in (0,1), \rho \in (-\infty, 1)$$

$$F^2(k^2, l^2) = \min\left\{l^2, \frac{k_2}{\gamma}\right\}, \gamma \in (0,1).$$

Notice that F^2 is not of class C^2. This choice was dictated by computability constraints. Problem (T) now is

$$Max \left[ak_1^\rho + (1-a)l_1^\rho\right]^{1/\rho}$$

$$\text{such that } y \le \min\left\{1 - l_1, \frac{k - k_1}{\gamma}\right\}.$$

The straightforward solution gives a PPF of the type

$$T(k,y) = \left[a(k - \gamma y)^\rho + (1-a)(1-y)^\rho\right]^{1/\rho}. \quad (9)$$

Problem (P) for our model economy is

$$\text{Max} \sum_{t=0}^{\infty} \left[a\left(k_t(1+\gamma\mu) - \gamma k_{t+1}\right)^\rho + (1-a)(1 - k_{t+1} + \mu k_t)^\rho\right]^{1/\rho} \delta^t$$

$$\text{such that } (k_t, k_{t+1}) \in D,$$

where

$$D = \left\{(x,y) \in [0,1] \times [0,1], \text{ such that } \mu x \le y \le \mu x + \frac{x}{\gamma}\right\}.$$

The competitive equilibrium sequences $\{c_t, y_t, q_t, r_t, w_t, l_t^1, l_t^2, k_t^1, k_t^2\}_{t=0}^{\infty}$ may then be derived from the sequence of optimal capital stocks $\{k_t\}_{t=0}^{\infty}$ that solve

(P)
$$W_\delta(k_0) = \text{Max} \sum_{t=0}^{\infty} V(k_t, k_{t+1})\, \delta^t$$

such that $(k_t, k_{t+1}) \in D$

$V(k_t, k_{t+1}) = T(k_t, k_{t+1} - \mu k_t)$, k_0 given in $[0, \bar{k}]$,

using the following relations, which hold either by definition or as a condition for equilibrium:

$$c_t = V(k_t, k_{t+1}); \quad (10a)$$

$$y_t = k_{t+1} - \mu k_t ; \tag{10b}$$

$$q_t = \delta W_\delta'(k_{t+1}) = -V_2(k_t, k_{t+1}); \tag{10c}$$

$$r_t = V_1(k_t, k_{t+1}); \tag{10d}$$

$$w_t = V(k_t, k_{t+1}) + q_t(k_{t+1} - \mu k_t) - r_t k_t ; \tag{10e}$$

$$l_t^i = l^i(k_t, k_{t+1} - \mu k_t), i = 1,2; \tag{10f}$$

$$k_t^i = k^i(k_t, k_{t+1} - \mu k_t), i = 1, 2; \tag{10g}$$

The functions $k^i(\cdot)$ and $l^i(\cdot)$ above are solutions to (T), and W_δ: $[0, \bar{k}] \to \Re$ is the value function associated with (P). As $W_\delta'(\cdot)$ denotes the derivative of this function, in Eq. 10c we are implicitly assuming that T is at least C^1 and applying the result obtained by Benveniste and Scheinkman *(12)*. In fact, we will sometimes make the stronger assumption that V: $D \to \Re_+$ is of class C^2 on $\underline{\text{Int}}(D)$.

Let $\tau_\delta(\cdot)$: $[0, \bar{k}] \to [0, \bar{k}]$ be the policy function associated with $W_\delta(\cdot)$, i.e.,

$$\tau_\delta(x) = \text{Arg}\max_y \left\{V(x,y) + \delta W(y); (x,y) \in D\right\}. \tag{11}$$

The optimal sequence $\{k_t\}$ is generated by the discrete dynamical system $k_{t+1} = \tau_\delta(k_t)$, starting at k_0. While $\tau_\delta(\cdot)$ and $W_\delta(\cdot)$ become analytically intractable as soon as V acquires even mild nonlinearities, a qualitative analysis can nevertheless be performed. We will show that, given any C^2 function θ: X → X, with X a closed interval of the real line, one can find a two-sector economy and a discount factor $0 < \delta < 1$ for which such a θ is the optimal policy function τ_δ. This result is an application to this specific case of a general theorem first proved in Boldrin and Montrucchio *(9)*.

Proposition 2. Let the feasible set $D \subset X x X$ be given, with X a closed interval of $\Re$. Let $\theta \in C^2(X;X)$ be such that graph(θ) ⊂ D. Then there exists a short-run return function V: D → $\Re$, continuous and strictly concave, and a discount factor, $\delta^* \in (0,1)$, such that $\theta \equiv \tau_\delta{}^*$, where $\tau_\delta{}^*$ is the optimal policy associated to the given (D,V, δ^*). Moreover, V(*x*, *y*) is increasing in *x* and decreasing in *y*, and one may recover from it two production functions F^1 and F^2 satisfying (T1), (T2).

Proof (Sketch). Let θ: X→X be given. Consider the function

$$U(x,y) = -\frac{1}{2}(y - \theta(x))^2 - \frac{L}{2}x^2 + m \cdot x$$

with L and *m* two positive real numbers. For L > 0 and large enough U is strictly concave in *x* and *y*. Also,

$$\max_y U(x,y) = U(x, \theta(x)) = W(x).$$

Let W, so defined, be the value function for the associated problem. A simple manipulation of Eq. 6 shows that the short-run return function $V(x, y) = U(x, y) - \delta W(y)$ in this case turns out to be

$$V(x,y) = -\frac{1}{2}(y - \theta(x))^2 - \frac{L}{2}x^2 + mx + \frac{\delta L}{2}y^2 - \delta my.$$

One needs to show that V is strictly concave and monotone in *x* and *y* for appropriately chosen L, *m*, and δ. It is not very difficult to see that this is always possible and that the following estimate for δ^* and the associated L^* are appropriate:

$$L^* = \eta\beta + \alpha\sqrt{\eta\beta}; \tag{12}$$

$$\delta^* < \left[2\alpha\sqrt{\eta\beta} + \eta\beta + \alpha^2\right]^{-1}, \tag{13}$$

where

$$\eta = \max_D |y - \theta(x)|,$$

$$\alpha = \max_X |\theta'(x)|, \text{ and}$$

$$\beta = \max_X |\theta''(x)|.$$

Now, in order to recover the "fictitious" production functions F^1 and F^2 that give rise to the return function V, one may proceed in the following way. Set the depreciation rate for capital $(1-\mu) = 1$ so that $k_{t+1} = y =$ output of $F^2(l^2, k^2)$. Then pick F^2 exactly as in our example, i.e., $y = \min\{1-l^1, (x-k^1)/\gamma\}$ with γ a parameter in (0,1) to be defined later. Such a choice for F^2 obviously satisfies (T1) and (T2). In order to recover $F^1(l^1, k^1)$, one has to repeatedly substitute for $y = 1-l^1$ and $x = \gamma y + k^1 = \gamma(1-l^1) + k^1$ in the definition of

$V(x, y)$, therefore obtaining

$$F^1(l^1, k^1) = -\frac{1}{2}\left[(1-l^1) - \theta(y-yl^1+k^1)\right]^2$$

$$+ m(y-yl^1+k^1) - \frac{L^*}{2}(y-yl^1+k^1)^2$$

$$+ \frac{\delta^* L^*}{2}(1-l^1) - \delta^* m(1-l^1). \quad (14)$$

Some tedious algebra will show that, given δ^* and L^* as defined in Eqs. 12 and 13, one can always pick $\gamma(\delta^*, L^*) \in (0,1)$ and $m(\delta^*, L^*) > 0$ such that F^1, as defined in Eq. 12, also satisfies (T1) and (T2). Q.E.D.

The formula given in Eq. 14 for the production function F^1 is, clearly, quite messy and does not seem to resemble any of the "typical" production functions that economists are keen to adopt in their applied exercises. Nevertheless, it suffices for our general existence purposes. On the other hand, as we will see, restricting the analysis to "typical" production functions does not rule out the emergence of complicated trajectories for $\left\{k_t\right\}_{t=0}^{\infty}$ as our example will prove.

Notice further that while both the F^1 and F^2 used in the example also satisfy homogeneity of degree one in (l^i, k^i) (i.e., returns to scale are constant) we have not made such an assumption for the general model. This was done on purpose as Proposition 2 would not be fully true otherwise. In fact, for a generic $\theta \in C^2(X;X)$ one can always find an F^2 that is homogeneous of degree one, but the resulting F^1 is typically not homogeneous.

In any case, homogeneity of degree one is of no harm to our search for complicated dynamics in the general two-section model, and it is therefore worth introducing it here.

(T3)
$F^i: \Re^2_+ \to \Re_+, i = 1,2,$

is homogeneous of degree one in k^i, l^i.

We can now collect a few results about the two-sector model that are relevant for our analysis.

Fact 1. Let $(x, y) \in \text{Int}(D)$ be a point on the policy function, i.e., $y = \tau_\delta(x)$. Assume $V \in C^2$ on Int(D). Then, if $V_{12}(x, y) > 0$ (< 0), the policy function is locally increasing (decreasing) at (x, y). Furthermore, if $(y, \tau_\delta(y)) \in \text{Int}(D)$, then $V_{12}(x, y) > 0$ $(<0, =0)$ implies the policy function is strictly increasing (strictly increasing, constant) at (x, y).

Proof. See Benhabib and Nishimura *(13)*.

Fact 2 (Turnpike). If $V_{12}(x, y) \geq 0$ for all $(x, y) \in \text{Int}(D)$, then any optimal sequence k_t converges to some k^* as $t \to \infty$, where $k^* \in \text{Fix}(\tau_\delta)$. Furthermore, for every given strictly concave V (and associated feasible set D), there exists a $\bar{\delta} < 1$ such that for all $\delta \in [\bar{\delta},1)$, the dynamical system $k_{t+1} = \tau_\delta(k_t)$ is globally asymptotically stable, i.e., there exists a unique k^* such that $k_t(k_0) \to k^*$ as $t \to \infty$, for every $k_0 \neq 0$.

Proof. See McKenzie *(7)*, Scheinkman *(14)*, Dechert and Nishimura *(15)*, and Deneckere and Pelikan *(16)*.

Fact 3. Assume $V_{12}(x, y) < 0$ for $(x, y) \in \tilde{D} \subset \text{int}(D)$. Let $(x^*(\delta), x^*(\delta)) \in \tilde{D}$ satisfy $\tau_\delta(x^*(\delta)) = x^*(\delta)$ for $\delta \in [\delta^-, \delta^+] \subset (0,1)$. Assume there exists $\delta^0 \in (\delta^-, \delta^+)$ such that:

(a)
$$V_{22}(x^*(\delta^0), x^*(\delta^0)) + \delta^0 V_{11}(x^*(\delta^0), x^*(\delta^0)) - (1+\delta^0)V_{12}(x^*(\delta^0), x^*(\delta^0)) = 0;$$

(b)
$$V_{22}(x^*(\delta), x^*(\delta)) + V_{11}(x^*(\delta), x^*(\delta)) - (1+\delta)V_{12}(x^*(\delta), x^*(\delta)) > 0, \text{ for } \delta \in [\delta^-, \delta^0);$$

(c)
$$V_{22}(x^*(\delta), x^*(\delta)) + V_{11}(x^*(\delta), x^*(\delta)) - (1+\delta)V_{12}(x^*(\delta), x^*(\delta)) < 0, \text{ for } \delta \in (\delta^0, \delta+].$$

Then, there exists a period-two orbit for τ_δ for all δ in some (right or left) neighborhood of δ^0.

Proof. See Benhabib and Nishimura *(13)*.

Fact 4. Under hypotheses (T1) and (T3), we have

(a)
$T_{12}(x, y) > 0$ for all (x, y)

such that $(k^1/l^1)(x, y) < (k^2/l^2)(x, y)$;

(b)
$T_{12}(x, y) < 0$ for all (x, y)

such that $(k^1/l^1)(x, y) > (k^2/l^2)(x, y)$.

Furthermore, if $(\bar{x}, \bar{y}) \in D$ is a feasible pair such that $T_{12}(\bar{x}, \bar{y}) = 0$, then $T_{12}(\bar{x}, y) = 0$ for all y feasible from $\bar{x}$.

Proof. See Benhabib and Nishimura *(13)*

and Boldrin *(10)*.

As pointed out in the Introduction, we will give conditions under which our simple two-sector competitive economy displays persistent and endogenous oscillations. We will appeal to the mathematical notion of "chaos" to describe such phenomena.

Definition 2. We say that $\tau_\delta: [0, \bar{k}] \rightarrow [0, \bar{k}]$ has *topological chaos* when there exists an orbit of period three for τ_δ, i.e., $\exists\, x \in (0, \bar{k}]$: $\tau_\delta^3(x) = x$, and $x \notin \text{Fix}(\tau_\delta)$. By Sarkovskii's theorem [see, for example Guckenheimer and Holmes *(17)*, p. 311], this implies that τ_δ has orbits of period n for any natural number n. It also implies [see Li and Yorke *(18)*], that there exists a nondenumerable set $S \subset [0, \bar{k}]$ such that all orbits of τ_δ with initial conditions $k_0 \in S$ exhibit aperiodic behavior. More formally this means there exists $\varepsilon > 0$ such that for every pair of points x and y in S with $x \neq y$

$$\limsup_{n\to\infty} \left|\tau_\delta^n(x) - \tau_\delta^n(y)\right| \geq \varepsilon$$

$$\liminf_{n\to\infty} \left|\tau_\delta^n(x) - \tau_\delta^n(y)\right| = 0$$

and for every $y \in \text{Per}(\tau_\delta)$ and $x \in S$,

$$\limsup \left|\tau_\delta^n(x) - \tau_\delta^n(y)\right| \geq \varepsilon.$$

The last formal result we need gives a set of computable sufficient conditions for the existence of topological chaos in a two-sector economy.

Fact 5. Assume there exists a $k^* \in (0, \bar{k})$ such that $V_{12}(k^*, \cdot) = 0$. Then τ_δ has topological chaos for all $\delta \in (0,1)$ that satisfy the following conditions:

(a)
$V_2(x, k^*) + \delta V_1(k^*, \cdot) = 0$, has two roots, $k_1 \in (0, k^*)$ and $k_2 \in (k^*, \bar{k}]$;

(b)
$V_2(x, k_1) + \delta V_1(k_1, k^*) = 0$, has a root $k_3 \in [k^*, \bar{k}]$;

(c)
$V_2(x, k_3) + \delta V_1(k_3, k_1) = 0$, has at least one real root.

Proof. See Boldrin *(10)*.

Fact 2 is the classical turnpike theorem; we will use it just to note that there exists an upper bound on the set of δ that may produce oscillating behavior. Fact 3 shows that when τ_δ is downward sloping around an optimal steady state (OSS), then a cycle of period two may bifurcate from the OSS when it loses stability. Since the information necessary to verify the hypothesis of this result is local in nature, we can use it to detect orbits of period two. This will turn out to be the first (or last) step of a bifurcation cascade leading to chaos in our specific model. A simple generalization of Fact 3 will also allow us to detect the existence of orbits of period four and, potentially, of any orbit with period 2^n. Facts 1 and 4 link the slope of τ_δ to factor-intensity conditions. For the case in which $\mu = 0$ (i.e., the capital stock lasts only one period) the causal relation is clear: τ_δ is increasing when the investment sector has a higher capital/labor ratio than the consumption sector, decreasing in the opposite case, and flat at the reversal points. When $\mu \neq 0$ we can easily see that

$$V_{12}(k_t, k_{t+1}) = T_{12}(k_t, k_{t+1} - \mu k_t) - \mu T_{22}(k_t, k_{t+1} - \mu k_t). \tag{15}$$

The slope of τ_δ also depends, therefore, on μ and the sensitivity of the price of capital to variations in the output of the investment sector. The critical point of τ_δ, when it exists, will not necessarily coincide with a factor-intensity reversal, and will not be independent of δ, as in the case when $\mu = 0$. On the other hand, note that τ_δ may now be nonmonotonic even in the absence of a capital-intensity reversal; this is true if T_{12} is negative everywhere (i.e., the consumption sector is always more capital intensive) and if both μ and T_{22} are "large enough" for small values of k_t. Having clarified the theoretical aspect of the problem, we will now proceed to apply these abstract results to our simple example.

Example. If we set also $\mu = 0$, then our $V(k_t, k_{t+1})$ becomes

$$\left[a(k_t - \gamma k_{t+1})^\rho + (1-a)(1-k_{t+1})^\rho\right]^{1/\rho}.$$

The second derivative V_{12} is zero for $k_{t+1} = 1$, or for $k_t = \gamma$, positive on the interior of D for all $k_t \in [0, \gamma)$ and negative for all $k_t \in (\gamma, 1]$. Simple manipulations of the Euler equation,

$$-\left[a(k_{t-1} - \gamma k_t)^\rho + (1-a)(1-k_t)^\rho\right]^{1/\rho - 1}$$

$$\cdot \left\{(1-a)(1-k_t)^{\rho - 1} + a\gamma(k_{t-1} - \gamma k_t)^{\rho - 1}\right\}$$

$$+ \delta a\left[a(k_t - \gamma k_{t+1})^{\rho} + (1-a)(1-k_{t+1})^{\rho}\right]^{1/\rho-1} \cdot (k_t - \gamma k_{t-1})^{\rho-1} = 0 \tag{16}$$

will show that the unique, interior steady state k(δ) can be expressed as

$$k(\delta) = \left\{1+(1-\gamma)\left[\frac{1-a}{a(\delta-\gamma)}\right]^{\frac{1}{1-\rho}}\right\}^{-1}. \tag{17}$$

Therefore, for $\delta \in [0, \gamma]$, we have no interior state, and for $\delta \in (\gamma,1)$, we have a unique interior steady state, which is on the upward sloping branch of τ_δ for $\delta < \gamma\ (1 + (1-a)/a\ \gamma^{\rho})$ and on the downward sloping one for $\delta > \gamma\ (1 + (1-a)/a \cdot \gamma^{\rho})$. Let us simplify the algebra by setting $\rho = 0$ in the definition of F^1. This is the Cobb-Douglas case, especially loved by macroeconomists. The function V(x, y) is now $(1-y)^{\alpha}(x-\gamma y)^{1-\alpha}$ with $\alpha = 1 - a$.

The proofs of the following list of results may be found in Boldrin and Deneckere *(11)*. Scheinkman *(19)* had previously suggested this model as a candidate for chaos.

Proposition 3. For $0 \leq \delta \leq \gamma$, the optimal path k_t converges to zero for any initial condition k_0 in [0,1], and no interior OSS exists. For $\gamma < \delta \leq \gamma/(1-\alpha)$, there exists a unique interior OSS k^* defined by Eq. 17 and the optimal path converges to k^* for any k_0 in (0,1].

Observe that Proposition 3 implies global asymptotic stability if $\alpha \geq 1-\gamma$ independently of δ! From Eq. 17, we may also derive some comparative statics results.

Proposition 4. The OSS level of the capital stock defined by Eq. 17:

(a)
increases with the discount factor δ;

(b)
decreases with the labor productivity factor α;

(c)
decreases with γ, the capital/labor ratio in the investment sector.

Let us now turn to the case where $\delta > \gamma/(1-\alpha)$. It is clear from the literature on optimal growth theory [for example, McKenzie *(6)* and Scheinkman *(14)*], that when k^* is locally stable for τ_δ, i.e., $|\partial\tau_\delta(k)/\partial k|_{k=k*} < 1$, the second-order system produced by the Euler equation (16) has a local saddle point structure at k^*. This means that, of the two eigenvalues of the characteristic polynomial

$$\delta V_{12}(k^*, k^*)\lambda^2 + [V_{22}(k^*, k^*) + \delta V_{11}(k^*, k^*)]\lambda + V_{12}(k^*, k^*) = 0 \tag{18}$$

associated with the linearization of Eq. 17, one lies inside and one lies outside the unit circle. In fact, the smaller eigenvalue corresponds to $\tau'_\delta(k^*)$ when the latter exists. For our example, Eq. 18 reduces to

$$\delta\,[(\gamma-k^*)/(1-k^*)]\,\lambda^2 - [\delta + (\gamma-k^*)^2/(1-k^*)^2]\,\lambda + (\gamma-k^*)/(1-k^*) = 0 \tag{19}$$

from which we may compute

$$\lambda_1 = (1-k^*)/(\gamma-k^*) < 0$$
$$\lambda_2 = [(\gamma-k^*)/(1-k^*)]\,\delta^{-1} < 0. \tag{20}$$

The signs of expressions in Eq. 20 follow from the fact that $k^* > \gamma$ whenever $\delta > \gamma/(1-\alpha)$. We can easily reduce Eq. 20 to

$$\lambda_1 = \alpha/[\gamma - (1-\alpha)\delta] < 0,$$
$$\lambda_2 = (\gamma - (1-\alpha)\delta]/\alpha\delta < 0. \tag{21}$$

Thus,

$$\lambda_1 \in (-\infty,-1) \text{ for } \delta \in (\gamma/(1-\alpha), (\alpha+\gamma)/(1-\alpha)), \text{ and}$$
$$\lambda_1 \in (-1,0) \text{ for } \delta > (\alpha+\gamma)/(1-\alpha). \tag{22}$$

$$\lambda_2 \in (-1,0) \text{ for } \delta \in (\gamma/(1-\alpha), \gamma(1-2\alpha)), \text{ and}$$
$$\lambda_2 \in (-\infty,-1) \text{ for } \delta > (\gamma/(1-2\alpha), \text{ if } \alpha < 1/2, \text{ and}$$
$$\lambda_2 \in (-1,0) \text{ for } \delta \in (\gamma/(1-\alpha),1) \text{ if } \alpha \geq 1/2. \tag{23}$$

Hence we have proven the following:

Proposition 5. The OSS k^* is locally asymptotically stable when $\alpha \geq 1/2$, and, when $\alpha < 1/2$, it is stable for parameter values δ in $(\gamma/(1-\alpha), \gamma/(1-2\alpha)) \cup ((\alpha+\gamma)/(1-\alpha),1)$.

The following is an immediate corollary to Proposition 5:

Corollary. The OSS k^* is locally asymptotically stable for all $\delta \in (\gamma/(1-\alpha),1)$ when $\alpha \geq (1-\gamma)/2$.

One should note that Proposition 5 states a local result only. Intuition suggests that k^* may in fact be globally, asymptotically stable, but a proof of this requires additional analysis [see Boldrin and Deneckere *(11)*, Section 4].

A natural question now arises: What happens when $\delta \in [\gamma/(1-2\alpha), (\alpha+\gamma)(1-\alpha)]$? A partial answer is the following proposition:

Proposition 6. Let $\alpha < (1-\gamma)/2$. Then the

policy function τ_δ has a cycle of period two in a neighborhood of $\delta^- = \gamma(1-2\alpha)$ and $\delta^+ = (\alpha+\gamma)/(1-\alpha)$. These cycles are locally stable when they exist for $\delta \in (\delta^-, \delta^+)$ and unstable in the other cases.

The reader should note that, given $\delta < 1$, it is possible to find α and γ in $(0,(1-\gamma)/2)$ and $(0,1)$, respectively, such that $\gamma/(1-2\alpha) = \delta$. This means that, at every level of discounting, we can always find some technology that has optimal two cycles. In fact, the dynamic behavior of this economy for $\delta \in (\delta^-, \delta^+)$ may become very complicated. Our contention is that, for suitable a, ρ, and γ, there exists an interval $(\delta^*, \delta^{**}) \subset (\delta^-, \delta^+)$ at which τ_δ has (at least) topological chaos. This was proven in Boldrin and Deneckere *(11)*. It was also shown there that a second "flip bifurcation" may lead to an orbit of period four. We use the same logic behind Fact 3 and Proposition 5.

The following is a corollary:

Corollary. Let $(x(\delta), y(\delta))$ be a period-two point for our model, and suppose it exists for all δ in (δ^-, δ^+) with $x(\delta) < \gamma$ and $y(\delta) > \gamma$. Then a cycle of period four exists for all values of $\delta^0 \in (\delta^-, \delta^+)$ at which either one of the following two equations hold

(a)

$$(x(\delta^0) - \gamma)/(1 - x(\delta^0)) = -(\delta^0)^2(1 - y(\delta^0))/(y(\delta^0) - \gamma)$$

(b)

$$(x(\delta^0) - \gamma)/(1 - x(\delta^0)) =$$

$$-(1 - y(\delta^0))/y(\delta^0))/(y(\delta^0) - \gamma).$$

To verify the presence of such bifurcations in our model, consider the example $\alpha = 0.03$, $\gamma = 0.09$. Proposition 5 implies that the steady state k^* is locally stable when δ lies in the interval [0.0928, 0.0957]. For discount factors in [0.0957, 0.0974], stable period-two orbits are present, verifying Proposition 6. At $\delta = 0.0974$, the period-two orbit $x^* = 0.0738$, $y^* = 0.3980$ bifurcates into a stable period-four orbit, which exists for $\delta \in [0.0974, 0.0978]$. In fact, our simulation results reveal that successive bifurcations eventually lead to chaos when δ reaches the value 0.099. This chaos exists for $\delta \in [0.099, 0.112]$, as can be checked directly by applying Fact 6 above to our model. Figure 1 describes the evolution of the policy function τ_δ for $\alpha = 0.03$ and $\gamma = 0.09$, as δ moves in (0,1) and Figure 2 depicts the set of (α, γ) parameters for which topological chaos is present for some $\delta \in (0,1)$. Both firgures are from Boldrin and Deneckere *(11)*.

One might suspect that part of the reason that such extreme values of the parameters are needed stems from the fact that the elasticity of substitution between capital and labor in the consumption good sector is fairly large (i.e., $\rho = 0$). For $\rho < 0$, the elasticity of substitution decreases. This brings us back to the initial formulation. The elasticity of substitution, σ, for the CES is equal to $1/(1-\rho)$; negative values of ρ thus permit much smaller values of σ. Our simulations reveal that chaotic optimal paths do arise for this model as well and that when σ is fairly low, chaos appears for values of the discount factor roughly three times larger than those found for the Cobb-Douglas model. A typical example has $\alpha = 0.2$, $\gamma = 0.2$, $\delta = 0.25$, and $\rho = -0.5$. Since the values for the discount factor at which chaos appears in the Cobb-Douglas model are themselves approximately 100 times larger than the ones found in the artificial economies constructed by Boldrin and Montrucchio *(9)* and Deneckere and Pelikan *(16)*, no definite conclusion can be drawn at this stage as to whether a model of this type could produce chaotic dynamics at more reasonable values of the discount factor.

Conclusion

In this chapter, we analyzed simple versions of aggregate general equilibrium models, which produce unique but sometimes cyclical and even chaotic paths for macroeconomic variables such as output, consumption, and investment. The reader should consult Boldrin and Woodford *(20)* for a more complete survey. Despite the fact that an analytical expression for the policy function is often unavailable, we are able to characterize the dynamic behavior of our economy in terms of its basic parameters: α, the labor share of income in the consumption sector; γ, the capital/labor ratio in the investment sector; and δ, the discount factor. For many values of the parameters, the

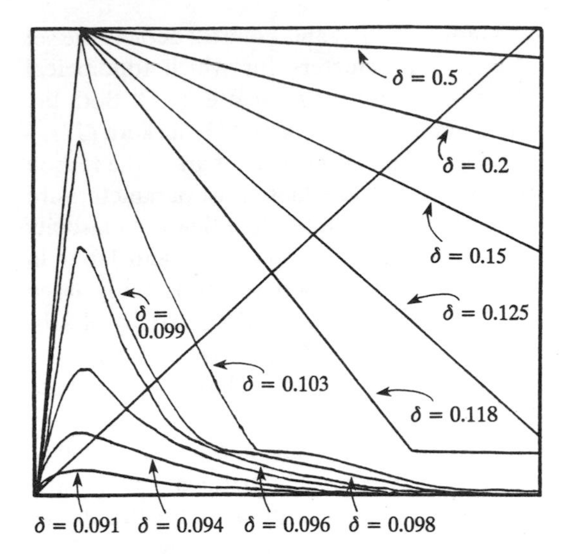

Fig. 1. Evolution of the policy function $\tau\delta$ as a function of the discount parameter δ (α = 0.03, γ = 0.09). Reprinted from *(11)*, with permission.

unique steady state was shown to be globally, asymptotically stable. For other values of the parameters, we obtained a unique period-two point, which was globally attractive. Successive bifurcations then led to a chaotic regime but only for rather unrealistic values of the parameters.

This should not be too surprising. As the analysis above indicated, the analytical complexity of the nonlinear model greatly exceeds that associated with linear stochastic models. The primitive stage of our research technology forced us to work with a rather rudimentary and rigid model. Nevertheless, our study casts some doubt on the notion that, in one-dimensional capital goods models, chaos is a useful way to model the apparently self-sustained nature of the trade cycle. The highly nonlinear, unimodal form for the policy function that is necessary in order to produce complex dynamics forces one to resort to rather unrealistic parameter values. A similar neoclassical multisector model, allowing for a higher dimensional state space is bound to be much more successful in this regard, since even slight departures from linearity may then produce strange attractors.

While still in its infancy, the study of nonlinearities in economic models is likely to provide insights into the forces behind observed economic fluctuations. In our model, we underlined the importance of *intersectoral substitution effects* (induced by different degrees of profitability in different sectors) as well as intertemporal substitution effects in determining factor allocation decisions, investment activities, and so on. Bypassing the nonlinearities with first-order approximations neglects the important contribution of these factors in amplifying and sustaining oscillations.

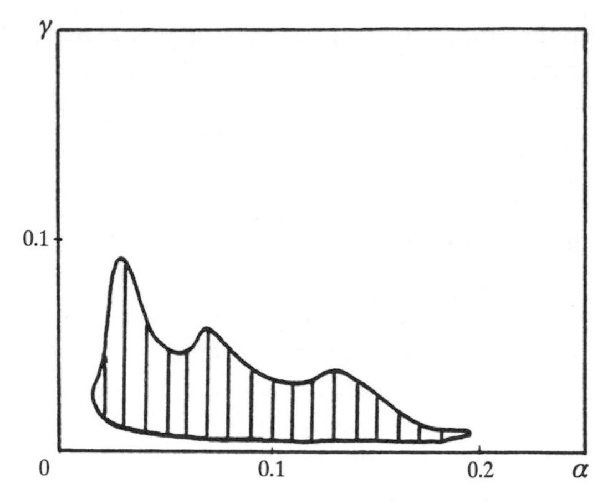

Fig. 2. Values of the parameters α and γ for which chaos is present for some δ in (0,1). Reprinted from *(11)*, with permission.

References

1. Day, R., *Am. Econ. Rev.* **72**, no. 3, 406 (1982).
2. Cass, D., *Rev. Econ. Stud.* **32**, 233 (1965).
3. Dechert, W. D., *Nationalökonomie* **44**, 57 (1984).
4. Bewley, T., *J. Math. Econ.* **10**, 233 (1982).
5. Boldrin, M., and L. Montrucchio, *Dynamic Complexities of Intertemporal Competitive Equilibria* (Oxford University Press, London-New York), forthcoming, 1991.
6. McKenzie, L. W., in *Handbook of Mathematical Economics*, vol. 3, K. J. Arrow and M. D. Intriligator, Eds. (North Holland, Amsterdam, 1986), pp. 1281–1355.
7. McKenzie, L. W., in *The New Palgrave* (Stockton Press, New York, 1987), pp. 712–720.
8. Boldrin, M., and L. Montrucchio, *Math. Model.* **8**, 627 (1986).
9. Boldrin, M., and L. Montrucchio, *J. Econ. Theory* **40**, 26 (1986).
10. Boldrin, M., in *Economic Complexity: Chaos,*

Sunspots, Bubbles and Nonlinearity, W. Barnett, J. Geweke, K. Shell, Eds. (Cambridge University Press, Cambridge, UK, 1988).
11. Boldrin, M., and R. Deneckere, *J. Econ. Dyn. & Control* **12**, 108 (1990).
12. Benveniste, L., and J. A. Scheinkman, *Econometrica* **47**, 727 (1979).
13. Benhabib, J., and K. Nishimura, *J. Econ. Theory* **35**, 284 (1985).
14. Scheinkman, J. A., *J. Econ. Theory* **12**, 11 (1976).
15. Dechert, W. D., and K. Nishimura, *J. Econ. Theory* **31**, 332 (1983).
16. Deneckere, R., and S. Pelikan, *J. Econ. Theory* **40**, 13 (1986).
17. Guckenheimer, J., and P. Holmes, *Nonlinear Oscillations, Dynamical Systems and Bifurcation of Vector Fields* (Springer-Verlag, New York, 1983).
18. Li, T. Y., and J. A. Yorke, *Am. Math. Mon.* **82**, 985 (1975).
19. Scheinkman, J. A. "General Equilibrium Models of Economic Fluctuations: A Survey" (mimeo, Department of Economics, University of Chicago, September 1984).
20. Boldrin, M., and M. Woodford, "Equilibrium Models Displaying Endogenous Fluctuations and Chaos," *J. Monetary Econ.* **25**, 193 (1990).

13

The Case for Chaos in Childhood Epidemics

William M. Schaffer, Lars F. Olsen
Greg L. Truty, Sandra L. Fulmer

Abstract

In the absence of mass immunization, childhood diseases such as chicken pox, measles, mumps, and rubella exhibit recurrent epidemics in large, First World cities. In chicken pox, the basic pattern is a yearly oscillation. In measles and rubella, apparently chaotic fluctuations are superposed on an annual cycle. Both patterns are observed in mumps. The case for chaos in childhood epidemics is rendered credible by comparison of real-world epidemics with differential equations that categorize the host population into Susceptible, Exposed, Infectious, and Recovered groupings (SEIR models) and Monte Carlo simulations that reproduce many of the qualitative and quantitative properties of real-world epidemics. Certain anomalies persist, suggesting avenues for future investigation.

Introduction

Childhood diseases — chicken pox, measles, mumps, and rubella — have long attracted the attention of mathematicians and epidemiologists whose interests tend toward mathematics and time-series analysis *(1–16)*. This continuing interest reflects a variety of factors, including the following:

(i) The biology is reasonably well understood and lends itself to mathematical representation.

(ii) Many of these diseases are, or at times have been, reportable. Hence there is a substantial body of data that can be studied either for its own sake or for the purpose of comparison with hypothetical models.

(iii) Often, the observed dynamics are neither so regular as to indicate simple oscillations with some small admixture of noise, nor so erratic as to suggest essentially random variations. In other words, there seem to be patterns for which mathematics might provide an accounting.

It would thus appear that childhood diseases constitute a series of well-defined epidemiological systems for which real-world behavior and mathematical models can be profitably compared. In addition, it should be noted that the success of mass immunization programs in western nations notwithstanding, some childhood diseases, notably measles, are major sources of mortality in the developing world *(17)*. Even in the First World, congenital rubella syndrome remains a continuing tragedy both for the children it afflicts and their families.

Phenomenology

Childhood diseases are characterized by recurrent epidemics (Fig. 1). Often there is a yearly cycle, which may reflect enhanced transmission rates in schools *(4, 18)* as well as climatic factors that operate to the same effect

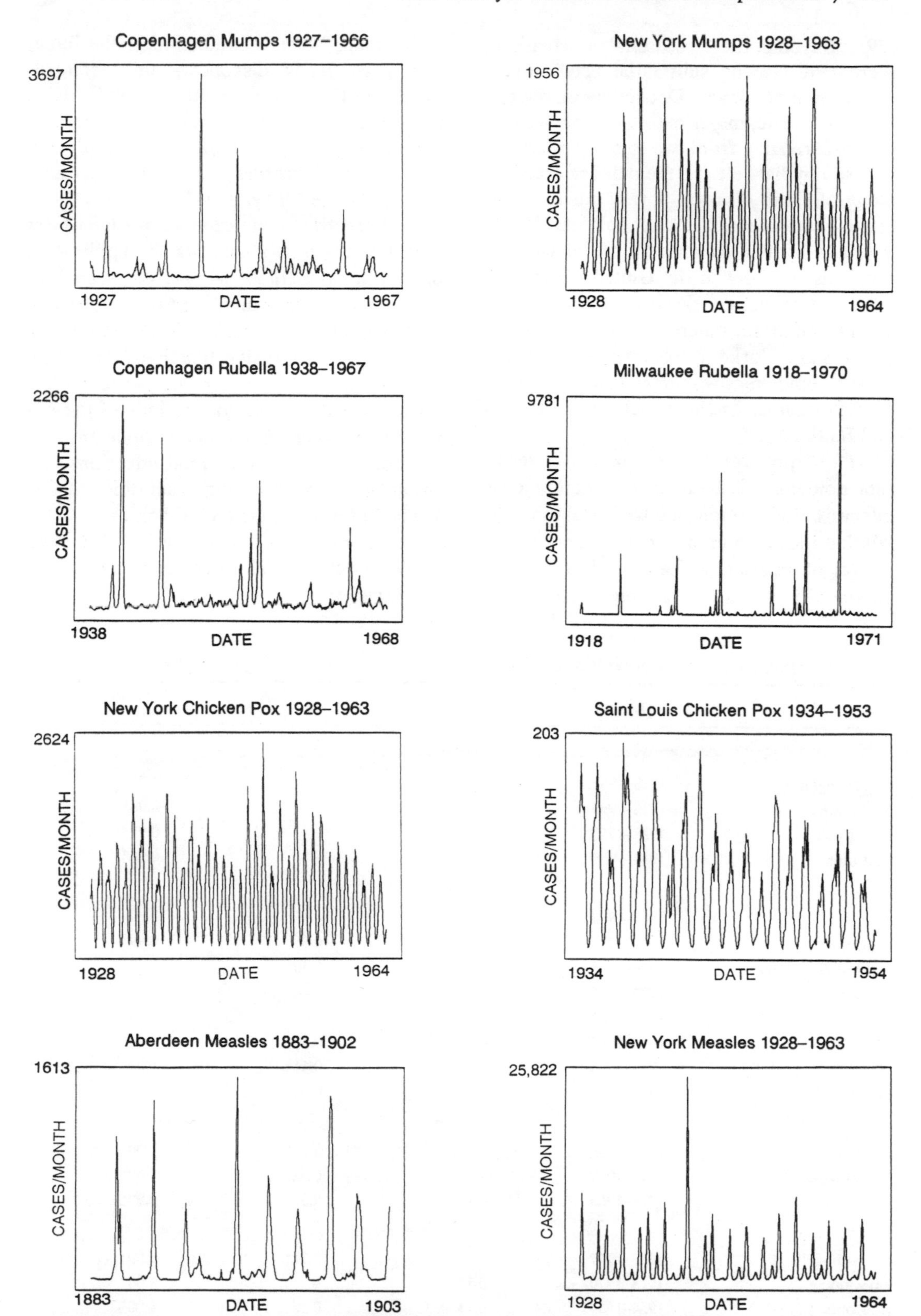

Fig. 1. Representative time series for childhood infections in large, First World cities prior to mass immunization. Numbers of monthly cases reports are plotted against time.

(19). Beyond these seasonal fluctuations, there may also be substantial between-year variation in incidence. Depending on the disease and locale, major outbreaks can occur at intervals ranging from two to seven years. As stressed by Bartlett *(5)*, the detailed epidemiology of these diseases is very much a local affair, with centers of population bound together by the movements of infected individuals. At least in the United States, synchronization breaks down over distances of several hundred miles. Thus, for example, London and Yorke *(6)* observed synchrony of measles epidemics within the five boroughs of New York, but asynchrony between New York and Baltimore.

One approach to the problem of recurrent epidemics is to focus on interepidemic intervals. For example, it is well established *(4, 20)* that the mean interval between major outbreaks of measles increases with decreasing community size and increasing isolation. That is, in small, isolated communities, the infection periodically disappears until reintroduced by the arrival of infected individuals from without. In such cases, one observes a rise in the mean age of infection as indicated by serological profiles *(21)*. [For a detailed case study, see Cliff *et al.* *(22)*.]

Alternatively, one can look for long-term periodicities in incidence via the application of traditional methods of time-series analysis *(23)*. This entails computing the power spectrum whereby a time series is represented as a superposition of cosine functions. Then one inquires as to whether there are particular frequencies that account for the bulk of the observed variance. Systematic application of such techniques to childhood infections *(14, 24)* suggests the following generalities (Table 1; Fig. 2) for large, First World cities:

(i) For chicken pox, one always observes a single spectral peak corresponding to the yearly cycle.

Table 1. Spectral analysis of childhood diseases.

City	Dates	Spectral peaks (cpy)		
		Chicken pox		
Copenhagen	1938–1967	–	–	1.00*
Milwaukee	1916–1965	–	–	0.98*
New York	1928–1963	–	–	0.99*
St. Louis	1934–1953	–	–	0.98*
		Measles		
Aberdeen	1883–1902	–	0.46*	0.89
Baltimore	1900–1927	–	0.45	0.99*
Baltimore County	1928–1963	–	0.40*	0.99
Copenhagen	1927–1966	–	0.40*	0.99
Detroit	1920–1962	–	0.46	0.99*
Milwaukee	1916–1965	–	0.40	1.00*
New York	1928–1963	–	0.41	1.01*
St. Louis	1934–1954	–	0.34	1.01*
		Mumps		
Copenhagen	1927–1966	–	0.25*	1.00
Milwaukee	1922–1970	–	0.32	1.00*
New York	1928–1963	–	0.37	0.99*
		Rubella		
Copenhagen	1938–1967	–	0.21*	1.00*
Milwaukee	1918–1970	0.14	–	1.00*
St. Louis	1934–1954	0.14	0.49	1.00*

*Asterisks denote major spectral features.

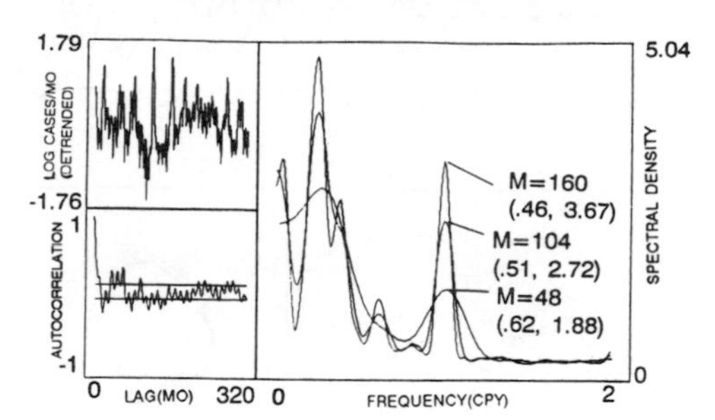

New York Mumps 1928–1963

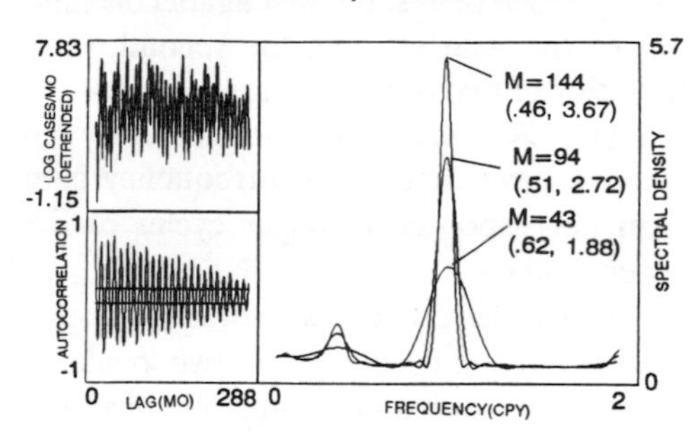

Copenhagen Rubella 1938–1967

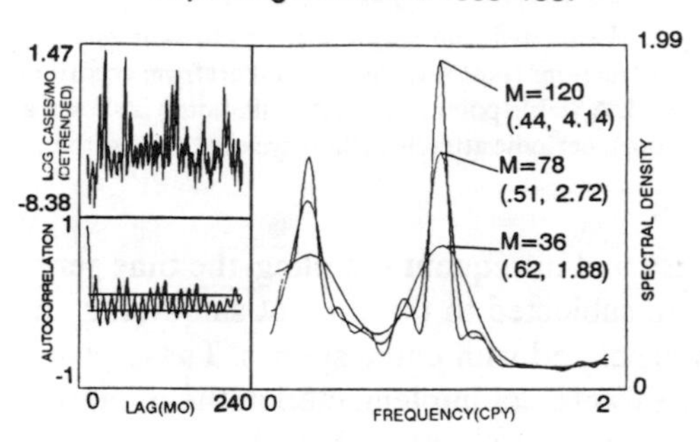

Milwaukee Rubella 1918–1970

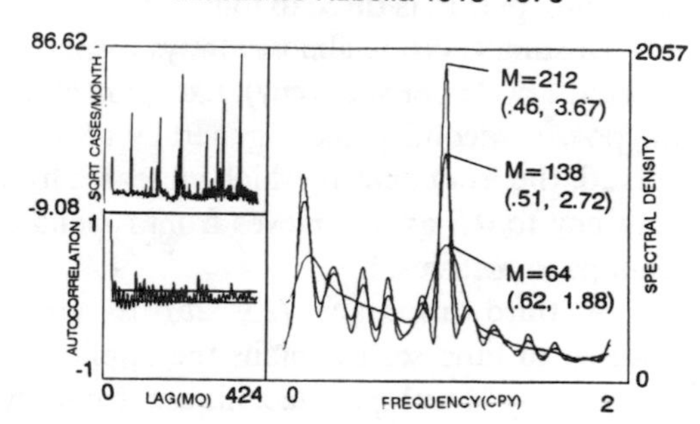

New York Chicken Pox 1928–1963

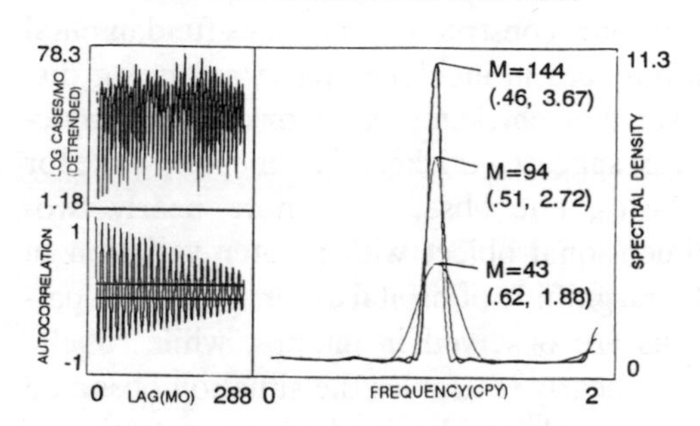

Saint Louis Chicken Pox 1934–1953

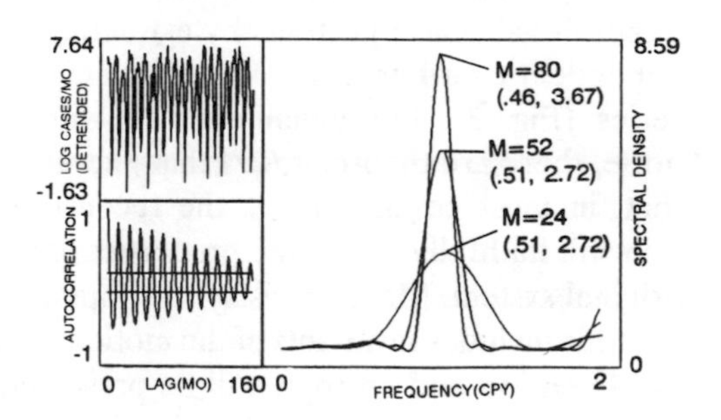

Aberdeen Measles 1883–1902

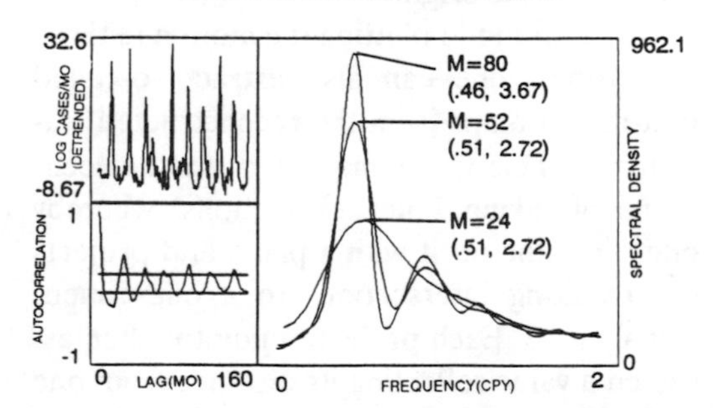

New York Measles 1928–1963

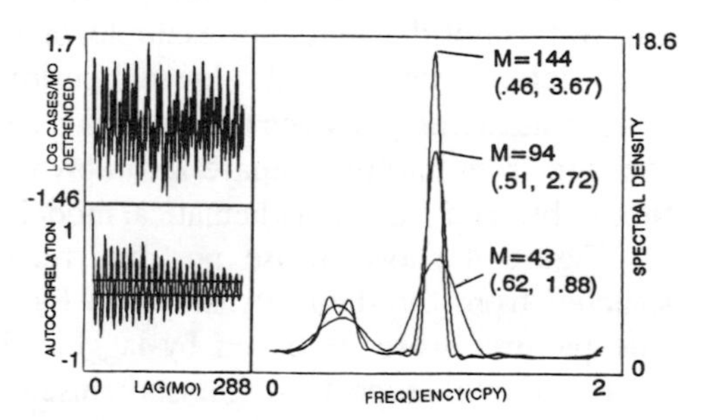

Fig. 2. Time series analysis of childhood epidemics. Each picture gives the time series (**upper left**) after log or square root transformation and detrending, the autocorrelation function (**lower left** – includes 95% confidence interval about zero correlation), and three smoothed spectra. The latter were computed using a Tukey window with the lag set equal to N/3, (N/3 + N/10)/2, and N/10, where N is the sample size.

(ii) In measles, one generally observed two spectral peaks, the first again corresponding to the yearly cycle; the second, to periodicities of two to three years.

(iii) A similar situation is observed in rubella, save that the low-frequency component corresponds to longer cycles of five to seven years.

(iv) Both the chicken pox (one spectral peak) and measles/rubella (two peaks) patterns are observed in mumps. When a low-frequency component is present, the corresponding period is three to four years.

In sum, there is almost always a spectral peak at 1 cycle per year (cpy). Except in chicken pox, a second peak is generally also observed, the frequency of which increases from 0.14 cpy to 0.5 as one moves from rubella to mumps to measles.

A third approach *(25, 26)* to epidemiological time series entails the application of recently developed techniques from dynamical systems theory. These procedures *(27, 28)* involve the construction of multidimensional phase portraits by extracting independent coordinates from a univariate time series (Fig. 3). For unlimited data without noise, there are theorems *(27)* that guarantee that, in most circumstances, the reconstruction will faithfully reflect the properties of the original system. More precisely, the topology, and this includes invariants of the motion such as dimension and entropy, will be preserved. Phase portrait reconstruction is especially useful when the resulting trajectory suggests a known form of dynamics, for example, one of the better understood low-dimensional chaotic attractors, or when trajectories reconstructed from nature compare favorably with those obtained from a mathematical model.

Figure 4 shows phase portraits reconstructed from the data sets shown in Fig. 1. The pictures were produced by lagging the data – monthly reports – against themselves. That is, three-dimensional trajectories were generated by plotting for all times, *t*,

$$x(t) \text{ vs. } x(t+T) \text{ vs. } x(t+2T). \quad (1)$$

Here, $x(t)$ is the number of cases reported, and T is the lag, generally equal to three months. To reduce the effects of observational

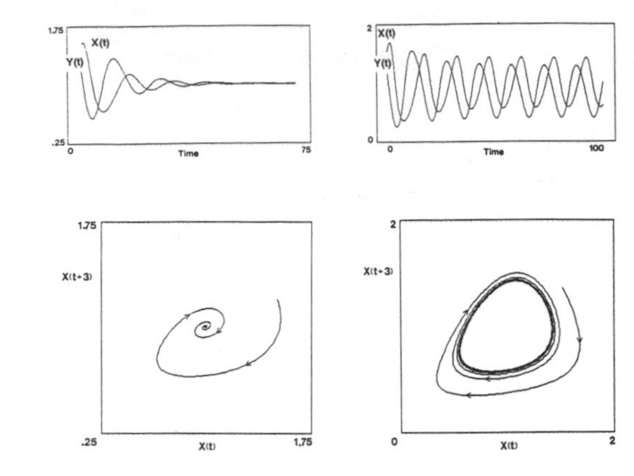

Fig. 3. Univariate time series (**top**) and two-dimensional reconstructions (**bottom**) computed therefrom according to Eq. 1. A stable point (damped oscillations) is shown at the top; a periodic attractor (limit cycle) at the bottom.

noise and infrequent sampling, the time series were subjected to three-point smoothing and interpolated with cubic splines. These procedures were not implemented in the numerical computations discussed below.

As in the case of spectral analysis, phase portrait reconstruction indicates fundamental differences in incidence patterns among diseases. For chicken pox, reconstructed trajectories suggest a thickened limit cycle, while for measles, one observes a more nearly two-dimensional object with greater variation in the magnitude of orbital excursions. Both patterns are observed in mumps, while rubella more nearly resembles the situation observed in measles but with a marked concentration of orbits near the origin.

In addition to plotting the motion in three dimensions, one can also extract so-called first-return maps from the reconstructed trajectories. This is accomplished via the technique of taking Poincaré sections, whereby one slices the orbit with a plane and projects the resulting intersections to a one-dimensional curve. Each projected point is then assigned a value reflecting its distance from one end of the section. Finally one plots the assigned values in temporal sequence, i.e., if $v(i)$ is the ith value, one plots

$$v(i+1) \text{ vs. } v(i)$$

for all pairs i and $i+1$. Such a construction gives the rule, if one exists, by which successive

Copenhagen Mumps

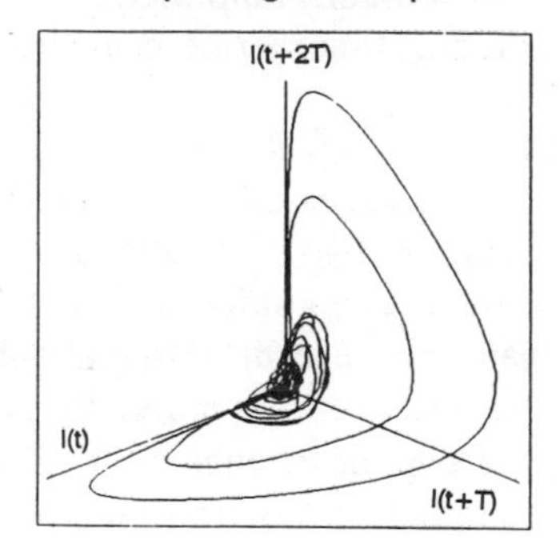

New York Mumps

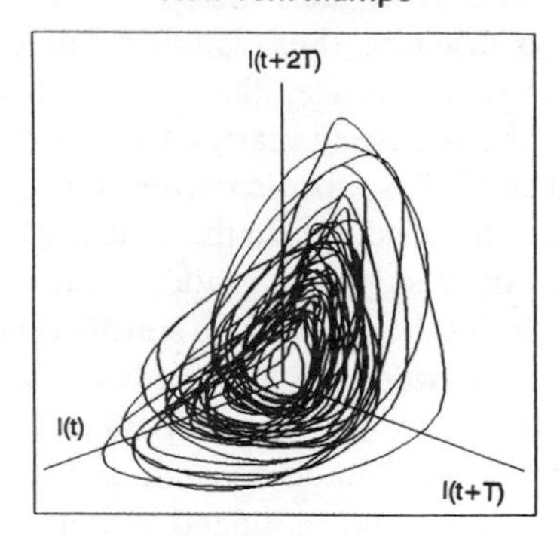

Copenhagen Rubella

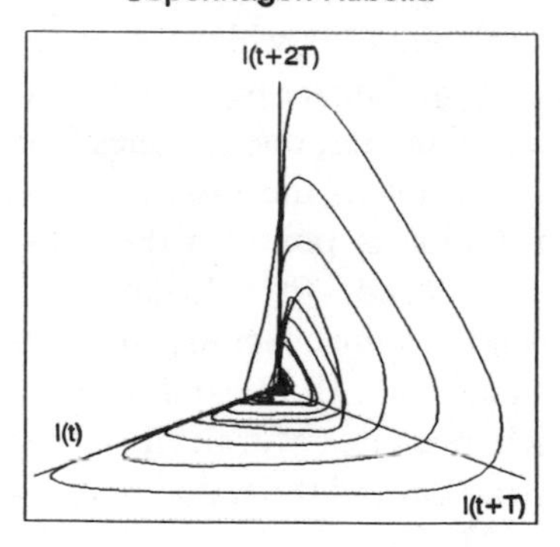

Milwaukee Rubella

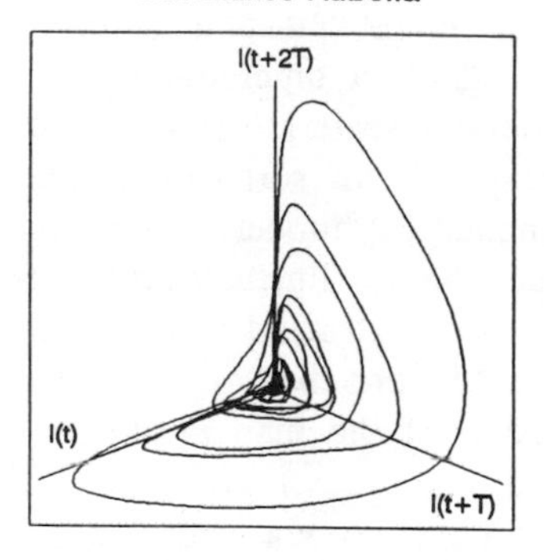

New York Chicken Pox

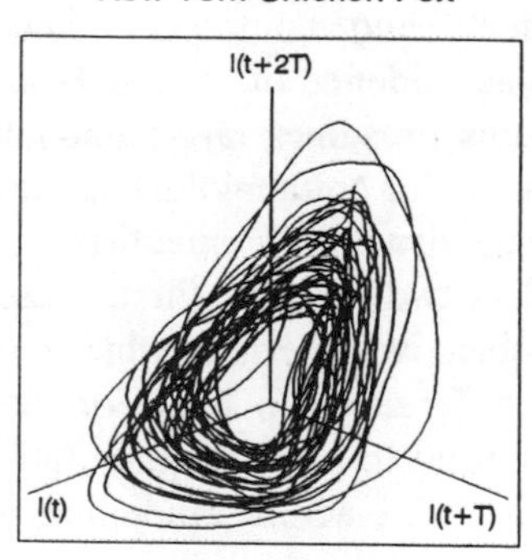

Saint Louis Chicken Pox

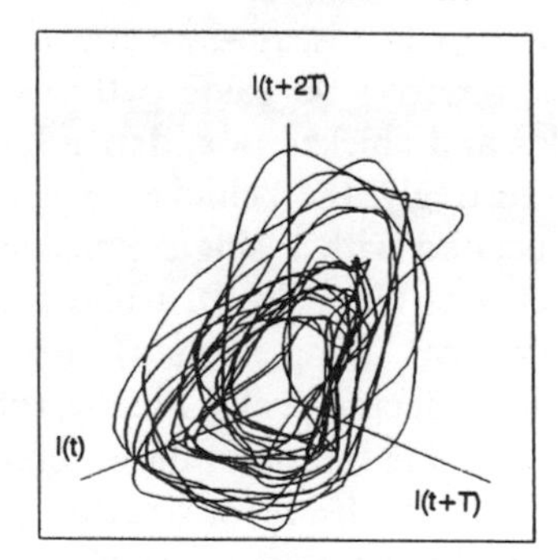

Aberdeen Measles

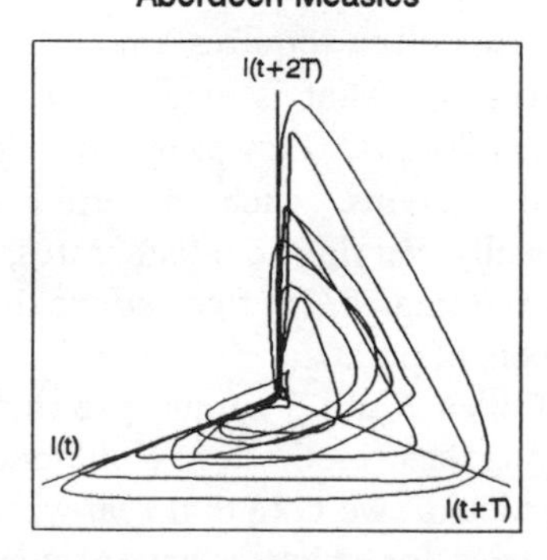

New York Measles

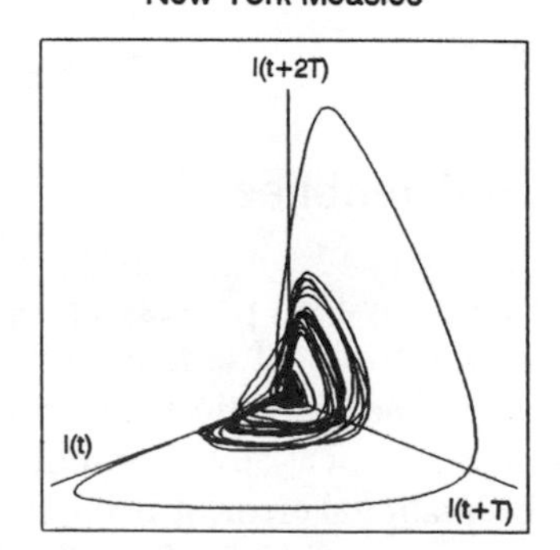

Fig. 4. Reconstructed phase portraits for childhood diseases. Orbits were reconstructed from the numbers of infective individuals reported monthly as prescribed by Takens *(27)* with a delay of three months. Prior to reconstruction, the data were subjected to three-point smoothing and interpolation with cubic splines.

orbital excursions may be predicted. For childhood diseases, there is often one excursion per year. Hence, the map effectively specifies the sequence yearly case rates.

Figure 5 gives representative return maps for each of the diseases. In the case of chicken pox, the points suggest a random scatter, and one concludes that there is no simple relationship either between the amplitudes of successive orbits or between the numbers of cases in succeeding years. Such a result is to be expected if the motion is, indeed, a simple limit cycle subject to chance perturbations *(29)*. By contrast, the maps obtained from measles epidemics suggest a unimodal curve, albeit with considerable scatter. It is well known *(30, 31)* that maps of this sort can be extracted from mathematical models exhibiting low-dimensional chaotic fluctuations. Note that the return maps extracted from chaotic systems need not be one dimensional. Often *(32)* the dimension of the map is higher. Then, provided one computes a sufficient number of points, one observes not one, but a whole series of curves with an obvious fractal *(33, 34)* structure.

Returning to the diseases, we note that for mumps, one observes the same patterns seen with measles and chicken pox, depending on locale. Interestingly, the "chicken pox" pattern is associated with a single peak in the power spectrum (New York), while for the "measles pattern" (Copenhagen), one observes multiple spectral peaks. Finally, rubella epidemics resemble the situation observed in measles, but, as in the reconstructed phase portraits, with a greater concentration of points near the origin.

Dynamical Quantities

Chaotic dynamical systems possess a property called "sensitivity to initial conditions" *(35)*, which means that nearby trajectories diverge on the average exponentially. Consequently, given any observational error, it is impossible to make long-term predictions. Specifically, if one performs two or more experiments, unresolvable differences in the system's initial state are eventually amplified to the point that the resulting time series cease to be correlated.

Figure 6 (top) illustrates an SEIR model for measles (discussed below). Here we depict the results of two simulations in which the initial numbers of infectives were varied by less than one percent. After about 15 years, the time series diverge markedly. At the same time, the overall frequency distributions and time-one maps are quite similar (Fig. 6, middle and bottom).

This example serves to emphasize a well-known point *(36)*, which is that descriptions of chaotic motion are necessarily statistical. As a practical matter, one assumes (sometimes this can be proved) the existence of an invariant distribution of points in the phase space and then proceeds to calculate quantities that depend thereon. Among the quantities often computed are the Lyapunov exponents *(37)*, which give rates of trajectorial expansion and contraction and the trajectory's fractal dimension *(33, 34)*.

Generally speaking, positive Lyapunov exponents and noninteger dimensions are taken as evidence of chaotic behavior. However, this presumes large amounts of noise-free data. For epidemiological time series, estimating dynamical quantities is primarily useful in two regards. First, these quantities offer objective criteria by which one can document differences in behavior. Second, they can be used to provide quantitative comparisons between actual epidemics and mathematical models. In passing, we note that computing fractal dimensions and Lyapunov exponents often requires amounts of data far in excess of what is available to practicing epidemiologists. However, for periodically forced systems, such as epidemics with seasonally varying contact rates, one can devise strategies *(24)* that reduce the required numbers of points.

Tables 2 and 3 summarize the results of applying these methods to childhood diseases. In particular, we note that real-world measles epidemics, for which we have accumulated the greatest amount of information, have a fractal dimension of about 2.5 and a maximum Lya-

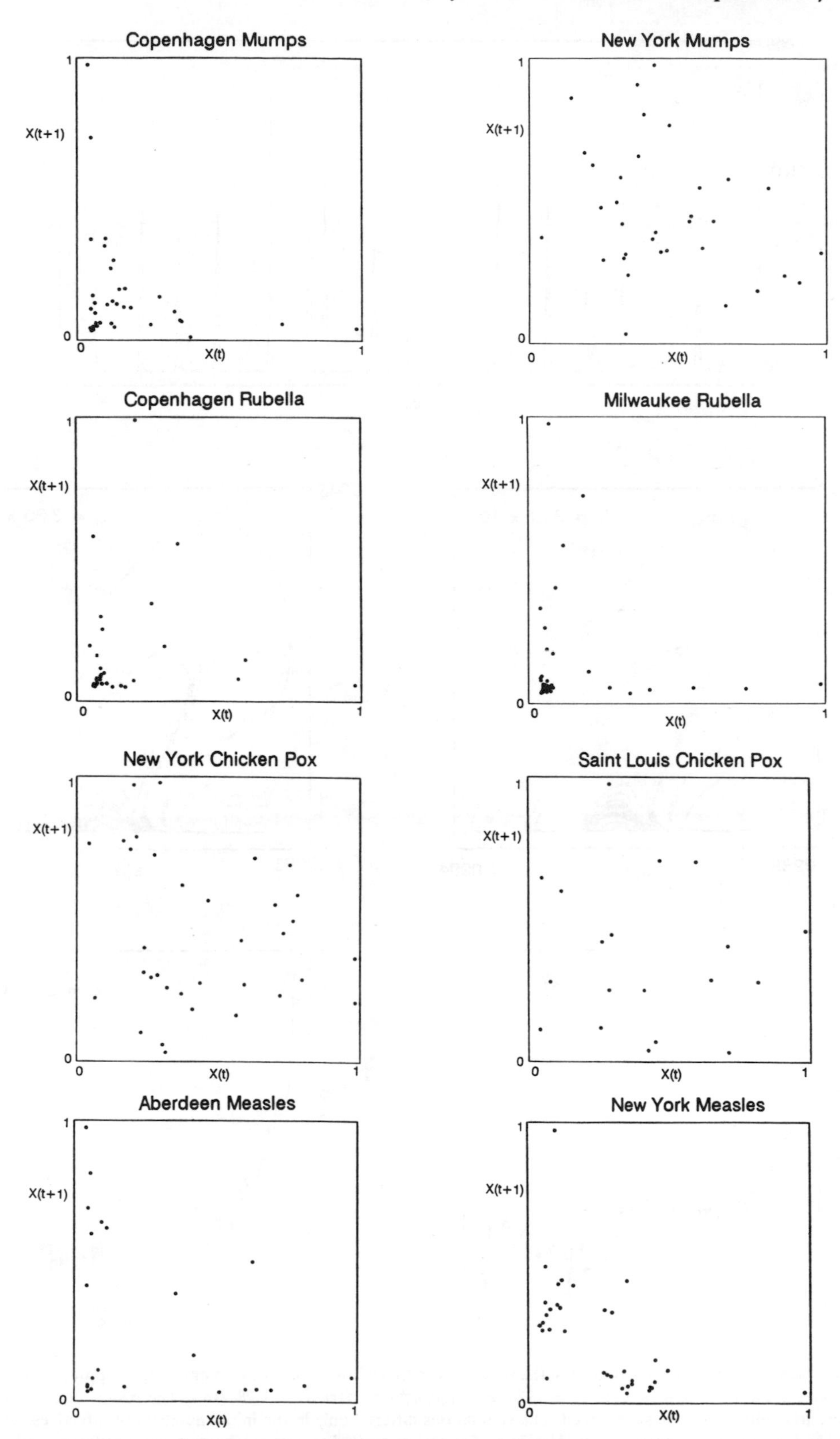

Fig. 5. Return maps extracted from the reconstructed phase portraits in Fig. 4.

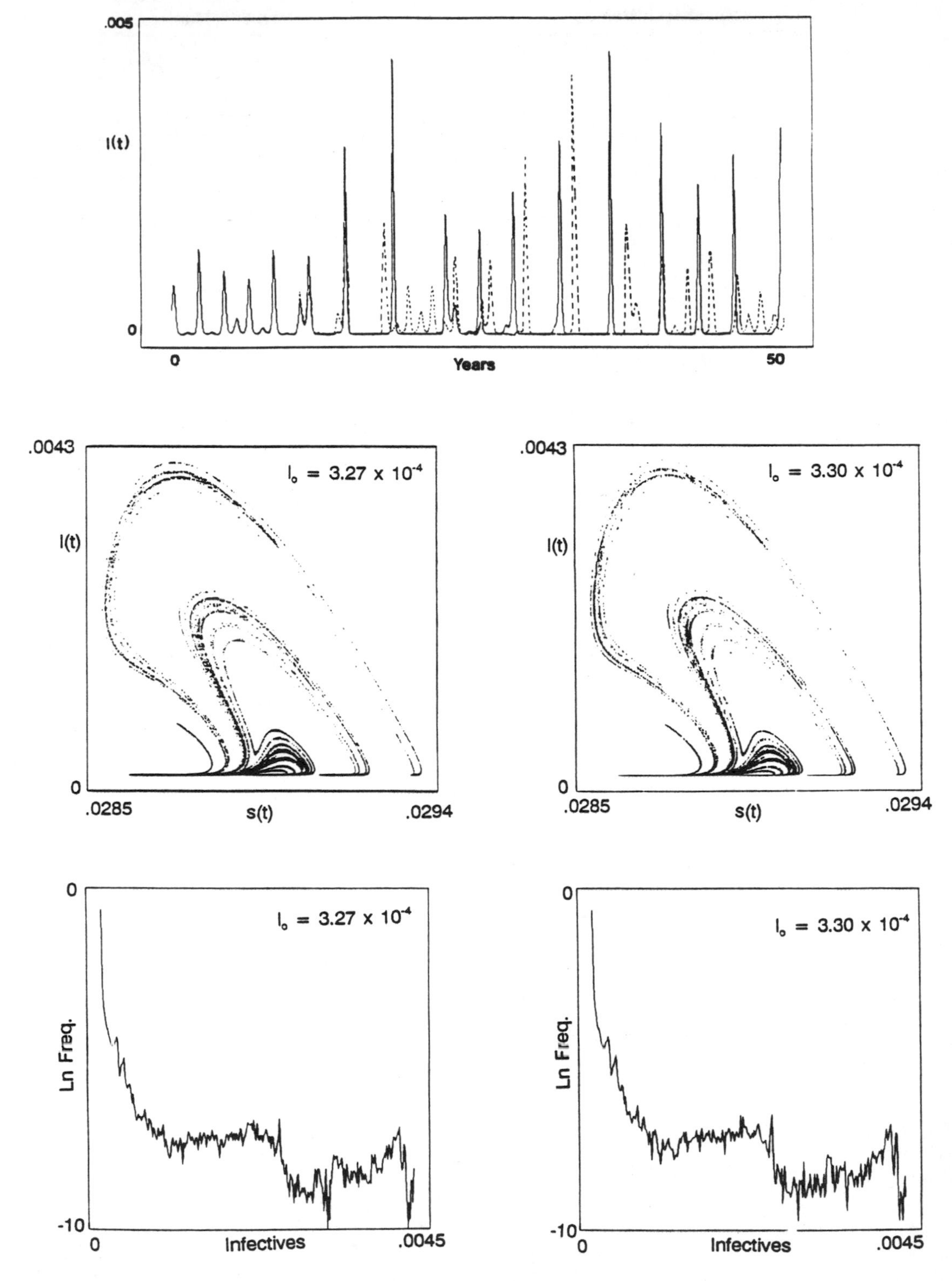

Fig. 6. Sensitivity to initial conditions in the seasonally forced SEIR equations with parameters appropriate to measles epidemics in large cities: $m = 0.02$, $a = 35.84$, $g = 100.0$, $\beta_0 = 1800$, $\beta_1 = 0.28$. **(top)** The results of two calculations (differential equations) are superposed. The simulations differed only in the initial numbers of infectives, which were varied by 1%. **(center)** Time-one maps. Numbers of infectives plotted against numbers of susceptibles at yearly intervals. **(bottom)** Frequency distributions for the numbers of infectives.

Table 2. Dimensional analysis of time-one maps.*

City	Embedding dimension				Mean**
	2	3	4	5	
			Chicken pox		
Copenhagen	1.89	2.18	2.31	2.35	2.28
Milwaukee	1.93	1.94	2.33	2.47	2.25
New York	2.21	2.03	2.01	2.96	2.33
St. Louis	1.37	1.88	1.92	1.49	1.76
			Measles		
Aberdeen	1.34	1.60	1.71	1.37	1.56
Baltimore	1.17	1.19	1.28	1.54	1.34
Baltimore County	1.10	1.20	1.27	1.78	1.42
Copenhagen	1.72	1.96	1.93	2.32	2.07
Detroit	1.37	1.43	1.37	1.51	1.43
Milwaukee	1.60	1.71	1.30	1.69	1.57
New York	1.23	1.55	1.75	1.76	1.68
St. Louis	1.08	1.39	1.14	1.14	1.22
			Mumps		
Copenhagen	2.29	1.87	2.14	1.88	1.96
Milwaukee	1.98	1.92	2.59	2.07	2.19
New York	1.99	2.23	2.45	2.81	2.50
			Rubella		
Copenhagen	1.81	2.20	1.86	1.82	1.96
Milwaukee	1.88	2.01	1.68	1.50	1.73
St. Louis	2.21	1.45	1.62	1.15	1.41
			Noise		
Uniform	2.41	2.63	3.51	4.03	–
Gaussian	2.39	3.04	3.52	4.86	–

*Correlation dimensions (*52*) computed for successive embeddings of all possible time-one maps (*24*). To obtain dimensions for the original time series, add 1.0 to the numbers in this table. **Average of embedding dimensions 3–5.

Table 3. Lyapunov exponents (bits per year) for disease.*

City	Disease			
	Chicken pox	Measles	Mumps	Rubella
Aberdeen	–	0.23 ± 0.05	–	–
Baltimore (early)	–	0.40 ± 0.06	–	–
Baltimore (late)	–	0.55 ± 0.04	–	–
Copenhagen	0.30 ± 0.02	0.60 ± 0.03	0.52 ± 0.05	0.25 ± 0.06
Detroit	–	0.59 ± 0.02	–	–
Milwaukee	0.11 ± 0.02**	0.71 ± 0.03	0.15 ± 0.05	0.23 ± 0.05**
New York	0.20 ± 0.03	0.44 ± 0.02	0.26 ± 0.03	–
St. Louis	0.27 ± 0.03	0.50 ± 0.02	–	***

*Maximum Lyapunov exponents estimated according to the method of Wolf *et al.* (*37*) as modified by Schaffer *et al.* (*24*) for periodically forced systems. **Embedded in three dimensions. ***Insufficient data.

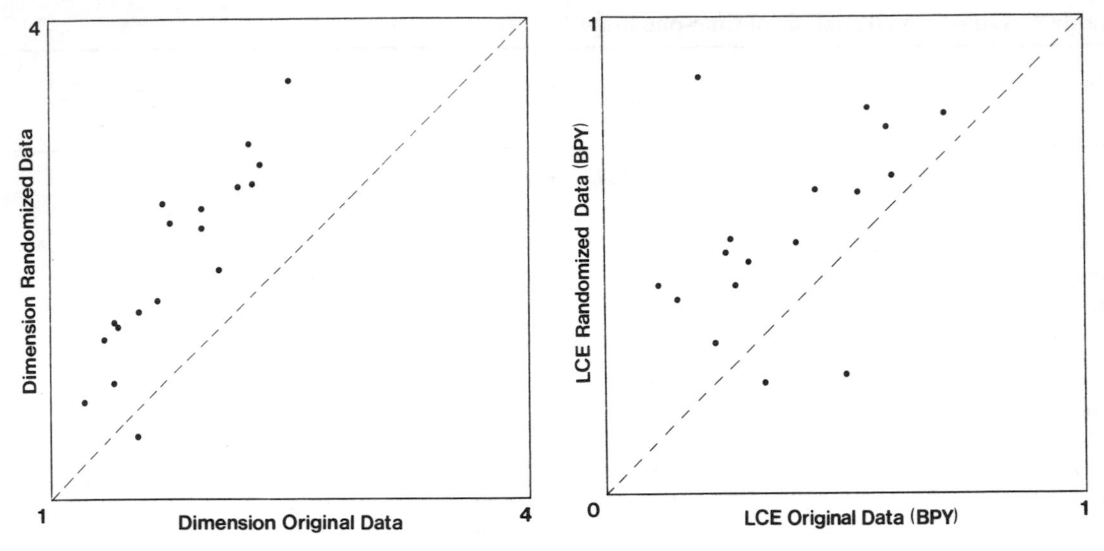

Fig. 7. Estimated dimensions (**left**) and Lyapunov exponents (**right**) for randomized epidemic time series exceed those computed from the original data. The randomized data were treated as the original time series with regard to embedding dimensions (Table 2) and time delays (see text).

punov exponent about 0.5 bits per year. These numbers can be compared with those obtained for chicken pox, for which the dimension ranges from 2.8 to 3.3, and the Lyapunov exponent from 0.1 to 0.3 bits per year. For mumps and rubella, the data are more variable, and, at this writing, insufficient to make meaningful generalizations.

Because available epidemiological time series are so short, the legitimacy of these calculations must be checked. In particular, one wants to be sure that the results are not simply artifacts of small sample sizes. A crude check is to generate random data sets of comparable length. In theory (i.e., for very long time series), such data should yield infinitely large dimensions and (positive) exponents. But for real- time series, the calculated dimensions and exponents will be finite. This gives a null hypothesis with which data can be compared. As Table 2 illustrates, 50-year time series of random numbers (Gaussian and uniform) yield higher dimensions than actual diseases.

A stronger test is to retain the distribution of the reported case rates but to shuffle their temporal sequence. For artifactual results, one expects the original and randomized data to yield comparable dimensions and exponents. But, as shown in Fig. 7, this is not the case. The randomized time series almost always yield higher dimensions and exponents than the originals.

Finally, one can study the behavior of mathematical models, for which we can calculate dimensions and exponents to greater precision. Table 4 gives Lyapunov exponents and the so-called Lyapunov dimension[1] for the seasonally forced SEIR model for parameter values appropriate for measles epidemics in large cities (see below). These calculations yield a maximum exponent equal to 0.4–0.5 bits per year and dimensions of about 2.5, both of which are in agreement with the values calculated for actual epidemics.

Mathematical Models

As noted by Dietz and Schenzle *(9)*, the history of mathematical epidemiology is a long one with numerous contributions of many kinds. Attempts to better comprehend epi-

1 If we equate the term "fractal dimension" with the information dimension, the Lyapunov dimension gives an upper bound *(25)*.

Table 4. Maximum Lyapunov exponents for SEIR equations.*

β_1	Equations	Data*
	"Measles"	
0.27	0.12	0.11 ± 0.003
0.28	0.46	0.42 ± 0.01
0.29	0.49	0.53 ± 0.02
	"Chicken pox"	
0.28	–	0.19 ± 0.09**

*Exponents measured in bits per year. Entries in the middle column computed directly from the equations as prescribed by Shimada and Nagashima *(53)* and Wolf *et al.* *(37)*. Entries in the right-hand column computed from the time series, $I(t)$, according to the method of Wolf *et al.* *(37)* as modified by Schaffer *et al.* *(24)*. **The state variables were subjected to perturbation by 2% multiplicative noise, i.e., $x(t)$ is replaced by $x(t)\ (1+Z(t))$, where $Z(t)$ is normally distributed with mean 0 and standard deviation 0.02. Perturbations were introduced at random intervals with the mean interval between perturbations being one year.

demiological phenomena using mathematical models date to the turn of the century, for example, Ross's *(38)* pioneering studies on malaria, and have continued ever since. Noteworthy in this regard were a series of papers by Anderson and May *(11–13)*. Other important contributions to the mathematical study of childhood diseases include papers by Bartlett *(3, 4)* and London and Yorke *(6, 7, 10)* and Dietz *(8)*. [For recent reviews, see Dietz and Schenzle *(9)* and Hethcote and Levin *(16)*.]

Our own concern is with SEIR models *(8, 39)* (Fig. 8). Note that the SEIR scheme is a general one that can be implemented in varying degrees of detail. One can, for example, include the complexities of age structure *(8, 40)*, geographic locale *(22)*, or even socioeconomic status. Or one can proceed in a simpler fashion, writing down coupled differential equations as follows:

$$\begin{aligned} dS(t)/dt &= m\,[1 - S(t)] - \beta\, S(t)\, I(t), \\ dE(t)/dt &= \beta\, S(t)\, I(t) - (m+a)\, E(t), \\ dI(t)/dt &= a\, E(t) - (m+g)\, I(t), \end{aligned} \qquad (2)$$

where S = susceptible; E = exposed; I = infectious. Here, there are four parameters

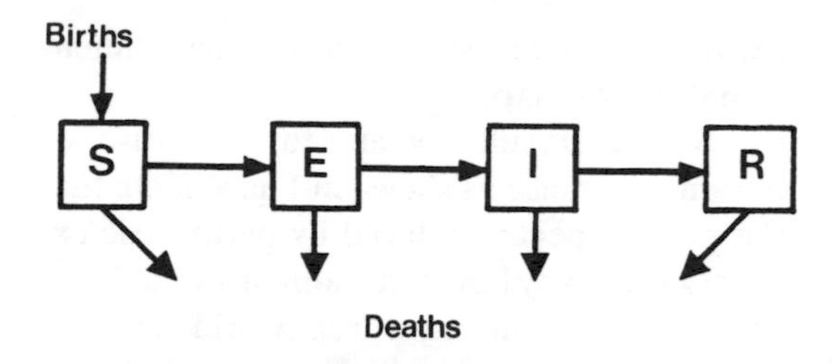

Fig. 8. The SEIR model for microparasitic diseases divides populations into Susceptible, Exposed, Infective, and Recovered categories. For childhood diseases, immunity is permanent, so there is no feedback from **R** to S. Specific realizations of the basic scheme differ in ancillary assumptions and mathematical implementation.

$(1/m)$ is the average life expectancy,

$(1/a)$ is the average latency period,

$(1/g)$ is the average infectious period,

β is the mean contact rate, i.e., average number of susceptibles contacted yearly per infective.

Of these, the first three can be measured independently of the actual time series. The final parameter, β, is generally estimated indirectly following a prescription given by Dietz *(8)*.

Equations 2 assume a constant population and exponential transfer rates between categories, and ignore the complexities of age structure and geography. Note that because natural immunity in childhood diseases is permanent, one can dispense with a fourth equation for the numbers of recovered individuals and concentrate on the three variables above.

As written above, the SEIR model has very simple dynamics. Either the disease dies out, or there is a single stable equilibrium about which one observes damped oscillations. Thus, the model does not account for the recurrent outbreaks actually observed. Two potential resolutions of this discrepancy were discussed some years ago by Bartlett *(3)*. The first is that demographic stochasticity (sampling errors) and external perturbations sustain the oscillations and prevent the system from ever reaching equilibrium. In this regard, it has been observed *(25)* that imposing multiplicative noise on the state variables for parameters appropriate for measles does, in fact, prevent the trajectory from settling down. However, the resulting orbit does not much resemble real-world epidemics. Nor can one extract from it any suggestion of a one-dimensional return map.

Similar results obtain for the case of parametric noise as shown in Fig. 9. Note that the power spectra induced by parametric excitation lack any hint of the annual cycle that is so prominent in most real-world measles epidemics. At the same time, one does observe a spectral peak corresponding to long-term periodicities of 2–3 years. Thus, as pointed out, for example, by Anderson and May *(11)*, the damping rate computed from Eq. 1 gives important information regarding the long-term dynamics.

An alternative mechanism for inducing recurrent epidemics is seasonal forcing. Here, one presumes that the force of infection varies over the course of the year, either due to physical factors — temperature, humidity, etc. — or because of enhanced transmission rates during the school year *(41)*. Investigations of the effect of seasonal forcing include those by Bartlett *(3)*, Yorke and London *(7)*, Dietz *(8)*, and Grossman *(42)*. Here, the mean transmission rate, β, is replaced by a time-dependent function such as

$$\beta(t) = \beta_0 (1 + \beta_1 \cos 2\pi w t) \qquad (3)$$

where $1/w$ is the period of the forcing, which is taken to equal one year. Other forcing functions, designed to more accurately model the school year, have been studied by Kot *et al.* *(43)* and Schenzle *(15)*.

Of particular interest is work by Aron and Schwartz *(39)*. These authors demonstrated that for parameters appropriate for measles, increasing the variation between summer and winter induces a series of period-doubling bifurcations, whereby a simple yearly cycle gives way to biennial dynamics and so forth, on up to chaos. Also relevant are studies by Schwartz and Smith *(44–46)* and by Kot *et al.* *(43)* indicating that for certain ranges of parameter values there can be more than one attracting set. Then the observed dynamics will depend on the initial conditions. These alternative attractors arise by so-called "saddle-node" bifurcations and undergo their own period-doubling routes to chaos (Fig. 10). Of interest is the fact that the resulting oscillations lack an identifiable seasonal component, in which regard they differ from the "main" period-doubling sequence described by Aron and Schwartz *(39)*.

In Fig. 11, we display bifurcation diagrams for chicken pox, measles, mumps, and rubella, using the more realistic forcing function studied by Kot *et al.* *(43)*. That is, for each disease, we calculate the long-term dynamics as evidenced by the numbers of infectives occurring yearly for a wide range of values of β_1. The remaining parameter values were held constant. Note that because we plot the number of infectives at yearly intervals, an annual cycle gives a single point on the bifurcation diagram; a biennial cycle, two points; etc. In addition, one can make the following observations:

(i) Each disease exhibits a very different sequence of changes. For example, the chicken pox parameters yield a simple limit cycle for all values of β_1. By way of contrast, for measles, one sees period doubling to chaos as discussed above.

(ii) In measles, mumps, and rubella, there is more than one sequence of dynamical states.[2] Thus, in the mumps simulations, an annual cycle can coexist with three- and five-year cycles. The real-world implication of this situation is that one expects to see a mixture of these possibilities either within or between cities. That is, chance perturbations should periodically bounce the system from one attracting set to another. In this regard, it is worth noting that epidemics of mumps in New York City exhibit the chicken pox pattern (noisy annual cycles), while those in Copenhagen resemble the situation observed in measles.

(iii) For measles, we observe the period

2 The diagrams were constructed as follows. Start with an initial value of the parameter value and arbitrary initial conditions. Integrate Eq. 1 for several hundred years and plot the results for the final hundred. Increment the parameter value and repeat using the final values of S, E, and I as initial conditions for the next experiment. This procedure allows one to "track" an attractor through parameter space and, when there are coexisting attractors, to demonstrate this fact by first increasing and then decreasing the parameter.

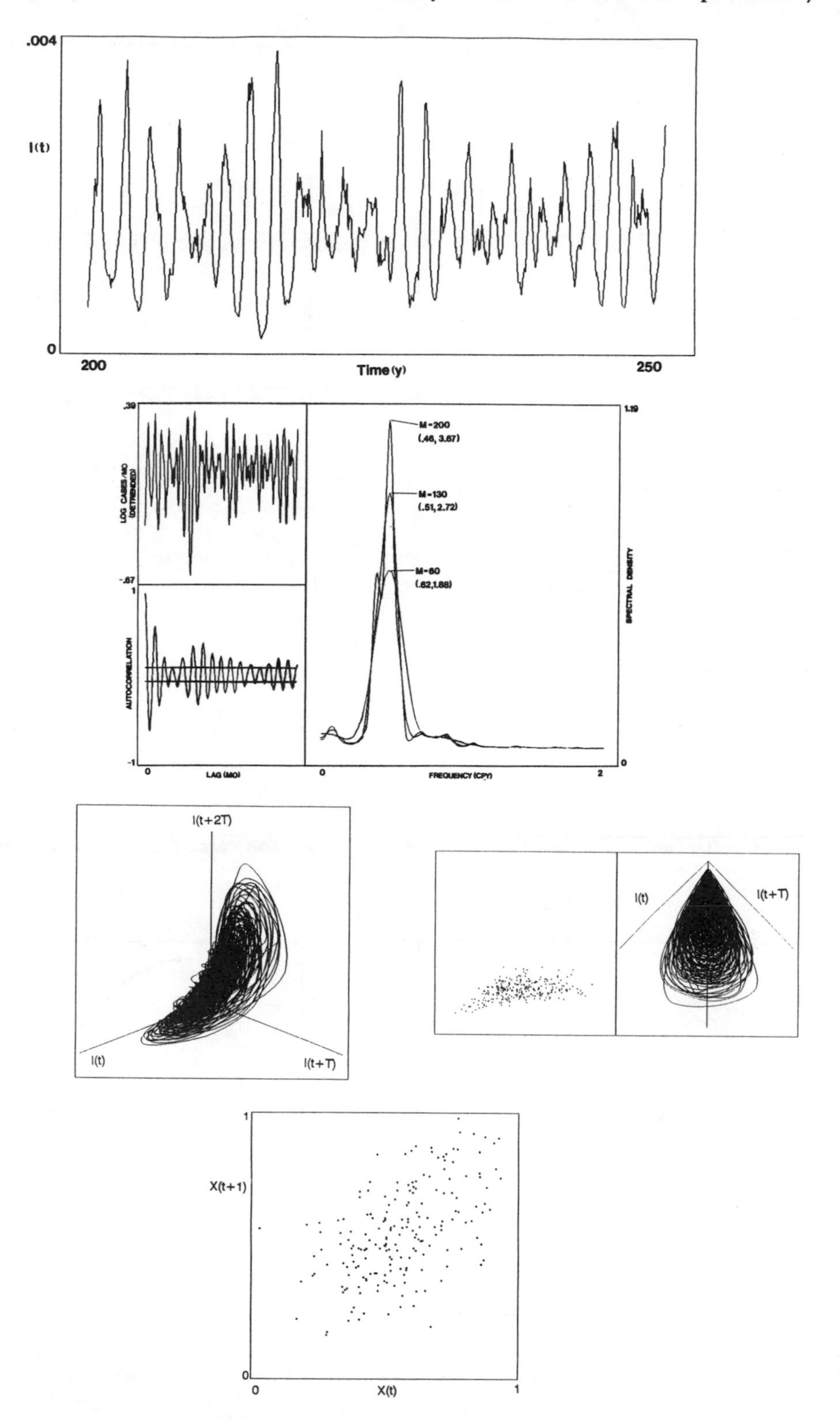

Fig. 9. Parametric forcing of the SEIR model (measles parameters with 5% multiplicative noise added to the reciprocal of the infectious period, g) without seasonality induces recurrent epidemics that lack the qualitative attributes of real-world epidemics. In particular, extracting a one-dimensional map yields an apparently random scatter of points.

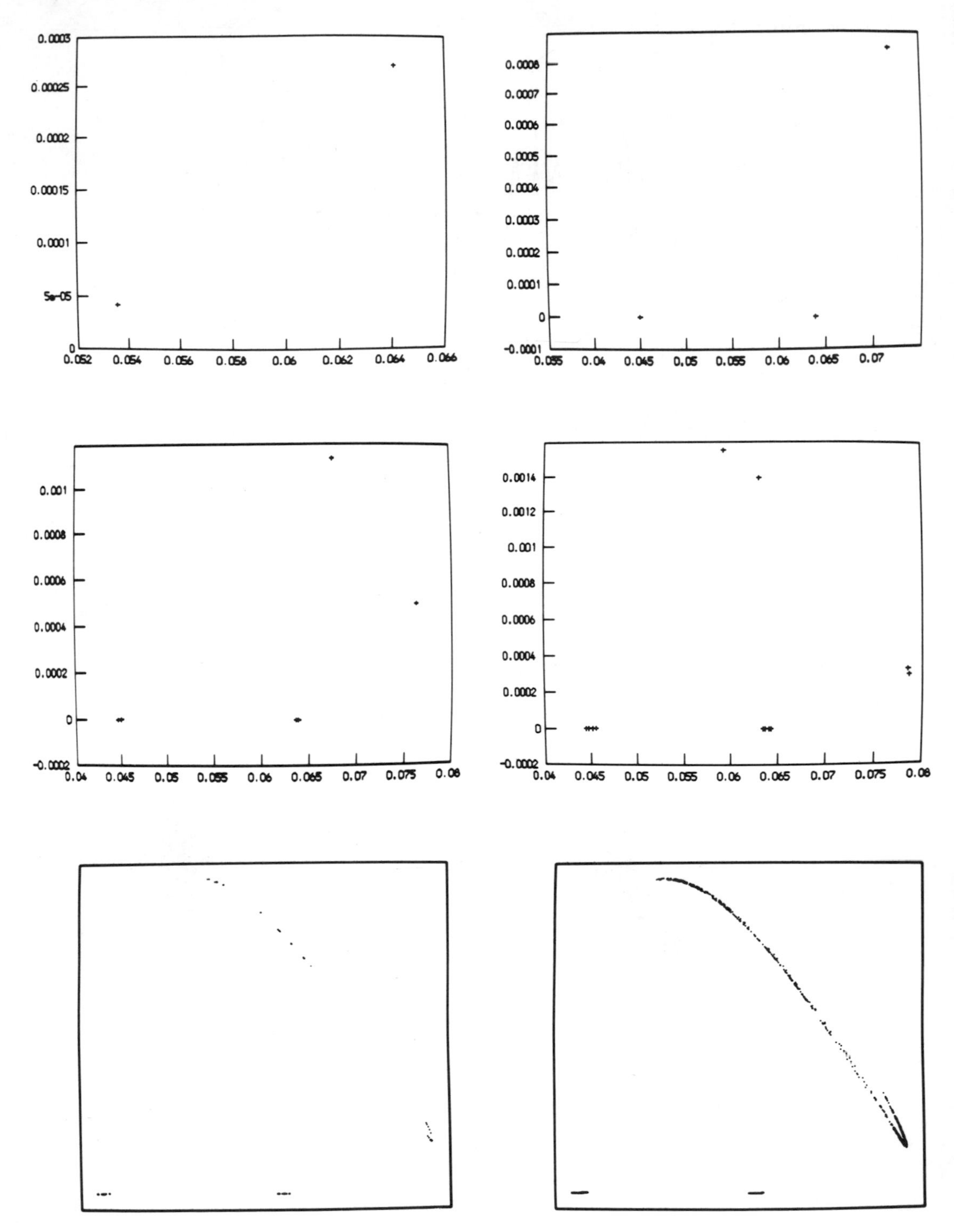

Fig. 10. Coexisting attractors (time-one maps) in the SEIR model for measles but with the intensity of seasonal forcing equal to 0.13 to 0.1535. The period-three cycle (**upper right**) and its descendents (**middle, bottom**) manifest large amplitude fluctuations and lack the obvious seasonality observed in the coexisting period-two cycle (**upper left**). [Reproduced with permission from *(25)*, copyright 1985 by Oxford University Press.]

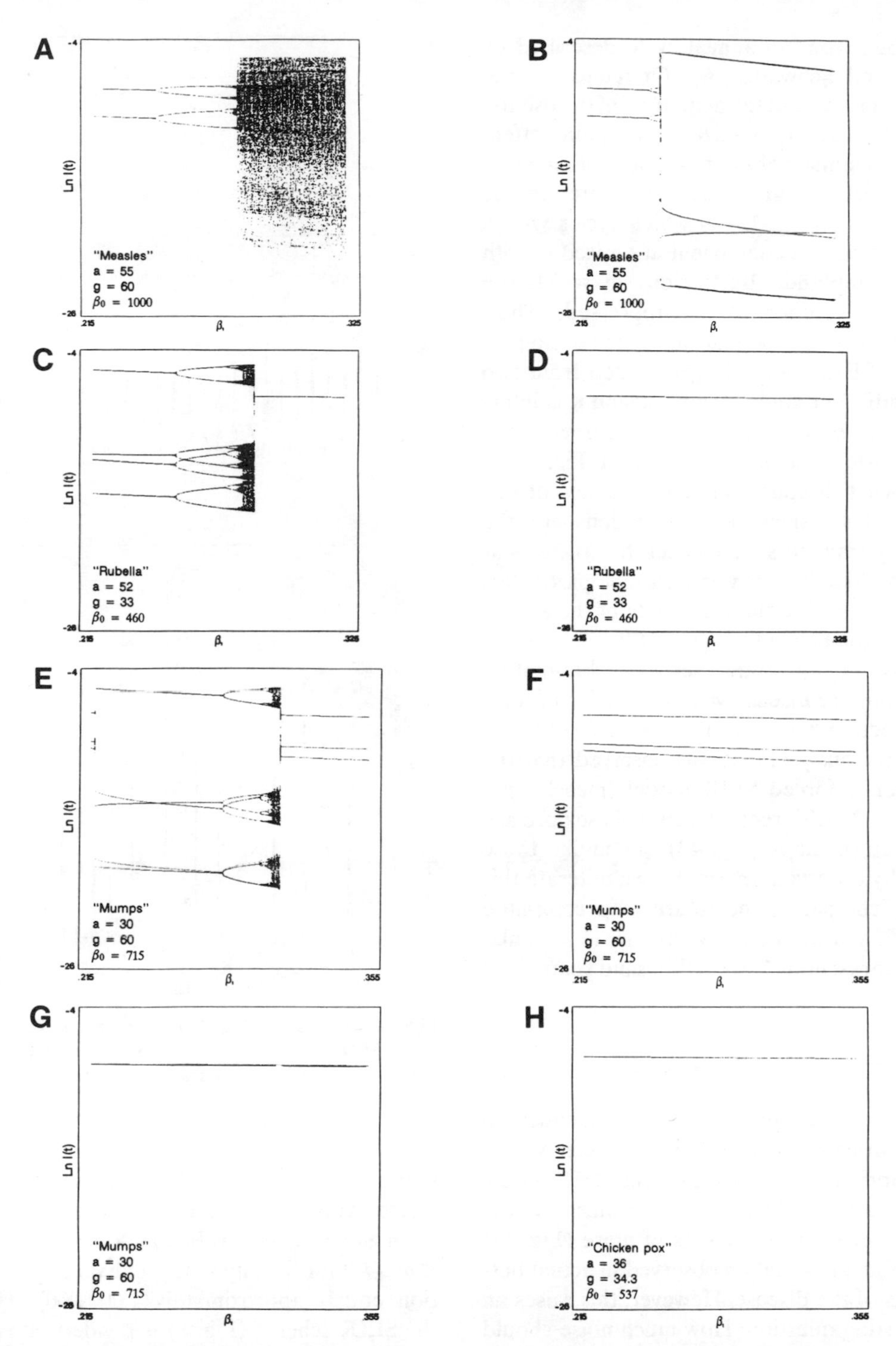

Fig. 11. Bifurcation diagrams for the SEIR equations using the modified forcing function proposed by Kot *et al.* *(43)*. The long-term dynamics (logarithm of the numbers of infectives) is plotted for different levels of seasonal variation in transmission. **(A)** The main period-doubling sequence with parameters appropriate for measles is shown at the left. Note the abrupt transition from a four-piece attractor to large-amplitude chaotic fluctuations. **(B)** Coexisting with the main bifurcation sequence is a four-year cycle of even larger amplitude. **(C-D)** Bifurcation diagrams for rubella. An annual cycle coexists with a four-year cycle that period-doubles to chaos. **(E-H)** Bifurcation diagrams for mumps (three coexisting attractors) and chicken pox (a single periodic orbit). While the details of the bifurcation diagrams depend on the choice of forcing function, the existence of sharp transitions does not. [Reproduced from *(43)*, with permission. Copyright 1988 by Elsevier Science Publishers.]

doubling from an annual cycle described by Aron and Schwartz *(39)*. Of interest is the abrupt increase in the amplitude of the oscillations at about $\beta_1 = 0.28$. At this point, effectively biennial behavior – alternating high-low years – gives way to more erratic fluctuations (Fig. 12) in which cycles with a strong seasonal component are mixed in with larger amplitude fluctuations in which the seasonal component is suppressed. Thus, beyond a certain point, chaos in the seasonally forced SEIR model is synthesized from two very different kinds of motion, and it is interesting to observe that both are observed in the real-world time series shown in Fig. 1. A somewhat disconcerting consequence of this fact is that a single set of equations with the same parameters can generate 20–30 year runs of data that are very different, particularly with regard to the magnitude of the annual peak in the power spectrum *(47)*.

(iv) In a qualitative sense, the bifurcation diagrams are insensitive to the choice of forcing function. However, the details do differ.

Previously *(25)* it was observed that the seasonally forced SEIR model (measles parameters) yields reconstructed phase portraits and return maps (Fig. 13) not unlike those actually observed. In Fig. 14, we reiterate this point, comparing the return map computed from 63 years of data in Baltimore to a simulation representing several thousand years.

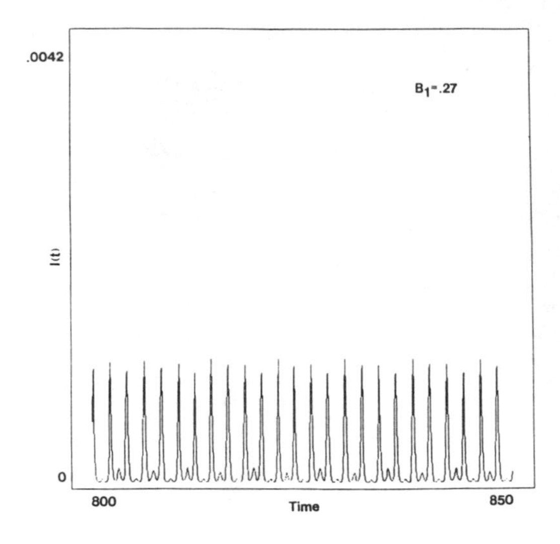

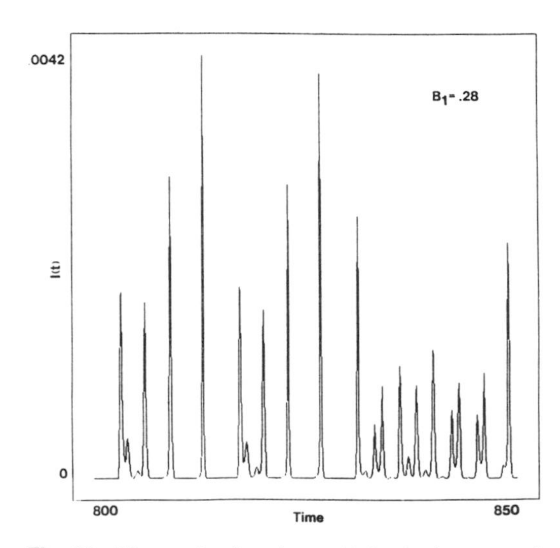

Fig. 12. Time series (numbers of infectives) generated by the SEIR model with parameters as given in Fig. 11A before and after the abrupt transition to large-amplitude fluctuations.

Monte Carlo Simulations

The foregoing approach can be extended to other diseases as well. For parameter values appropriate for chicken pox, the SEIR model yields simple limit cycles, which in the presence of small amounts of noise (Fig. 15) resemble the situation observed in actual outbreaks of the disease. However, this raises an interesting question: How much noise should one add? The question is not an idle one, since adding sufficient noise to chicken pox simulations can induce behavior unlike that observed in large cities (Fig. 16).

One approach to this problem is to assume that the noise is primarily stochastic, i.e., that one is dealing with finite populations and hence with unique individuals and events. Such an approach has been adopted by Olsen *et al. (48)* for the city of Copenhagen (population equals approximately 1,000,000). Here the SEIR scheme (Fig. 8) is divided into discrete transitions, e.g., an individual shifts from the Susceptible class to the Exposed. To each transition, one assigns probabilities that depend on the disease as well as the choice of time step and population size. Finally, because even in large populations, random extinction

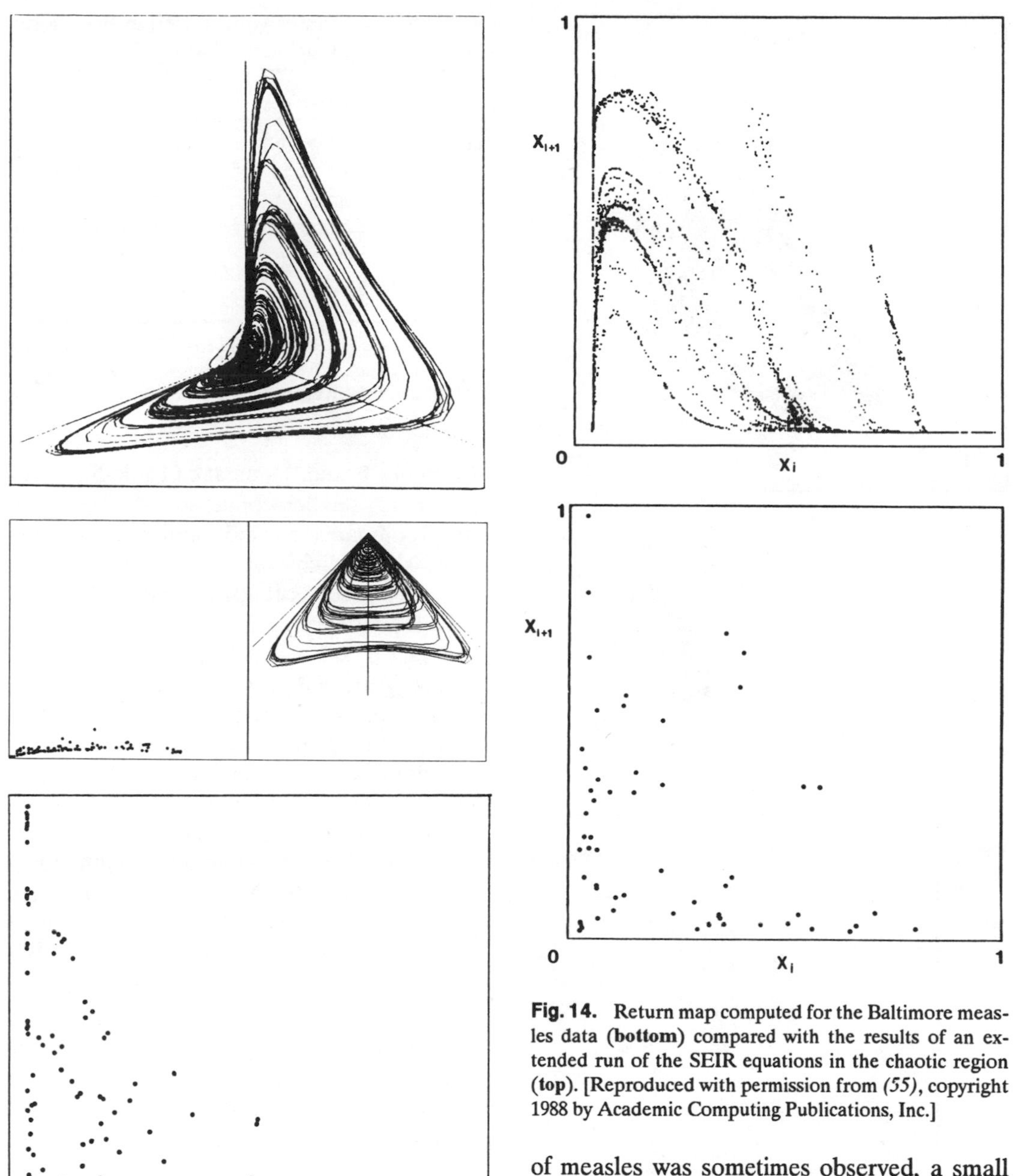

Fig. 13. Behavior of the seasonally forced SEIR model in the chaotic region. The trajectory was reconstructed by lagging the numbers of infectives, the coordinate axes being $I(t)$, $I(t+T)$, and $I(t+2T)$, with $T = 3$ months. The return map suggests a fattened version of a unimodal curve. Parameter values as in Fig. 6.

Fig. 14. Return map computed for the Baltimore measles data (**bottom**) compared with the results of an extended run of the SEIR equations in the chaotic region (**top**). [Reproduced with permission from *(55)*, copyright 1988 by Academic Computing Publications, Inc.]

of measles was sometimes observed, a small trickle of infectives was allowed to enter the population with constant probability.

Figure 17 summarizes their findings. Each plate contrasts actual data (left) with the results of a simulation (right). Once again, the computations are in generally good agreement with observation. The notable exception is rubella, where the simulations fail to reproduce the observed concentration of points

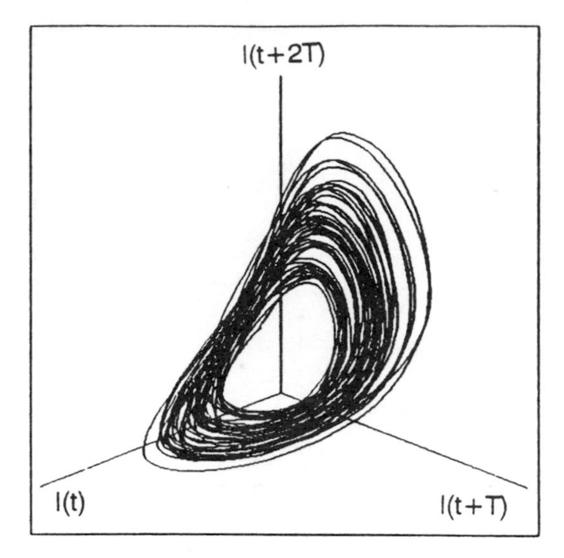

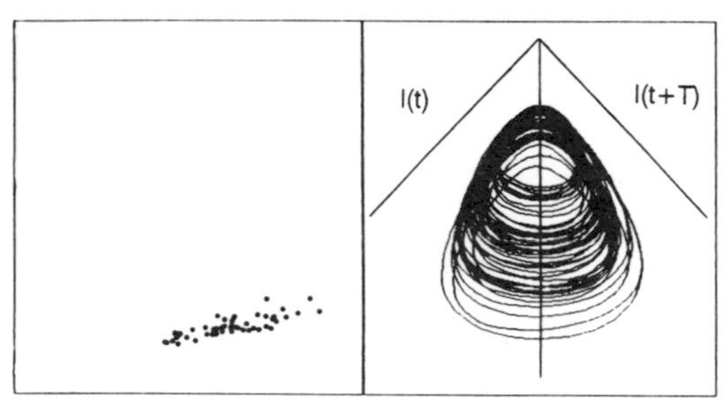

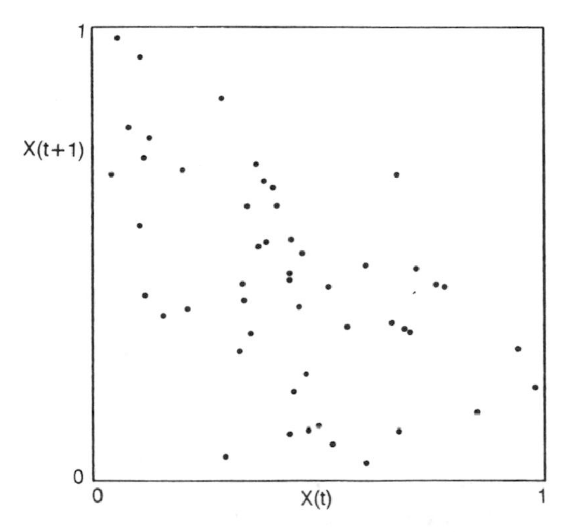

Fig. 15. Behavior of the SEIR model in the region of a yearly limit cycle. Parameters appropriate for chicken pox; $a = 36.0, g = 34.3, \beta_0 = 308.7, \beta_1 = 0.28$. Two percent multiplicative noise was added as described in the footnote to Table 4.

near the origin in the phase portrait and map.

Perhaps even more compelling than the Copenhagen data is the situation observed on

Table 5. Parameter values used in Bornholm Island Monte Carlo simulations.*

	Chicken pox	Measles
m	3.858×10^{-8}	3.858×10^{-8}
v**	4.0×10^{-6}	4.0×10^{-6}
a	5.787×10^{-5}	9.259×10^{-5}
g	6.617×10^{-5}	1.069×10^{-4}
0	2.072×10^{-8}	3.846×10^{-8}
β_1	0.24	0.28

*All constants scaled to a time step of one minute. We assume a "standard" year of 360 days = 518,400 minutes.
**Influx of infectives.

Bornholm Island, Denmark (Table 5), which has a much smaller population of 50,000. Figure 18 compares real and simulated epidemics of measles and chicken pox. For measles, the computer predicts that the magnitude of the annual spectral peak should be much reduced, as in fact it is. For chicken pox, only 11 consecutive years of data were available, but these revealed marked deviations from the annual cycle observed in larger communities. Such deviations were also observed in the simulations. An interesting implication of these observations is that population size can be viewed as a bifurcation parameter, with reductions in this quantity having roughly the same effect as adding noise (Fig. 19). In simulation experiments, the maps obtained for chicken pox with small populations are comparable to those observed for measles with large populations. In other words, small population size (noise) induces behavior that is reminiscent of the true chaos produced by the SEIR model for other parameter values.

Chaos Versus Periodicity Plus Noise

Much of the interest in chaos hinges on the possibility that heretofore imperfectly understood phenomena might be better comprehended in light of the new dynamics. In particular, one hopes that this may apply to natural systems outside the carefully controlled conditions of the laboratory. The fore-

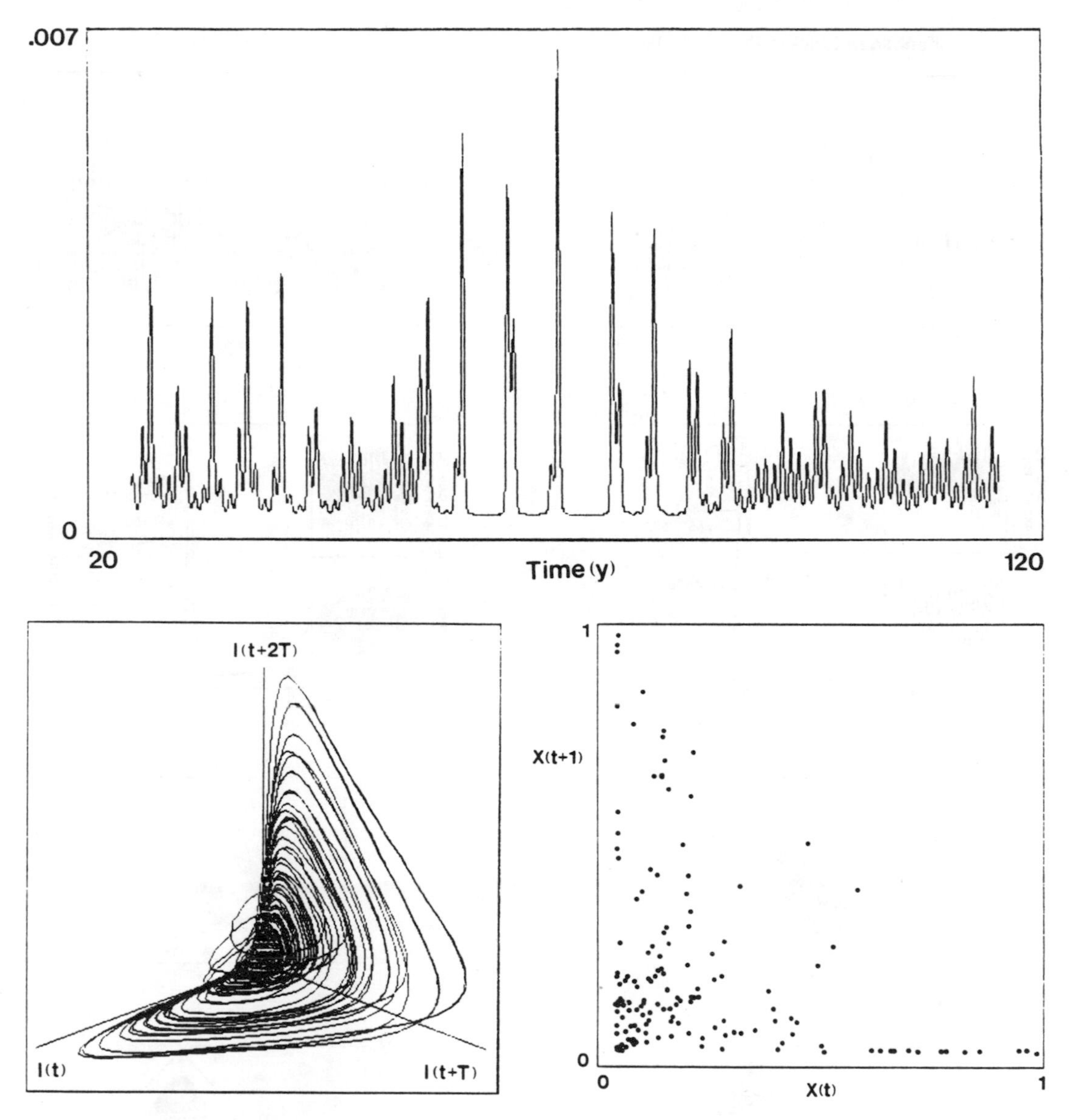

Fig. 16. Behavior of the SEIR model in the region of limit cycle behavior (chicken pox) but with 5% multiplicative noise added. Other parameters as in Fig. 15.

going studies suggest that childhood diseases may provide an example of just this circumstance. But do they?

Recently, Schwartz *(46)* reiterated his belief that measles epidemics are better understood as biennial cycles corrupted by noise. In particular, he maintains that the intensity of seasonality, β_1 in Eqs. 2, is not sufficient to place the system in the chaotic region. In fact, published estimates *(6)* of this parameter for New York City are somewhere in the vicinity of 0.25, which is close to the value at which essentially biennial dynamics give way to erratic fluctuations that are obviously chaotic. Moreover, neither β_1 nor any of the other parameters in Eqs. 2 can be estimated with sufficient accuracy to say with certitude that the system is to the right or the left of the point at which the dynamics become chaotic. Finally, Eqs. 2 and 3 are certainly not the "real" equations that determine the course of actual epidemics. Too many factors are omitted.

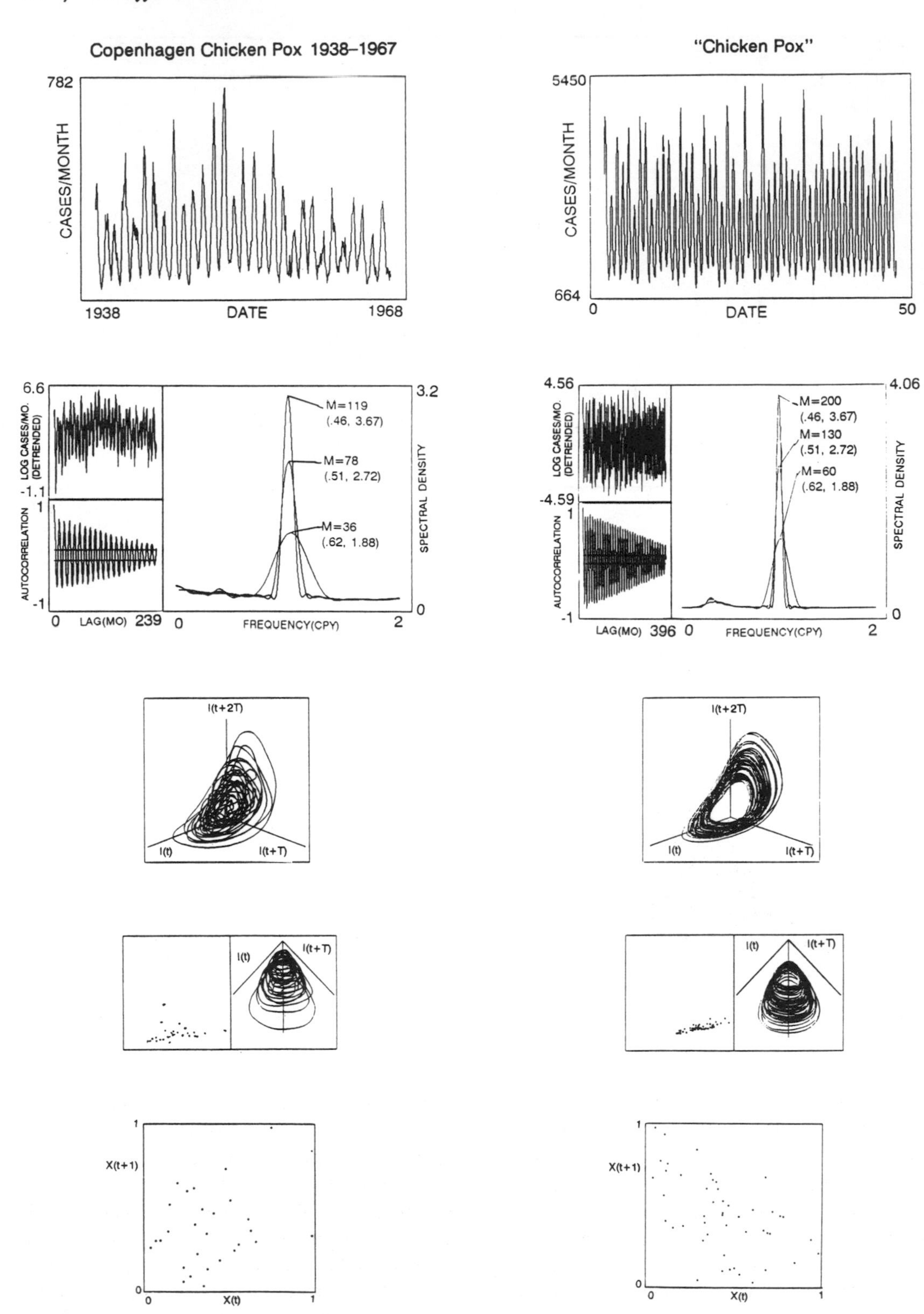

Fig. 17. Actual and simulated epidemics of chicken pox, measles, mumps, and rubella for Copenhagen, Denmark. Simulations represent Monte-Carlo implementations of the basic SEIR model with seasonal forcing for a population size of 1,000,000. [Reproduced with permission from *(48)*, copyright 1988 by Academic Press.]

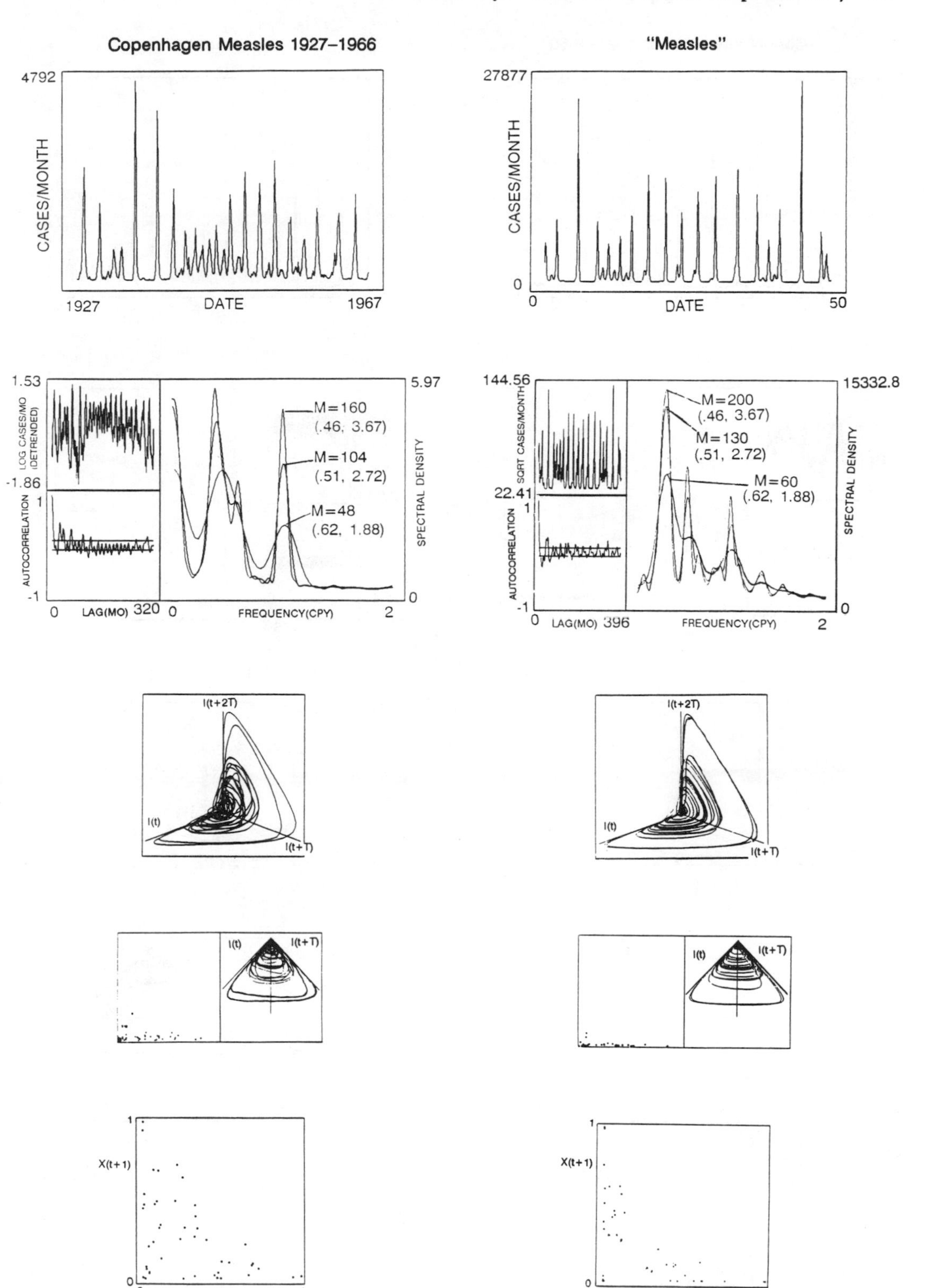

Fig. 17 *(continued)*

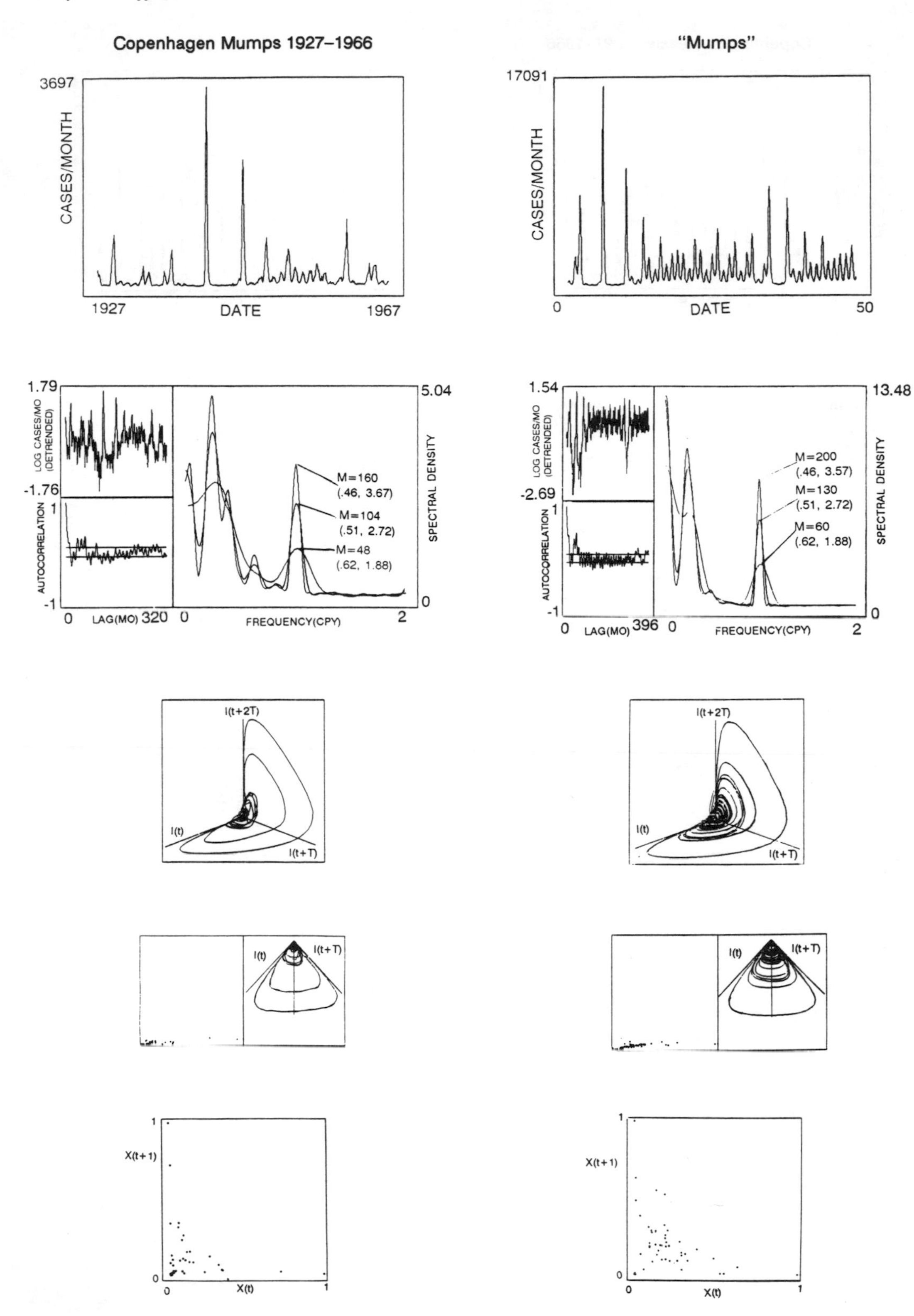

Fig. 17 *(continued)*

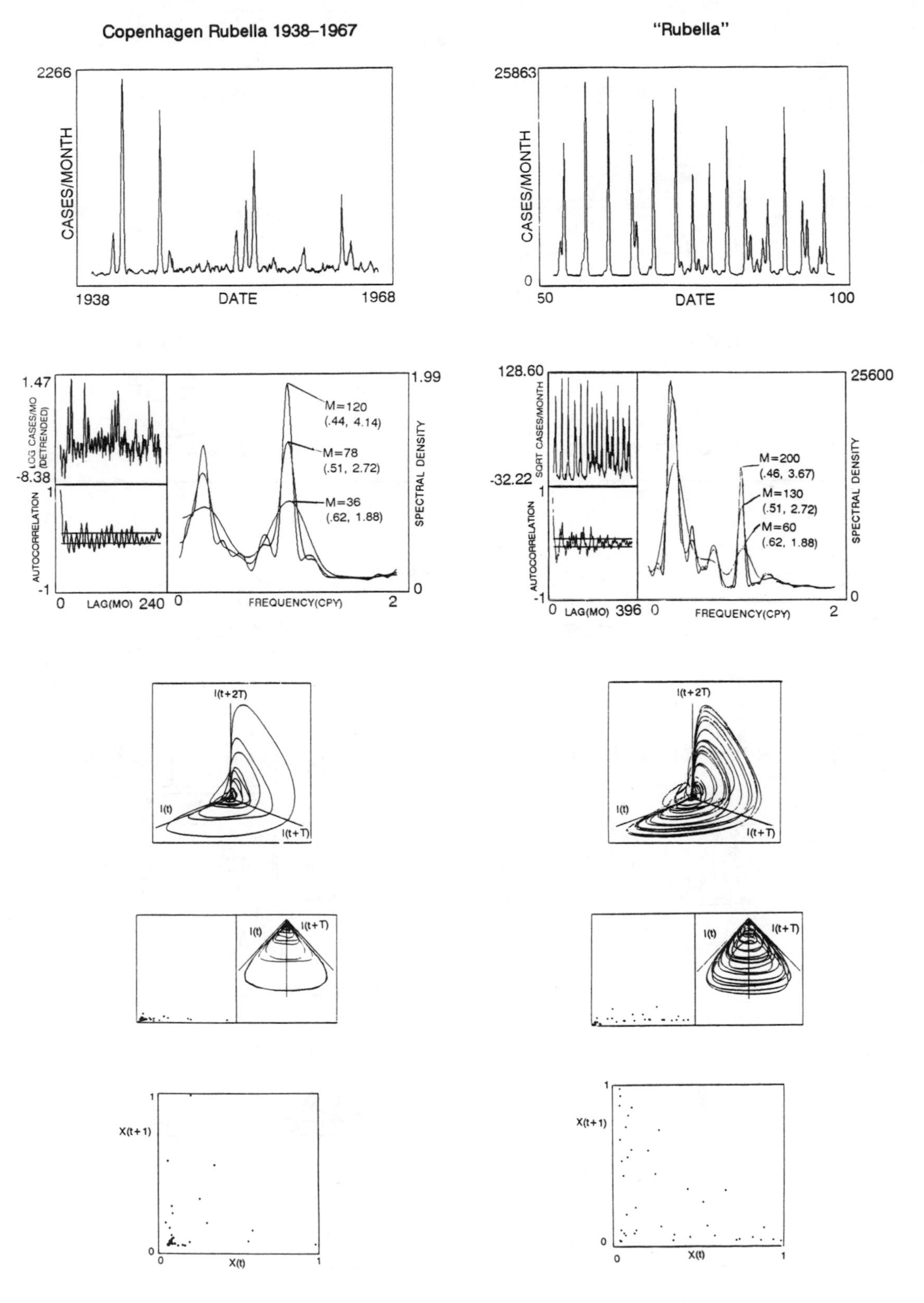

Fig. 17 *(continued)*

Measles

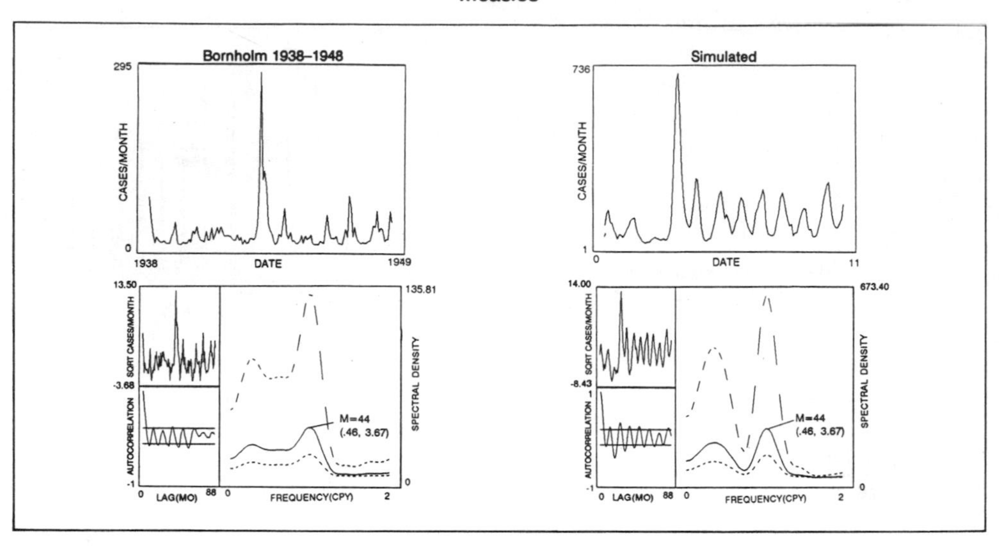

Chicken Pox

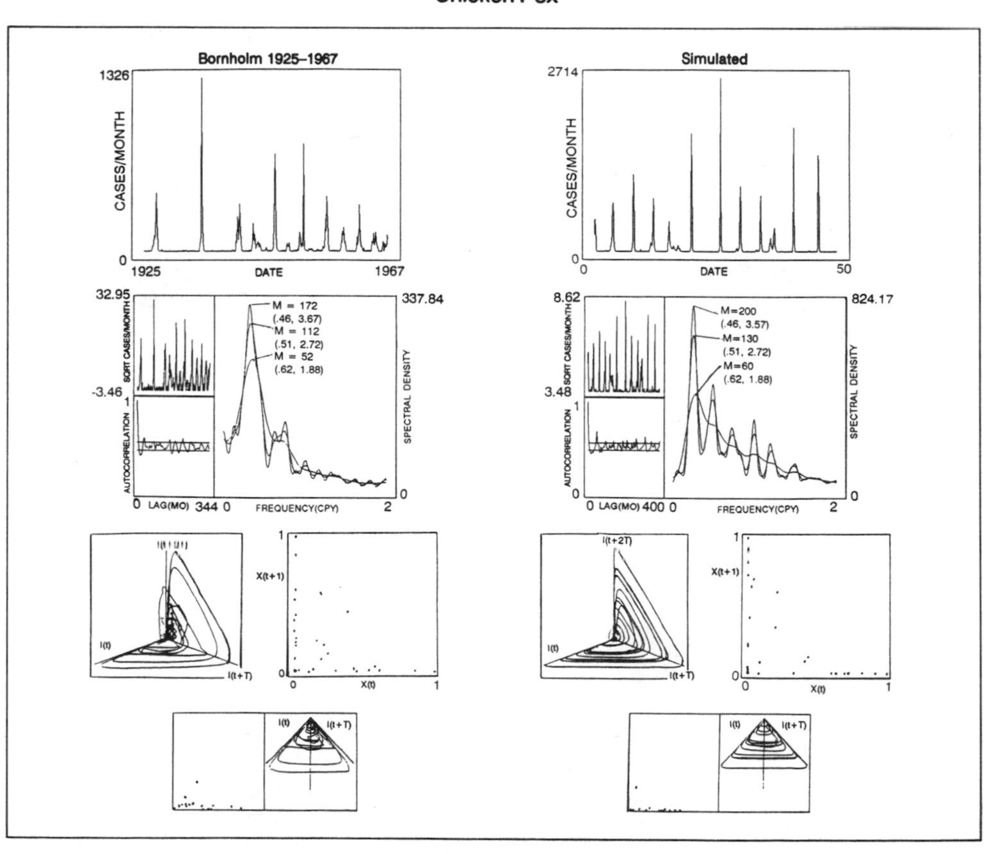

Fig. 18. Actual and simulated epidemics of chicken pox and measles for the Danish Island of Bornholm—population = 50,000. Monte Carlo simulations performed with parameter values given in Table 5.

What one *can* say is that the equations that govern the world probably exhibit behaviors similar to what is seen in the SEIR model, though perhaps for slightly different parameters. Consequently, to address the issue raised by Schwartz, one needs to consider the data. In particular, we pose the following question: Do the observed dynamics more nearly resemble those induced by the SEIR model in the twice-periodic region and in the presence of noise – or are they more suggestive of those induced by the model when it is placed squarely in the chaotic region? This is, of course, a difficult matter to resolve.

Having said so, we turn to the computer and estimate fractal dimensions and Lyapunov exponents for a series of values of β_1 in the presence of varying levels of noise. Each experiment, corresponding to 50 years of hypothetical data, was repeated ten times, thus allowing us to gain a handle on the expected variability. Figures 20 (fractal dimensions) and 21 (Lyapunov exponents) summarize our results. We note the following:

(i) In the chaotic region, estimated dimensions and exponents were relatively insensitive to the force of external perturbations.

(ii) For a given noise level, the estimated dimension declines and the Lyapunov exponent increases as one moves from the region of twice-periodic orbits to large-amplitude chaos. This is in keeping with the observation that the estimated dimensions of chicken pox epidemics exceed those computed for measles, while for the exponents, the converse holds. More generally, our results accord with those of Schaffer *et al. (26)* and Olsen *(49)*, who studied of the effects of noise on discrete mappings and differential equations.

(iii) Comparing the estimated dimensions of measles epidemics observed in eight First World cities with our calculations (Fig. 20), we conclude that real-world outbreaks of measles are somewhere in the vicinity of the critical value of β_1. Similar results obtain for the Lyapunov exponents (Fig. 21).

On the face of it, the last point would appear disappointing. In fact, we suggest that,

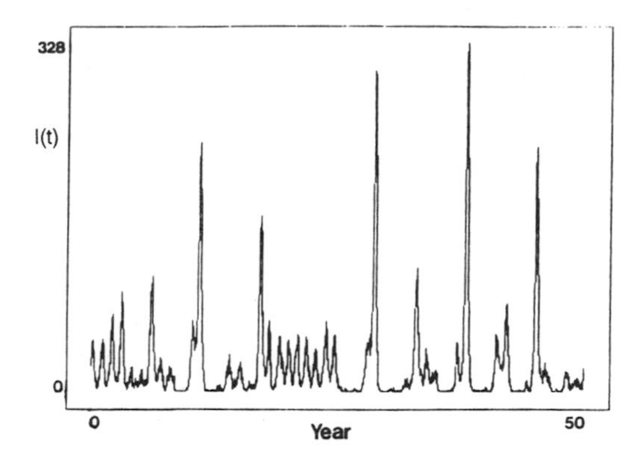

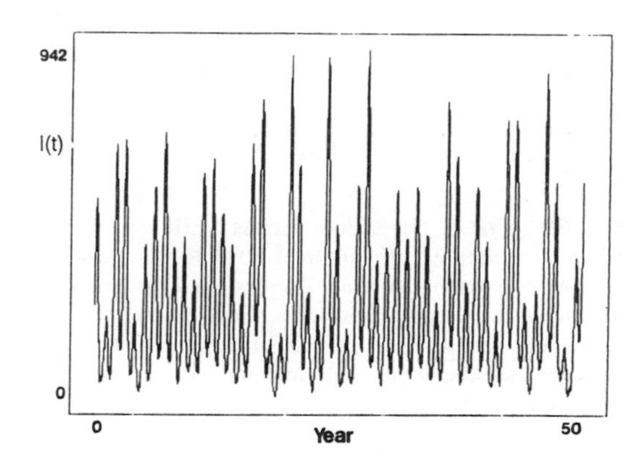

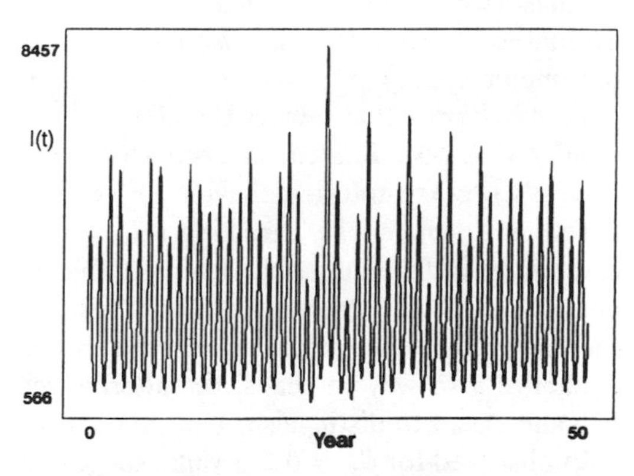

Fig. 19. Simulated epidemics of chickenpox using a Monte Carlo implementation of the seasonally forced SEIR model *(48)*. **(top)**N = 50,000. ($\beta_0 = 2.072 \times 10^{-8}$; $\nu = 4.0 \times 10^{-6}$). **(middle)**N = 500,000. ($\beta_0 = 2.072 \times 10^{-9}$; $\nu = 4.0 \times 10^{-5}$). **(bottom)**N = 5,000,000. ($\beta_0 = 2.072 \times 10^{-10}$; $\nu = 4.0 \times 10^{-4}$). Note the emergence of a regular annual cycle with increasing population size.

from the viewpoint of accounting for the observed phenomenology, it matters little whether or not measles epidemics are chaotic in the sense that chaos would be observed in the absence of noise and demographic

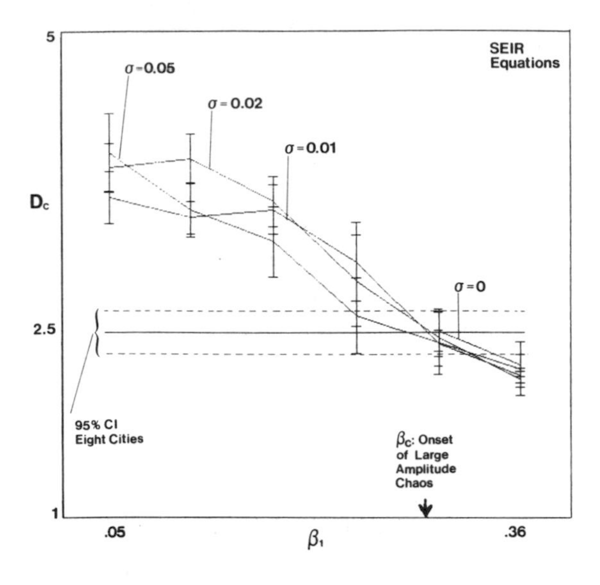

Fig. 20. Fractal dimensions for the SEIR model (differential equations) for different noise levels and magnitudes of seasonal forcing. Dimensions calculated for actual measles epidemics suggest that the degree of seasonality in nature more or less corresponds to the point at which essentially biennial dynamics give way to large-amplitude chaos.

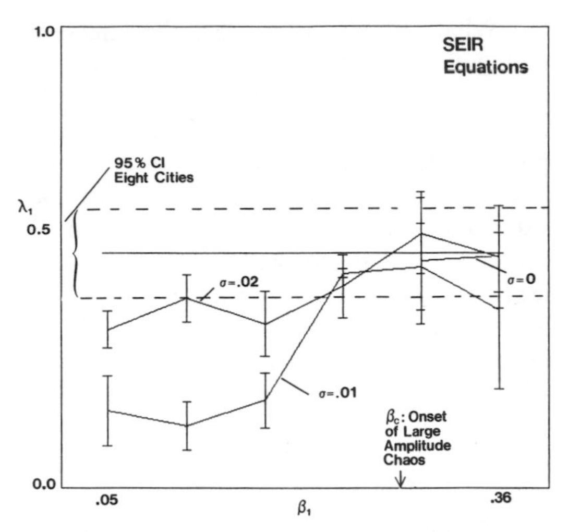

Fig. 21. Lyapunov exponents for the SEIR model (differential equations) for different noise levels and magnitudes of seasonal forcing. Compare with Fig. 20.

stochasticity. Such complications are unavoidable. Their effect on systems close (in parameter space) to a chaotic attractor is to induce behavior that mimics the actual chaos nearby *(29, 50)*. Indeed, in such cases, the system's pre-asymptotic behavior, i.e., before it settles down on the attracting set, can be essentially indistinguishable from full-blown chaos *(10, 51)*. Of course, as one moves further away from the point at which periodic dynamics give way to chaos, the alternatives become easier to distinguish. Thus, the return map observed for $\beta_1 = 0.2$, a value suggested by Schwartz as a reasonable maximum for real world epidemics, *does* appear to differ from both the SEIR model in the chaotic region and actual epidemics (Fig. 22). In sum, real world epidemics of measles are either truly chaotic or nearly so, and unless one is willing to study far more precise models, for example, of the sort described by Schenzle *(15)*, the question posed by Schwartz cannot be resolved by estimating a parameter.

Persistent Anomalies

Despite the overall agreement between the seasonally forced SEIR model and actual epidemics, certain discrepancies persist. In measles, the models predict peaks that are too sharp and troughs that are too deep. Moreover, there is often too much power associated with the low-frequency peak in the spectrum. And, in the case of rubella, neither the differential equations nor Monte Carlo simulations are able to account for the small amounts of yearly "jitter" that separates major outbreaks.

One possible basis for these anomalies is the fact that real cities are linked by a seasonally varying flow of infectives, which will tend to retard the long-term accumulation of infectives that provide the necessary kindling for widely spaced epidemics without a seasonal component. Explorations of this hypothesis are now in progress.

Acknowledgment

This work was supported by grants from the Danish Research Council, the United States

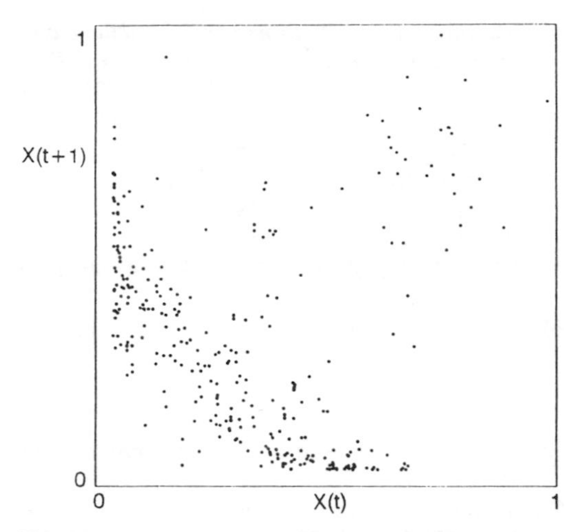

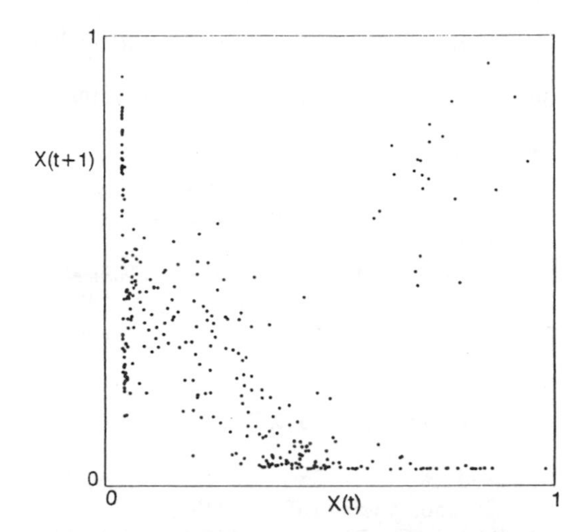

Fig. 22. Return maps computed for the SEIR model with parameters appropriate for measles but with reduced levels of seasonal forcing and with the addition of noise. (**Left**) $\beta_1 = 0.15$; (**Right**) $\beta_1 = 0.20$.

National Institutes of Health, and the North Atlantic Treaty Organization (NATO) Scientific Affairs Division.

References

1. Soper, H. E., *J. R. Stat. Soc.* **92,** 34 (1929).
2. Bailey, N. T. J., *The Mathematical Theory of Infectious Diseases And Its Applications* (Griffin, London, 1975).
3. Bartlett, M. S., *Proc. Third Berkeley Symp. Math. Stat. Probab.* **4,** 81 (1956).
4. Bartlett, M. S., *J. R. Stat. Soc.* **A 120,** 48 (1957).
5. Bartlett, M. S., *Stochastic Population Models in Ecology and Epidemiology* (Methuen, London, 1960).
6. London, W. P., and J. A. Yorke, *Am. J. Epidemiol.* **98,** 453 (1973).
7. Yorke, J. A., and W. P. London, *Am. J. Epidemiol.* **98,** 469 (1973).
8. Dietz, K., *Lect. Notes Biomath.* **11,** 1 (1976).
9. Dietz, K., and D. Schenzle, in *A Celebration of Statistics,* A. C. Atkinson and S. E. Fienberg, Eds. (Springer-Verlag, New York, 1985), pp. 167–204.
10. Yorke, J. A., N. Nathanson, G. Pianigiani, J. Martin, *Amer. J. Epidemiol.* **109,** 103 (1979).
11. Anderson, R. M., and R. M. May, *Nature* **280,** 361 (1979).
12. Anderson, R. M., and R. M. May, *Science* **215,** 1053 (1979).
13. May, R. M., and R. M. Anderson, *Nature* **28,** 455 (1979).
14. Anderson, R. M., B. T. Grenfell, R. M. May, *J. Hyg. Camb.* **93,** 587 (1984).
15. Schenzle, D., *IMA J. Math Appl. Med. Biol.* **1,** 169 (1984).
16. Hethcote, H. W., and S. A. Levin, in *Applied Mathematical Ecology,* S. A. Levin, J. G. Hallam, L. G. Gross, Eds. (Springer-Verlag, Berlin, 1989), pp. 193–211.
17. Callum, C., *World Health Stat. Q.* **36,** 80 (1983).
18. Fine, P. E. M., and J. A. Clarkson, *Int. J. Epidemiol.* **11,** 1 (1982).
19. Anderson, R. M., in *Population Dynamics of Infectious Diseases. Theory and Applications,* R. M. Anderson, Ed. (Chapman and Hall, New York, 1982), pp. 1–37.
20. Black, F. L., *J. Theor. Biol.* **11,** 207 (1966).
21. Anderson, R. M., and R. M. May, *Nature* **318,** 323 (1985).
22. Cliff, A., P. Haggett, J. Ord, G. Versey, *Spatial Diffusion: An Historical Geography of Epidemics in an Island Community* (Cambridge University Press, Cambridge, U. K., 1981).
23. Jenkins, G. M., and D. G. Watts, *Spectral Analysis and Its Applications* (Holden-Day, San Francisco, 1968).
24. Schaffer, W. M., L. F. Olsen, G. L. Truty, S. L. Fulmer, D. J. Graser, in *From Chemical to Biological Organization,* M. Markus, S. C. Müller, G. Nicolis, Eds. (Springer-Verlag, Berlin, 1988), p. 331.
25. Schaffer, W. M., *IMA J. Math. Appl. Med. Biol.* **2,** 221 (1985).
26. Schaffer, W. M., and M. Kot, *J. Theor. Biol.* **112,** 403 (1985).
27. Takens, F., in *Dynamical Systems and Turbulence,* D. A. Rand and L. S. Young, Eds. (Springer- Verlag, New York, 1981).
28. Packard, N. H., J. P. Crutchfield, J. D. Farmer, R. S. Shaw, *Phys. Rev. Lett.* **45,** 712 (1980).

29. Schaffer, W. M., S. Ellner, M. Kot, *J. Math. Biol.* **24,** 479 (1986).
30. Lorenz, E. N., *J. Atmos. Phys.* **20,** 130 (1963).
31. Rössler, O. E., *Z. Natürorsch.* **31a,** 259 (1976).
32. Olsen, L. F., in *Stochastic Phenomena and Chaotic Behavior in Complex Systems,* P. Schuster, Ed. (Springer-Verlag, Berlin, 1984), pp. 116–123.
33. Mandelbrot, B., *Fractals: Form, Chance and Dimension* (Freeman, San Francisco, 1977).
34. Farmer, J. D., E. Ott, J. A. Yorke, *Physica* **7D,** 153 (1983).
35. Ruelle, D., *Ann. N. Y. Acad. Sci.* **316,** 408 (1979).
36. Eckmann, J.-P., and D. Ruelle, *Rev. Mod. Phys.* **57,** 617 (1985).
37. Wolf, A., J. B. Swift, H. L. Swinney, J. A. Vastano, *Physica* **16D,** 285 (1985).
38. Ross, J., *The Prevention of Malaria* (John Murray, London, 1911).
39. Aron, J. L., and I. B. Schwartz, *J. Theor. Biol.* **110,** 665 (1984).
40. Hethcote, H. W., and J. W. Van Ark, *Math. Biosci.* **84,** 85 (1978).
41. Fine, P. E. M., and J. A. Clarkson, *Int. J. Epidemiol.* **11,** 5 (1982).
42. Grossman, Z., *Theor. Popul. Biol.* **18,** 204 (1980).
43. Kot, M., W. M. Schaffer, G. L. Truty, D. J. Graser, L. F. Olsen, *Ecol. Modell.,* **43,** 75 (1988).
44. Schwartz, I. B., *J. Math. Biol.* **18,** 233 (1983).
45. Smith, I. B., *J. Math Anal. Appl.* **64,** 467 (1983).
46. Schwartz, I. B., in *Biomedical Modeling and Simulation,* J. Eisenfeld and D. S. Levin, Eds. (J. C. Baltzer, A. G. Berlin, 1989), p. 201.
47. Schaffer, W. M., in *Chaos in Biological Systems,* H. Degn, A. V. Holden, L. F. Olsen, Eds. (Plenum Press, New York, 1987), pp. 223–248.
48. Olsen, L. F., G. L. Truty, W. M. Schaffer, *Theor. Popul. Biol.* **33,** 344 (1988).
49. Olsen, L. F., *Z. Natürforsch. Teil A.* **40,** 1283 (1985).
50. Crutchfield, J. P., J. D. Farmer, B. A. Huberman, *Phys. Rev.* **92,** 45 (1982).
51. Kantz, H., and P. Grassberger, *Physica* **17D,** 75 (1985).
52. Grassberger, P., and I. Procaccia, *Physica* **9D,** 189 (1983).
53. Shimada, I., and T. Nagashima, *Prog. Theor. Phys.* **61,** 1606 (1979).
54. Grassberger, P., in *Chaos*, A. V. Holden, Ed. (Manchester University Press, Manchester, UK, 1986), pp. 291–311.
55. Schaffer, W. M., and G. L. Truty, *Academic Computing* **March/April,** 34–35; 61–63 (1988).

Note added in proof. Versions of Figs. 14, 20, and 21 and portions of Figs. 2, 16, and 18 appear in Olsen, L. F., and W. M. Schaffer, *Science* **249,** 499 (1990).

14

Chaos and the Making of International Security Policy

Alvin M. Saperstein

Abstract

An important part of policy making is predicting the outcome of various options. Using mathematical models as a basis for prediction, as is often done, it is important to realize that, though deterministic, these models may lead to nonpredictable – chaotic – results. Policy makers must learn to predict the unpredictable – which may imply crisis instability and war – and react accordingly. As an example of using a model to make such predictions and draw appropriate policy conclusions, a procurement model for a bilateral Strategic Defense Initiative (SDI) is set up and numerically analyzed. The policy question is whether or not an implementation of SDI would transform an offensive defense into a defensive defense. The methodological question is whether such a transformation is crisis stable or unstable. The answer seems to be that, under present circumstances, a defensive-defense transformation is possible but that it is unstable.

Introduction

A fundamental aspect of science is its creation and use of models to *explain* observed phenomena. The policy maker (whether public or private) also uses models: to *predict* the future outcomes of present events and procedures and so ascertain which of the many possible present procedures to choose so as to gain the desired future outcome. Such policy-making models have varied greatly over the course of human history. They have included the reading of goat's entrails or the position of the stars, the extrapolation or replication of presumably similar past occurrences, and – in modern times – mathematical/scientific/technological models such as actuarial or systems theory models to fortell the future. Given the long history of battles and wars going significantly contrary to the most careful of plans, competent security policy making should also envision the worst-case-possibility that no prediction will be possible, that the outcome of a given policy will be so "chaotic" that no foretelling of the results of that policy is possible. The appropriate model must then be capable of *predicting unpredictability*.

I assume that the readership of this chapter consists of scientists who may be in the position to advise national security policy makers and citizens who are expected to select and control such policy makers. Hence the purpose of this chapter is to discuss some aspects of the scientific modeling of international security policy, using some of the new ideas developed by the field of deterministic chaos – long-term unpredictability in a completely deterministic model.

Most people – both citizens and policy makers – "think in terms of linear causality" *(1)* as a result of ordinary experience. Given the rapid response times now possible with modern strategic nuclear weapons (and even "conventional force" projection), the world of international events is now more tightly coupled than ever, casting doubt on the appropriateness of such linear intuition. Policy must be formulated in complex, nonlinear, strongly interacting situations. We need to experience comparable mathematical models so as to develop new "tools of thought," new intuitions, to enable us to anticipate policy outcomes and hence to choose policies wisely.

One important aspect of international relations that can be so modeled is the arms race between hostile nations or groups of nations. No attempt is made to explain the hostility. It is a "given," the foundation of a Richardson-like model *(2)* of arms procurements in which nations base their purchase and deployment of weapons upon the state of the weapons stocks of their hostile partners as well as on their own stocks and aspirations. Rational policy making, with such a reciprocal feedback model, is viewed as a dynamical flow from the present situation towards a desired future outcome and away from undesired outcomes. The parameters of the mathematical relationships characterizing the dynamical flow are characteristic of the different nations in the system of competing states and result from complex averages over the "microscopic" aspects of the interacting societies. Such parameters represent both inherent societal traits and explicit policy choices and are assumed to change relatively slowly, except as the result of deliberate policy choices that lead to different presumed flows.

In a "stable" model, one might envision the different possible outcomes as different fixed points (point "attractors") of the flow. To ensure a desireable outcome, we should choose our starting point to be within the basin of attraction of the desired fixed point (Fig. 1). Two types of policy choice are possible: if the model is fixed, choose the starting point S to be within the desired basin; if the starting point S is fixed, change the model parameters until the desired basin of attraction includes S. Of course, since no real-world determination is indefinitely precise, the starting point should be kept at least the error distance away from the boundaries between the basins to make sure that the desired outcome is attained. This approach assumes a nonchaotic flow with straightforward "bifurcations." Given such a model, policy making should be fairly straightforward, assuming that

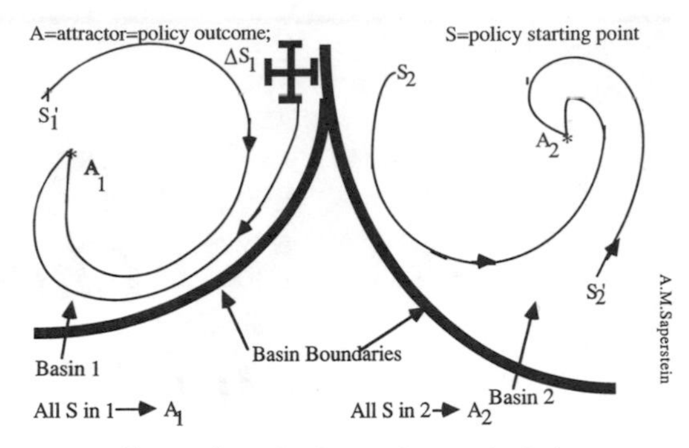

Fig. 1. "Basins of attraction" for policy choices.

the desired outcome is known (or at least that the undesired outcomes can be recognized).

Except for the new languages – deterministic or dynamical flow, fixed point, attractor, and basin – there is nothing very new in the above description of policy making. In the form of the Richardson arms-race model or the Lancaster equations of combat *(2, 3)*, such descriptions have been often used and abused *(4)*. Some may object to the use of deterministic, few-variable theories in situations that apparently rest upon very large numbers of (quasi-infinite!) degrees of freedom – the many diverse, complex, constantly shifting, individual people making up the populations of the competing nation system. They might expect completely random behavior-stochasticity or "old-fashioned chaos." But nations do act "smoothly" most of the time – just as do macroscopic physical bodies, though they too are made up of very many molecular degrees of freedom. When averaging over these many stochastic "microscopic" variables to obtain a few-variable, deterministic, "macroscopic" theory of nations or of physical objects, we expect the randomness to be averaged out, leaving a predictable situation. The randomness of the microlevel is not usually expected to be significant on the macrolevel. Therefore, smoothly varying macroscopic security models were perceived to be useful *(5)*.

However, we now know that such few-variable theories can also manifest chaos – extreme sensitivity to small changes in input or model parameters. That is, deterministic macrotheories (of materials or populations) can show macrolevel chaos above and beyond the expected intrinsic microlevel chaos. Indeed, the microfluctuations can be the small changes to which the macrosystem responds in an extremely sensitive manner. The question then becomes, How does such deterministic chaos in a model manifest itself in world affairs?

Chaos and International Security

With the suggestion *(6)* that chaos in a model may be related to the transition from "cold war" to "hot war" in the international system supposedly represented by the model, interest rose in the role of deterministic chaos in world affairs.

Some, interested in the continued dominance of military means in the relations between states, pursue the suggestion of "order within chaos." They recognize the general chaotic characteristic of the occurrence of war – the loss of predictability and control in the prewar juggling of competing states *(7)* – and the prevalence of chaos within the ensuing battles themselves – "the fog of battle" *(8)*. They believe that the successful general is the officer who recognizes, learns, and uses the "laws of chaos." Up to now, such victorious commanders have been recognized and selected retrospectively: they are the ones who have "won." It would be more efficient to recognize them prospectively – while still students at military academies. If the laws of chaos could be ascertained from mathematical models and taught to officer candidates, we might be able to deliberately create future Napoleons rather than wait for their occasional chance appearance. The goal here would be to use chaos rather than to avoid it.

Others, more interested in the de-emphasis of the military role in international relations – believing that war has lost its "utility" in a nuclear age – have focused on chaos as pure unpredictability. Given the prevalence of small disturbances in any international system, the possibility of chaos implies the loss of predictability and hence control by the actors within the system. Such a loss of control, given short-response time, long-range nuclear weapons, may mean war and catastrophe. Those holding this paradigm – of which the author is one – maintain that chaos cannot be used but must be avoided. In formulating policy, choices that may lead to chaos must be shunned. When using deterministic models to help formulate policy, those models and/or sets of input parameters that indicate the possibility of a loss of predictability should sound loud warnings of policy choices to be avoided.

A possible approach to modeling unpredictability in the context of Fig. 1 is to assume that the model implies "fractal boundaries" *(9)* between the basins of attraction. In this case, there would be a "random bifurcation" if the model's starting "point" overlaps the boundary region. (Remember that starting "point" is a misnomer: it is a actually a starting "region" since all parameters are only determined to within a finite experimental uncertainty.) Given the starting configuration, one would not know which of the possible outcomes would be attained and hence could not formulate policy rationally. However, assuming that the fractal boundary has a limited "width," as the flow progresses it will eventually end up definitely in a specific basin of attraction, after which time the flow and its attractor will be completely predictable. If at that time the perceived outcome is undesirable, there will still be opportunity (the attractor will not yet have been approached) to reverse course, change policy (change the model and/or its parameters) so as to head towards a more desirable attractor.

Overall, such a flow (and the international situation that it is presumed to model) is still predictable and controllable – hence "safe." (However, there is the possibility that the fractal boundary width is not limited – that it is an "area filling curve" which covers the entire region of interest in parameter space. Such a situation is illustrated by the Sierpinski gasket *(10)*, a Canter set in two dimensions, where the "included region" may be one basin of attraction, the "excluded region" may be the basin of attraction for the opposite outcome. Such a model implies a situation in which the outcome is completely unknown until it actually occurs; no control is possible.)

I have suggested that an indication that chaos could be the result of a given policy, that one will not be able to predict which of several possible outcomes (attractors) is the more likely, is a good reason to change that policy. Actually, that is too broad a statement. If the two possible outcomes are continued "peace" (even of a hostile nature) and war, with all its nuclear implications, most nations would try to avoid a chaos that made it impossible to choose between the two. If the indications are that a war, if initiated, will be won, with losses far less than gains, the only unpredictability being which specific individuals among the class of professional soldiers will actually be among the small number of battle fatalities, it is unlikely that any nation would draw back from such a policy and its implicit chaos. Thus, chaos itself is not a sufficient policy deterrent; the range of options obscured by the chaos (the range between the extreme attractors) must also be carefully considered.

There is, of course, no need to assume that the different outcomes, among which confusion is possible, are discrete, with separate (identifiable or nonidentifiable) basins of attraction. The set of outcomes (or a subset thereof) may be a continuum modeled by a "strange attractor" *(11)* in which case neighboring points in the single basin of attraction become exponentially separated when they reach the attractor. This is the usual definition of "deterministic chaos" *(12)*; it implies extreme sensitivity to fluctuations *anywhere* along the dynamical flow. Once in the basin of a strange attractor, there is no way of predicting how things will turn out; all possibility of future control is lost in such a basin. (Of course, since we are presuming a deterministic situation, the "immediate future" of the system at any stage is not shrouded in mystery; it is the final outcome – the attractor – that may be so shrouded.) If the range of variation of outcome along the strange attractor includes both states of peace and states of war, any policy leading to such an attractor implies a complete loss of any reasonable sense of security.

Thus we cannot say that chaos itself, in a model of hostile international relations, is equivalent to the prediction of war *(13)*. The chaos may not be fundamental to the system or subsystem in which it occurs; the subsystem may not be fundamental to the peace-war transition. If the amplitude of the chaotic fluctuations has the same order of magnitude as that of the total variation possible in the flow of the dynamical subsystem in question (i.e., is not just like a noise amplitude superimposed upon a predictively varying flow), it seems reasonable to say that the chaos is fundamental to the system (as it is in a fully developed turbulent flow, which is quite distinct from a laminar flow with some "noisy" ripples). If the fundamentally chaotic subsystem is something like violent fluctuations in the number of infantry troops in the National Guard contingents of various American States, it is unlikely that this augers the outbreak of international war. (There are many domestic political forces that would lead to the same result.)

To link the prediction of chaos to the prediction of war, the modeled subsystem must be a foundation of national security. In the nuclear age, the state of the strategic nuclear missiles of the two superpowers must be considered such a fundamental ground of international security. Chaos in such a system must be, at least heuristically, analogous to "crisis instability" *(14)* in the relationship between the super powers.

A "crisis unstable" situation regarding the two powers is one in which neither party is confident as to its ability to wait out a crisis (not striking out at its partner) because of the fear of losing its ability to respond to or protect itself from the actions of the opponent because of the prior actions of that opponent. Rather then wait out these prior actions, then, each is tempted to strike first. For example, suppose each side has 1000 intercontinental ballistic missiles (ICBMs), each having four warheads, and is thus capable of wiping out all of the opponent's missiles by firing 500 missiles, assigning two warheads to each target to ensure its destruction. (Each side is thus capable of a completely reliable "decapitating" strike.) The first to fire "wins" since it still has 500 missiles left to threaten the utter destruction of the civilization of its now helpless opponent. (I am assuming the absence of an efficient "fire under warning or attack" scheme or effective missile defense system.)

Thus, neither side can afford to wait out perceived threat or other crisis manifestations from the other.

> I didn't want to hit him though I had the means and motivation, which I thought he knew about, and so I thought that he thought that I was going to do so and therefore was planning to hit me, disabling me before I could hit him, because I know he had the means, and so – in order to prevent that – I hit him *(15)*.

Thus, the slightest hint of hostile action by an opponent in such a situation implies a massive preemptive attack, i.e., a small perturbation leads to a major system-changing response. This is the extreme sensitivity to small disturbances characteristic of deterministic chaos.

Crisis instability itself may not be the same as the outbreak of war in a hostile international system, just as physical instability may not give rise to an immediate change of state in a physical situation. A metastable liquid may require a dust grain to trigger the freezing transformation; so, some minimum-sized fluctuation may be needed to trigger an actual change of a "metastable" international state from peace to war. But there are always fluctuations in the real world of competing states, and if the situation is hostile enough, there will be no shortage of big fluctuations, e.g., insults.

Physical experience indicates that, once initiated, chaos easily procedes through a series of steps of ever-increasing complexibility to complete "turbulence" – a real change of state. Presumably, the same would be true of a realistic model of such a system, e.g., the Navier-Stokes equations for liquids. A comparably complex, nonlinear model of the interaction between the strategic nuclear systems of the superpowers would presumably also show such a change of state. Hence, it seems likely that – in a hostile system, such as has been prevalent for the past few decades – a crisis instability is equivalent to the outbreak of war; thus, the prediction of chaos in such a system is the prediction of the associated change of state, from cold war to hot war.

Illustration: A Procurement Model for SDI

To illustrate and explore these concepts further, we *(16)* have created a nonlinear procurement model for SDI – the "Strategic Defense Initiative" proposed by President Reagan in 1983 *(17)*. The threat of mutual strategic nuclear annihilation under which both "superpowers" – the United States and the Soviet Union – and the rest of the world have uncomfortably lived for the last few decades is certainly a foundation for current international relations. Any policy that purports to change the security postures of the major powers from the present "offensive defense" (deterrence, protecting oneself by the credible threat of doing enormous harm in retaliation for any damage done to oneself – MAD, "mutually assured destruction") to a "defensive defense" (having a shield which stops the missiles from falling upon you) is certainly addressing a fundamental aspect of the present international system.

SDI promises to shield the United States from the attack of Soviet ICBMs by a fleet of satellites in low-earth orbits capable of intercepting and destroying most of these attacking ICBMs in their initial boost phases. The "kills" would be accomplished over the launching territory via the rapid and long-range delivery of "exotic" energy (lasers, particle beams) or "conventional" energy (kinetic energy, e.g., guided missiles).

Presumably the Soviet Union would protect itself in a similar manner from our threatening ICBMs. Both nations have enormous investments (economic, material, social, psychological) in their offensive arms – their stocks of ICBMs. They are not likely to accept the degradation of these investments by the satellite based, defensive, anti-ICBM weapons. Hence satellite-killing weapons systems will be introduced that, because their purpose is to insure the penetration of the defensive satellite shield by the offensive ICBMs, must also be classified as offensive weapons. Our SDI model for the two competing-nations system thus consists of three components for each nation (Fig. 2): a defensive satellite system capable of destroying the opponent's ICBMs shortly after lift-off from the attacker's territory, the offensive ICBM system designed to land nuclear warheads upon the "soft" cities ("countervalue") or upon the "hardened" weapons and control centers ("counterforce") of the enemy, and the offensive anti-satellite weapons deployed to insure that the associated ICBMs will successfully penetrate the defensive shield.

Our model is a *procurement* model – a discrete, nonlinear extension of Richardson-like models *(2)* – not an attempt to describe the actual destructive interplay between these components that would occur should war actually break out, i.e., not a battle *(3)* or attrition *(18)* model. Each side procures and deploys numbers of the offensive and defensive components sufficient to carry out its preconceived purposes. But the number required to carry out a task depends upon the opponent's reactions. For example, in a MAD strategy, it may be sufficient for the U.S. to be confident that it can drop 500 nuclear warheads upon Soviet territory in retaliation for a prior Soviet attack. If these 500 warheads are mounted on reliable missiles, which are confidently immune to prior attack, and if the Soviets have no defense against such missiles, then the U.S. need only procure these 500 warheads. If the U.S. expects that the Soviets may intercept 100 warheads, then it will have to procure 600, etc. The result is that the procurement of each component is nonlinearly coupled to the procurement of the others.

Our analysis of this model is directed at two policy-oriented questions: (i) Are there parameter sets for which the time evolution of the model indicates a decrease in offensive weapons (with respect to the initial values) so that the attractor reached from the present configuration of the international system is a "defensive" end point? In such a case, the policy represented by these parameters should lead to a transition from the present offensive defense to the desired defensive defense. (ii) Given these desired parameters, is the resultant dynamical flow actually predictable, i.e., is the attractor an "ordinary" point or cycle attractor? Or, does the model

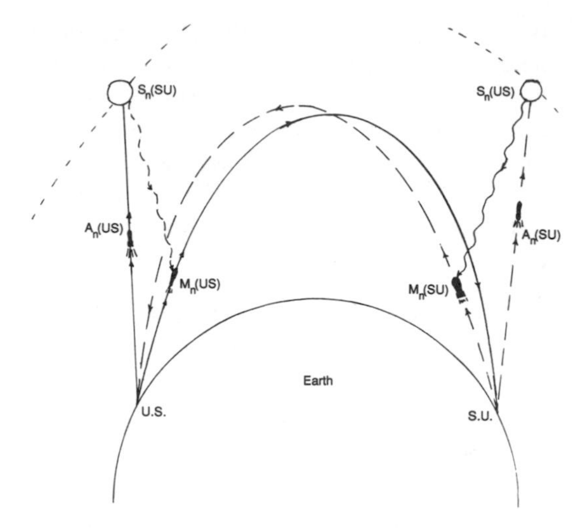

Fig. 2. Interactive SDI model including defensive ABM satellites (S) and offensive ICBMs (M) and satellite killers (A).

predict (and the policy imply) a "strange attractor" – chaos – in which case, the policy presumably leads to a crisis-unstable state of the international system and is hence to be avoided. The first question is answered by computing the time evolution of the model for many different parameter sets. The second is explored by adding noise to the system and noting the variation of the output.

Let $M_n(i)$ be the number of ICBMs deployed by nation i ($i = 1$ represents the U.S., $i = 2$, the Soviet Union) in the nth year (assuming yearly budget cycles, starting currently at $n = 0$). If μ_i is the complete cost (capital, entire life-cycle operating, and decomissioning) to side i of a single ICBM, then the total cost to i in year n for the procurement of ICBMs is $\mu_i\,[M_n(i)-M_{n-1}(i)]$. Similarly, let σ_i be the complete cost of each satellite anti-ICBM "battle station" (ABMs) whose numbers are $S_n(i)$, and let α_i be the cost per ground-based, satellite-killing missile (ASAT), which number $A_n(i)$. Then the total annual cost to side i to procure its strategic nuclear system is

$$D_n(i) = \mu_i \hat{\theta}\left(M_n(i) - M_{n-1}(i)\right) + \sigma_i \hat{\theta}\left(S_n(i) - S_{n-1}(i)\right) + \alpha_i \hat{\theta}\left(A_n(i) - A_{n-1}(i)\right), \tag{1}$$

where

$$\hat{\theta}(x) = x\,\theta(x) = \max(x, o) \qquad (2)$$

($\theta(x)$ is the Heaviside unit step function: $\theta(x) = 1$ for $x > 0, \theta(x) = 0$ for $x < 0$) is used since the cost of these weapons can never be negative. Aircraft, submarines, theatre missiles, and any other alternative means for threatening the homeland of the opponent with nuclear destruction are not included in this initial, purely illustrative model of strategic nuclear warfare. If D_i is the maximum of societal resources that i can devote to strategic nuclear warfare in any one year, then $D_i - D_n(i)$ can never be negative. In fact, we expect a smooth decrease in strategic nuclear weapon procurement as $D_n(i)$ approaches D_i from below, i.e., as $\hat{\theta}([D_i - D_n(i)]/D_i)$. This finite resource requirement, via Eq. 1, is the first nonlinear constraint among the variables of our model.

Side i procures ICBMs in order to carry out a definite offensive strategic purpose which requires the delivery of some minimum number, R_i, of nuclear warheads to the opposing side j's territory. If the strategy is one of deterrence, as assumed in this chapter, then R_i must be sufficient to produce "intolerable" damage to j when landed in retaliation; it is usually assumed *(19)* that $R_i = 500$. (An alternative first-strike strategy, in which the goal is to destroy all of the opponent's ICBMs and hence prevent them from striking back, would be described by $R_i = M_n(j)$.) If $R_n(i)$ is the number of warheads that i believes will actually penetrate j's defenses and destroy their designated targets (the belief being based upon the configuration of i and j forces in year n), then i will annually procure more offensive missiles until $R_n(i)$ equals R_i, after which no further ICBM procurement is necessary. If $R_n(i)$ exceeds R_i, then i may decommission missiles in a "build down" which might allow a transition to a defensive mode. Thus we have an "ICBM builder's recursion relation":

$$M_{n+1}(i) = M_n(i) + f_i\left[R_i - R_n(i)\right]\hat{\theta}\left(\frac{D_i - D_n(i)}{D_i}\right), \qquad (3)$$

which is a set of two coupled recursion relations specified by six free, positive model parameters: f_i, D_i, R_i. The determination of $R_n(i)$ requires further specification of the behavior of both sides.

Before we can calculate the number of warheads that should destroy their targets, we must first specify the number $\bar{\eta}_n(i)$ of i's warheads which penetrate j's satellite defense. If m_i is the number of warheads per ICBM for side i (mean multiple, independently targeted, reentry vehicle or "MIRV" number), $\Delta M_n(i)$ is the number of i's retaliatory missiles destroyed in their boost phase by j's satellites, and $Q_n(i)$ is the number of ICBMs that i does not fire in his first strike against j's missile silos, keeping them in reserve for a second strike, then

$$\bar{\eta}_n(i) = m_i\left[M_n(i) - Q_n(i) - \Delta M_n(i)\right]\theta(\bar{\eta}_n(i)) \qquad (4)$$

(where the unit step function is included to make sure that $\bar{\eta}_n$ is a positive number). The number held in reserve must be sufficient for carrying out their presumed final deterrent purpose in spite of the opponent's assumed ABM system; hence

$$Q_n(i) = \frac{1}{m_i}R_i + \Delta M_n(i). \qquad (5)$$

Note that ΔM, the loss of ICBMs to the ABM defense, is assumed to be the same whether the ICBM attack is a first or second strike, a small or large scale attack, (i.e., it is assumed that the defense system cannot be saturated by the attackers), a questionable assumption highly favorable to the SDI concept. Also favorable to this defensive concept is the functional form assumed for ΔM, that it be proportional to the square of the number of defensive elements *(20)*, rather than the linear relationship put forward by SDI's critics *(21)*.

The goal of this exercise is to test the policy of developing and deploying SDI. Thus, whenever there is uncertainty as to the structure or parameters of the model, the choice most favorable to SDI is made. If the modeling results are unfavorable to the SDI policy, significant doubt is cast upon the policy; if the results are favorable, one can only infer that further research is needed to see whether the basic core of the model favors SDI or whether the favorable results are attributable to these unduly optimistic, uncertain assumptions.

Thus we assume

$$\Delta M_n(i) = \left[\frac{1}{\beta_j} S_n^{eff}(j)\right]^2, \qquad (6)$$

where the "effective" number of satellite battle stations is still to be determined. The parameter β_i is given by Canavan *(20)* in terms of the ABM laser "brightness," the target ICBM "hardness," and geometric factors. The latter account for the fact that only a fraction (determined geometrically by Canavan) of the satellite battle stations can engage the rising ICBMs at any one time since they are all continuously orbiting the earth.

Let $\eta_n(i)$ be the number of *j*'s missiles that are destroyed in their silos by *i*'s ICBMs, and let p_i be the probability that a single *i* warhead destroys a single *j* missile. If $\bar{\eta}_n(i) \le M_n(j)$, then

$$\eta_n(i) = p_i \bar{\eta}_n(i). \qquad (7)$$

If

$$\bar{\eta}_n(i) > M_n(j),$$

then

$$\eta_n(i) = M_n(j)\left[1 - (1 - p_i)^{(\bar{\eta}_n(i)/M_n(j))}\right]. \qquad (8)$$

In view of the trend of increasing missile accuracies of the last few years (and the recent increases in U.S. warhead explosive yield), we assume $p_i = 1$ for the calculations of this chapter.

Then $\bar{\eta}_n(i) = \eta_n(i)$, and the number of *i*'s retaliatory warheads that land upon *j*'s territory, needed to complete Eq. 3 is

$$R_n(i) = m_i\left[M_n(i) - \eta_n(i) - \Delta M_n(i)\right]\theta\left(R_n(i)\right). \qquad (9)$$

In order to compute ΔM, we must know the "effective number of satellites," those that can actually destroy rising ICBMs via Eq. 6. This S^{eff} is less than the number of deployed ABM satellites since some of the latter may be destroyed by the aggressor's anti-satellite weapons at the moment the attack is initiated, before the satellites have a chance to focus upon the ICBM attack. We assume perfect accuracy for the ASATs, just as we have for the ICBMs and ABMs, which implies that the number of satellites destroyed is equal to the number of weapons fired at them.

Hence

$$S_n^{eff}(i) = \left[S_n(i) - A_n(j)\right]\theta\left(S_n(i) - A_n(j)\right), \qquad (10)$$

where the θ function is used to keep S^{eff} positive, as it should be. (Actually, this is an overestimate for S_n^{eff} since $S_n(i)$ are in orbits covering the entire globe, whereas the destroyed satellites are just those over the ICBM-firer's territory at the time of attack initiation. Again this overestimate is favorable to the SDI concept.) Thus, to complete the computation of Eq. 3, we must have relations for S and A (number of ABMs and ASATs).

The defensive-minded satellite builder will wish to have sufficient ABM satellites to destroy *all* of the prospective attacker's offensive missiles in the boost phase. Using Canavan's result, Eq. 6, the minimum number of satellites required to do this job occurs when $S_n^{eff}(i) = \beta_i [M_n(j)]^{1/2}$. If $S_n^{eff}(i)$ is less than this, some of the $M_n(j)$ ICBMs will penetrate *i*'s defense to land warheads on *i*'s territory; thus *i* will build more satellites if resources allow. If the number of ABM weapons exceeds this requirement, *i* will decommission some so as to save resources. Thus, we have a pair of "satellite builder's recursion relations"

$$S_{n+1}(i) = S_n(i)$$

$$+ g_i\left[\beta_i\sqrt{M_n(j)} - S_n^{eff}(i)\right]\hat{\theta}\left(\frac{D_i - D_n(i)}{D_i}\right) \qquad (11)$$

similar in structure to Eq. 3, where g_i is another pair of positive model parameters, corresponding to the rate at which ABM satellite battle stations will be built, which must be specified.

Similarly, the goal of the ASAT builder is to be able to destroy a sufficient number of defensive satellites so as to enable a predetermined number of its ICBMs to successfully penetrate the opponent's defensive shield. Let $Z_n(j)$ be the maximum number of *j* satellites that will allow *i* to penetrate *j*'s defense to the desired degree. If *j* has an effective number of ABM satellites greater than this number (i.e., if $S_n^{eff}(j) > Z_n(j)$), then *i* will have to destroy more, implying the further procurement of satellite-killing missiles if allowed by resource constraints. Hence the "satellite-killer-missile builder's recursion relations" are

$$A_{n+1}(i) = A_n(i)$$

$$+ h_i\left[S_n^{eff}(j) - Z_n(j)\right] \theta \left(\frac{D_i - D_n(i)}{D_i}\right), \tag{12}$$

where the parameters h_i determine the rate at which i can build and deploy ASAT systems.

All that remains, in order to have a closed set of coupled recursion relations capable of being iterated forward in time from given initial values, is to specify $Z_n(j)$ which, because of Eq. 6, implies that j can destroy $[Z_n(j)/\beta_j]^2$ of i's ICBMs. An additional $\eta_n(i)$ of i's potentially retaliatory ICBMs are destroyed in their silos by j's presumed prior strike. Hence, in order to be confident that it can deter j's first strike by being able to land R_i of its missiles upon j after that strike, $Z_n(j)$ must be limited by

$$R_i \geq m_i\left[M_n(i) - \eta_n(i) - \left(\frac{Z_n(j)}{\beta_j}\right)^2\right]. \tag{13}$$

Using Eqs. 4, 5, and 6 for $\eta_n(i)$ in Eq. 13, the limiting value for $Z_n(j)$ is given by

$$\left[\frac{Z_n(j)}{\beta_j}\right]^2 \geq M_n(i) - \frac{R_i}{m_i}$$

$$- m_j\left(M_n(j) - \frac{R_j}{m_i} - 2\left[\frac{S_n^{eff}(i)}{\beta_i}\right]^2\right). \tag{14}$$

Using the equality in Eq. 14 plus Eq. 10 for S^{eff}, $Z_n(j)$ is completely determined by previously defined quantities. Hence, given the 20 model parameters ($R_i, m_i, \beta_i, D_i, \mu_i, \sigma_i, \alpha_i, f_i, g_i, h_i$,) and the 12 initial conditions ($M_o(i)$, $S_o(i)$, $A_o(i)$, $M_{-1}(i)$, $S_{-1}(i)$, $A_{-1}(i)$ – note that we have to specify two initial conditions for each of the dynamical variables because of the inherent time delay of the system; the dynamical flow equations only hold for positive times), the six coupled nonlinear procurement recursion relations, Eqs. 3, 11, and 12, can be numerically solved for any positive n. The dynamical flow into the future is completely determined.

Iterating the procurement-recursion relations for many different parameter sets, each related in some "reasonable fashion" to the present U.S.–U.S.S.R. strategic nuclear weapon stocks, we may come up with answers to the first question: Can an SDI policy be implemented so as to bring about a transition from the current offensive-defense paradigm to a more desireable defensive defense? In terms of our model, can we find a flow in which the superpower ICBM stocks diminish while the defensive weapon stocks (ABMs) do not grow unreasonably? To approach the second question – can such a desirable SDI policy avoid an undesirable regime of crisis instability – we add a noise component to the model and test its output for variation with respect to the input fluctuations. "Sensitivity to small errors can be considered one of the defining features of chaos" *(22)*.

The dynamical flow model, Eqs. 3, 11, 12, can be symbolically represented as

$$x_{n+1} = F(x_n, \lambda), \tag{15}$$

where the dynamical variables of the system are compressed into the state vector $x_n = [M_n(1), M_n(2), S_n(1), S_n(2), A_n(1), A_n(2)]$, while the build-up parameters of the model are given by the vector $\lambda = [f_1, f_2, g_1, g_2, h_1, h_2]$. Perturbations and uncertainties in these parameters are simulated by having the vector λ vary randomly in time within a range determined by the "noise level" ε. To do this, Eq. 15 is modified to

$$x_{n+1} = F(x_n, \lambda_n) \tag{16}$$

and the six-component vector $\lambda_n = \lambda(1+\xi_n)$ where the six noise components ($\xi_{n,k}$, $k = 1,\ldots 6$) are random variables uniformly distributed in the interval $[-\varepsilon, \varepsilon]$. Starting with a set of initial conditions (x_o, x_{-1}), a given parameter set λ, and a noise level ε, the recursion relations, Eq. 16, are iterated for $n \leq 33$ timesteps (years). One such set of 33 iterations (plotting x_n against n) represents a single "orbit" of our model. We then consider an ensemble of 10^5 such orbits, each orbit arising from the same functional form F, the same starting values, and the same average parameters, differing only via the different realizations – at each value of n – of the six random variables ξ_n. For each value of n and each dynamical variable M, S, A, we compute the mean value of the variable and the upper and lower limits of its range over the ensemble of orbits; this range in outputs is indicated by error bars in our figures. This addition of noise to each point of the orbit is supposed to mimic a political reality in which the constantly shifting national and international political scene produces random changes in procurement decisions – and hence model parameters – each year. It also represents the uncertainty in determining the actual parameters due to the military secrecy policies of the two competing superpowers.

The size of the response range for a given input noise level is a measure of the "sensitivity to small errors" of our SDI model with a given set of parameters and initial values. If the range of response (the length of the error bar) is small compared to the mean value throughout the flow (i.e., if the percentage output range is comparable to the input noise level), we consider that dynamical flow to be stable and predictions made with its help to be credible. If the opposite is true – meaning that the output "noise" is greater than the input – we consider the flow to be unstable. If in addition the size of the output error bars increases noticeably faster than linearly with respect to time (perhaps exponentially over part of the flow), we infer the existence of positive Lyapunov exponents *(9, 12)* and, hence, deterministic chaos. The model is then predicting crisis instability.

Model Parameters and Results

At the present time, neither superpower has satellite ABMs or ASATs; the two sides have known numbers of ICBMs with known characteristics that can be taken from the open literature, e.g., Canavan *(20)*. Hence we assume

$$M_{-1}(1) = M_0(1) = 1000,\ M_{-1}(2) = M_0(2) = 1400$$

$$S_{-1} = S_0(i) = 0,\ A_{-1} = A_0(i) = 0$$

$$m_1 = 2, m_2 = 6 \qquad (17a)$$

It is usually assumed that 500 thermonuclear warheads are amply sufficient to do intolerable damage to a modern society *(19, 20)*; hence

$$R_i = 500. \qquad (17b)$$

The efficiency parameter β_i, of the not-yet-existing satellite battle stations is obtained *(20)* by extrapolating "reasonable" present-day physical and geometrical characteristics; the result is

$$\beta_i = 2.5. \qquad (17c)$$

The cost parameters are not as well known and have been taken *(16)* from estimates published by proponents of SDI (they do not include research and development costs); we use (B = 10^9).

$$\mu_i = \$0.1\text{ B}, \alpha_i = \$0.05\text{ B}$$

$$\sigma_i = \$3.0\text{ B}, D_i = \$40\text{ B} \qquad (17d)$$

The build-up parameters are completely unknown and have been estimated *(16)* as

$$f_i = 0.04, h_i = 0.04$$

$$g_1 = 0.03, g_2 = 0.04. \qquad (17e)$$

(These imply that each side can initially build about 20 new ICBMs and 3 ABM satellites per year.)

The set of initial and model parameters contained in Eqs. 17 will be referred to as the "standard model." If used with a 50% noise level (ε = 0.5), they will be designated as SP (50); the output for this model is shown in Fig. 3. (If the noise level is doubled, written as SP (100), the output variations roughly double, whereas the mean values change by no more than 5%.) The predicted variations of ICBM numbers (Fig. 3A) and ABM numbers (Fig. 3B) are less than 30%; the ASAT numbers (Fig. 3C) may fluctuate by somewhat more than 100%; there is no significant departure from linearity in the growth of the output ranges, which implies that there are no positive Lyapunov exponents. It appears that the SP model allows a meaningful prediction that the strategic arms race between the two powers will probably remain in the offensive mode with ever increasing numbers of ICBMs, ABMs, and ASATs on both sides. These results are unchanged with large changes in

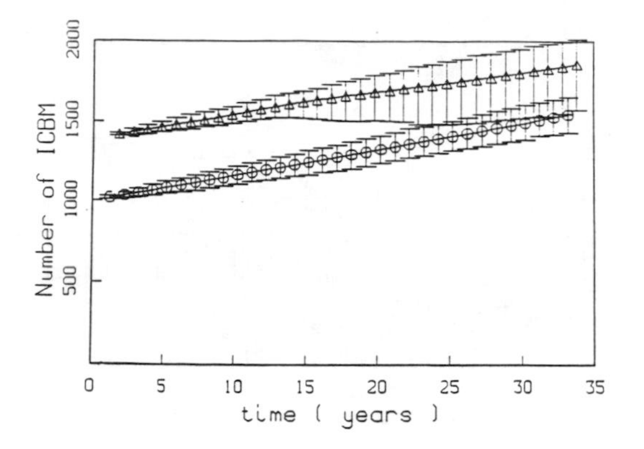

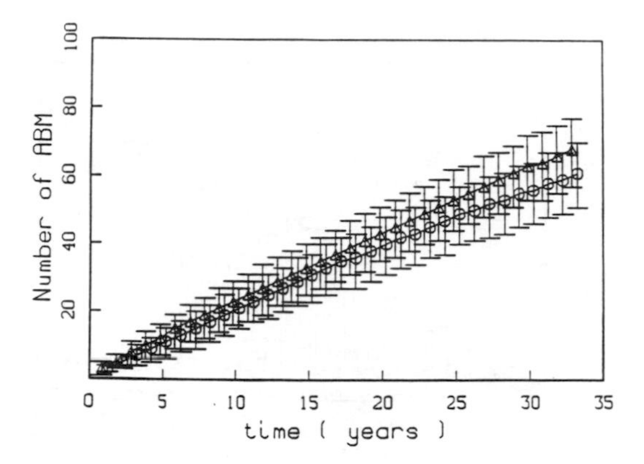

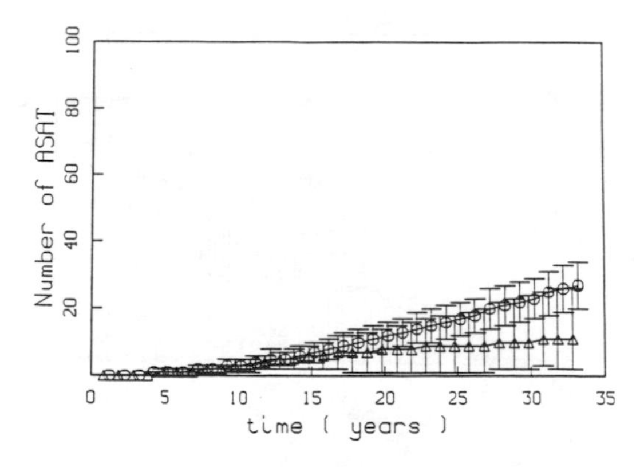

Fig. 3. M (ICBMs), S (ABMs), and A (ASATs) as a function of time: circles for the U.S., triangles for the Soviet Union, represent mean values. The associated error bars represent the output fluctuation stemming from input noise. The parameters are SP(50): 50% noise in the standard model.

the economic constraints, i.e., increased D_i. This implies that changes in the cost parameters will also be inconsequential as long as we do not depart by orders of magnitude from SP (which doesn't seem very likely).

Since the only significantly unknown parameters are the SDI build-up rates, we have explored large variations in f_i, g_i, and h_i. If the rate of production of offensive weapons is increased by a factor of 10 ($f_i = 0.4 = h_i$), i.e., so that each side is capable of building 200 ICBMs per year, the expected result is a vast build-up of these weapons, although qualitatively the flow remains similar to the SP case. We do not expect a change in the character of the arms race until there is more emphasis on defensive weapons. Such a qualitative change is seen (Fig. 4) when the defensive weapon production rate is increased ten-fold ($g_1 = 0.3$, $g_2 = 0.4$), which implies an initial ABM satellite deployment rate of 30 per year for each side. With this model flow, the Soviets lose the unambiguous offensive lead they had with the SP flow; now ICBM and ABM deployments are roughly the same (within the noise fluctuation band) although the U.S. lead in ASAT is now more definite. The output fluctuations are still comparable to input noise level, and there is no sign of an exponential growth in the output noise. There is thus no evidence for a loss of predictability, and the predicted flow shows no transition from an offensive-dominated mode to one dominated by the defense.

There is a clear transition to the desired defense-dominated mode if there is a ten-fold increase in both offensive and defensive weapon procurement rates and a very large increase in the economic resources available ($f_i = h_i = 0.4$, $g_1 = 0.3$, $g_2 = 0.4$, $D_i = \$10^5$ B). Figure 5 shows that, for the mean of this flow there is a marked decrease in the ICBMs on both sides, accompanied by reasonable numbers of ABMs and ASATs. Thus a quick build-up in both offensive and defensive strategic weapons would seem to be a desirable policy. However, note that in this case the range of output fluctuations is large with respect to both the input noise and the mean flow values. Furthermore, the rate of departure of the maximum output fluctuations from the means is significantly greater than linear and may very well be exponential, suggesting positive Lyapunov exponents. This indicates a prediction of chaos in the model and hence of crisis instability (war?) in the represented

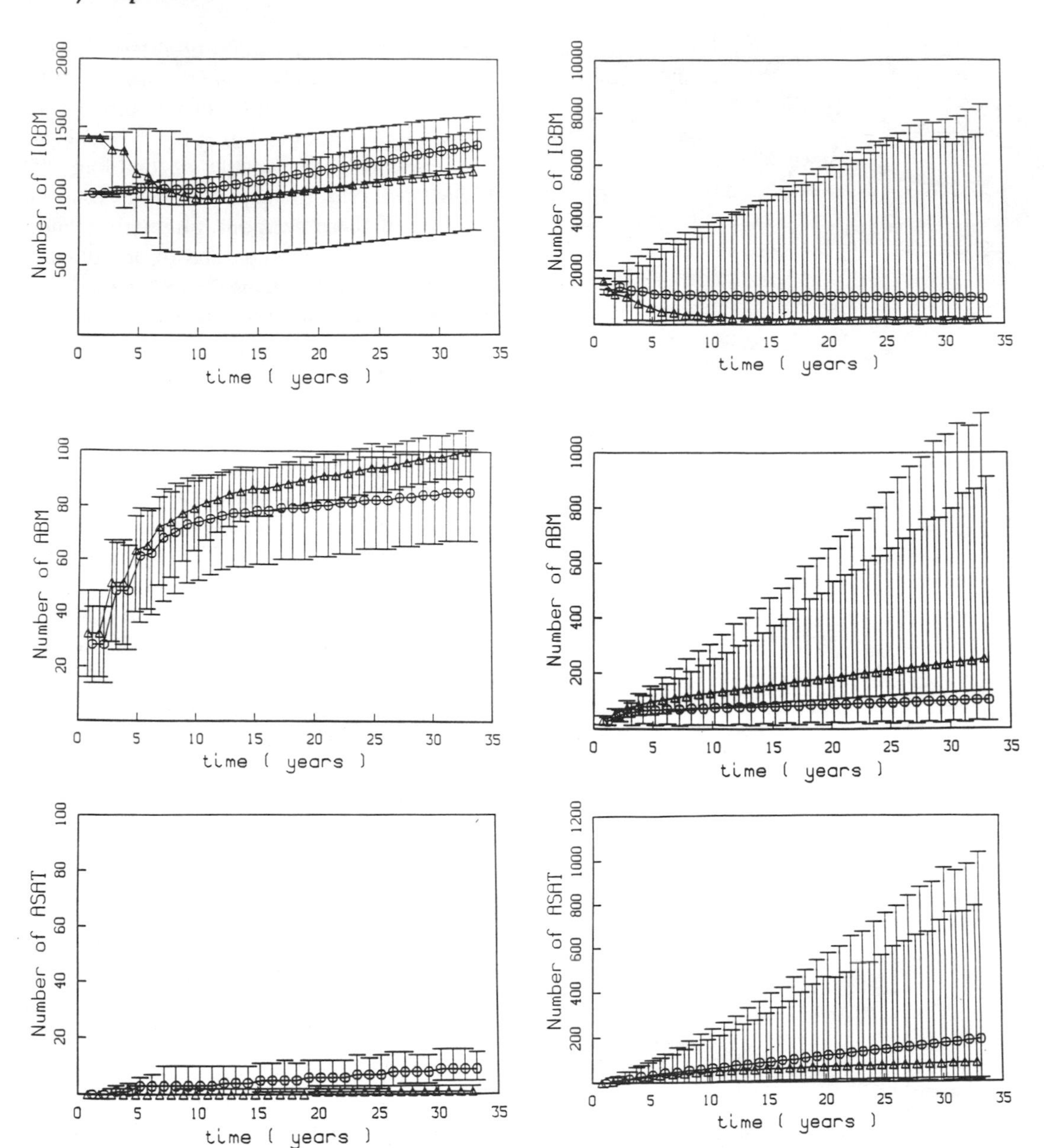

Fig. 4. The same as Fig. 3 except that the production rate of defensive weapons (ABM) has been increased ten-fold ($g_2 = 0.3, g_2 = 0.4$).

Fig. 5. The same as Fig. 3 except that the production rates of both defensive weapons (ABM) and offensive weapons (ICBM, ASAT) has been increased ten-fold, economic constraints relaxed ($f_i = h_i = 0.4, g_1 = 0.3, q_2 = 0.4$, D_i = $105B).

world system. So, such a policy is to be avoided rather than encouraged.

Conclusion

Only a few examples of parameter variations have been presented here [more are found in *(16)*] to give the reader an insight into the possibilities of the general approach and the specific model. The conclusion seems to be that an SDI transition to the defensive mode occurs in a crisis-unstable situation. Admittedly the model is incomplete. Many more variables — describing other aspects of the strategic nuclear posture, tactical nuclear weapon configurations, conventional force posture, and the roles of nations other than the two superpowers – would be needed before "practical" policy makers would be expected to pay much attention to the model and its specific results. But the preliminary model itself may be very useful to the general public considering SDI if its prediction of instability is meaningful in spite of its shortcomings.

Not having done an exhaustive search through the many-dimensional parameter space, we cannot say that our model "proves" that a strategic shift from offensive defense to defensive defense is always accompanied by chaos. (But, if a model is such that finding stable regions of parameter space is not relatively easy, it would not seem wise to base important policy upon that model.) Nor can we insist that the suggestions of our incomplete, heuristic model pertain to the real strategic world with its many more variables – bombers, submarines, tactical weapons, other nations, etc. Based upon political *(17, 25)* and physical intuition, our model and results thereof seem usefully suggestive. If a model has parameter regions of instability, adding variables is likely to increase the size of these regions, implying that the instabilities would manifest themselves earlier as the parameters are varied. "... the more complex a system is, the more numerous are the types of fluctuations that threaten its stability"*(23)*. A good example of this trend occurs in fluid dynamics: the critical Reynolds number, which determines the transition from laminar to turbulent flow, is lowered when thermal variables are added to the fluid model; thus fluid stability is decreased when heat effects become important to the fluid flow.

It follows that, if an incomplete model of international conflict such as ours shows instability in given regions of parameter space, a more complete, "realistic" model is also likely to be unstable – in fact, to be unstable in larger regions of the parameter space, i.e., to be harder to stabilize. The converse is not true: if a given model, representing a system, is stable, a more complex, more realistic model of the same system may be unstable.

The least that should be learned by policy makers from the simplified modeling of international relations is to be extremely cautious when there exists a possibility of a chaotic transition, especially if the location of the critical parameters (i.e., policies leading to such a transition) are unknown. One cannot be confident of "pushing" an opponent safely around the "policy field" when there may be a precipice in that field over which each might drag the other. Good illustrations of the need and application of such caution are numerous in technology. For example, new aircraft (especially military ones) are introduced to push out the boundaries of "performance space" (lift, speed, agility, endurance, etc.) as much as possible. Absent a complete theory of aircraft performance (which was certainly true prior to the advent of massive computers and wind tunnel facilities), the line between stable handling and disasterous turbulent transition could not be precisely predicted. Test pilots would be warned of the possibility of such a transition and told to explore the new regions of performance space very gingerly. With the use of radio and "black boxes" to bring back the information when test pilots did not themselves survive, much could be learned about the possibilities of aircraft performance – which could thus be extended from generation to generation. There seem to be no such substitutes for extreme caution in the realm of potential nuclear war catastrophe, no possibity of generations learning from the errors of previous generations.

Thus, in a modern, rapidly changing, potentially catastrophic, nuclear-armed world, policy makers can no longer depend upon a "common sense" *(24)* formed by previous, unrelated experience. They must rely upon modeling, and both policy maker and citizen should acquire experience in making, criticizing, and interpreting such models. A fundamental task in modeling for security policy making should be the checking of the model for the stability of its predictions. The possibility of chaos in the dynamical flows, which are the core of most such models, is much greater than is commonly supposed.

As for SDI itself, the lesson of our model

and of many other analyses *(17, 25)*, seems to be that we should all be very leery of any policy furthering the deployment of SDI until after there have been major reductions in the strategic weapons stocks of all parties.

Acknowledgments

The author is grateful to the Aspen Center for Physics where the collaboration with Gottfried Mayer-Kress that culminated in this work was initiated. He is also indebted to the Los Alamos National Laboratory where the computations that led to the figures of this chapter were carried out by Dr. Mayer-Kress.

References

1. Prigogine, I., and I. Stengers, *Order Out of Chaos* (Bantam, New York, 1984), p. 203.
2. Richardson, L. F., *Arms and Insecurity* (Boxwood, Pittsburgh, 1960); Richardson, L. F. *Statistics of Deadly Quarrels* (Boxwood, Pittsburg, 1960); Saaty, T. L., *Mathematical Models of Arms Control and Disarmament* (Wiley, New York, 1968).
3. Lanchester, F. W., in *The World of Mathematics* J. R. Newman, Ed., vol. 4 (Simon and Schuster, New York, 1956), pp. 2138-2157; Epstein, J. M., *The Calculus of Conventional War* (Brookings, Washington, DC, 1985).
4. Kaufman, W. W., in *Alliance Security: NATO and the No-First Use Question*, J. D. Steinbruner and L. V. Sigal, Eds. (Brookings, Washington, DC, 1983), pp. 43-90.
5. Saperstein, A. M., *Am. J. Phys.* **54**, no. 7, 607 (1986).
6. Saperstein, A. M., *Nature* **309**, 303 (1984).
7. Tuchman, B., *The Guns of August* (Dell, New York, 1962).
8. von Clausiwitz, K., *On War*, A. Rapoport, Ed. (Penguin, New York, 1968).
9. Schuster, H. G., *Deterministic Chaos,* 2nd revised ed. (Weinheim, FRG, VCH, 1988), Plates VIII-XV.
10. Gleick, J. *Chaos: Making a New Science* (Penguin, New York, 1987), p. 100.
11. Ott, E., *Rev. Mod. Phys.* **53**, 655 (1981).
12. Berge, P., Y. Pomeau, T, C. Vidal, *Order Within Chaos*, (Wiley, New York, 1986); Eckmann, J. P., *Rev. Mod. Phys.* **53**, 643 (1981); Thompson, J. M. T. and H. B. Stewart, *Non Linear Dynamics and Chaos* (Wiley, New York, 1986).
13. Grossman, S., and G. Mayer-Kress, *Nature* **337**, 701 (1989).
14. Smoke, R., *National Security and the Nuclear Dilemma*, (Addison-Wesley, Reading, MA, 1984); Beckman, P. R., L. Campbel, P. W. Crumlish, M. N. Dobkowski, S. P. Lee, *The Nuclear Predicament* (Prentice Hall, Englewood Cliffs, NJ, 1989).
15. Saperstein, A. M., in *Challenges and Opportunities From Now to 2001*, H. F. Didsbury, Jr., Ed. (World Future Society, Bethesda, MD, 1986). pp. 108 – 113.
16. Saperstein, A. M. and G. J. Mayer-Kress, "A systematic procedure for evaluating the impact of new security policies upon the maintenance of peace: SDI vs. Star Wars," *Proc. of the 30th Annual Meeting of the Society of General Syst. Research*, J.A. Dillon, Ed. (Philadelphia, Intersystems Publications, Salinas, CA, 1986), pp. L-100 – L-109; Saperstein, A. M. and G. Mayer-Kress, *J. of Conflict Resolution* **32**, 4, 636 (1988).
17. U.S. Congress, Office of Technology Assessment, *Ballistic Missile Defense Technologies*, UTA-15c-254 (Government Printing Office, Washington, DC, 1985); Garwin, R., K. Gottfried, H. W. Kendall, *The Fallacy of Star Wars* (Vintage, New York, 1984); Graham, D., *High Frontier* (TOR/Tom Doherty Associates, New York, 1983).
18. Saperstein, A. M., *J. Peace Res.* **24**, no. 1, 47 (1987).
19. Schroeer, D., *Science, Technology and the Nuclear Arms Race* (Wiley, New York, 1984).
20. Canavan, G., "Simple Discussion of the Stability of Strategic Defense" (Los Alamos National Laboratory Report P/AC-85-81); Canavan, G., H. Flicker, O. Judd, K. Taggert, "Comparison of analyses of strategic defense" (Los Alamos National Laboratory Report P/AC-85-27, 1985).
21. Garwin, R., *Nature* **315**, 286 (1985).
22. Crutchfield, J. P., J. D. Farmer, B. Huberman, *Phys. Rep.* **92**, no. 2, 45 (1985), p. 52.
23. Prigogine, I., and I. Stengers, *Order Out of Chaos* (Bantam, New York, 1984), p. 188.
24. Gray, C. S., and K. Payne, *Foreign Policy* **Summer**, 14 (1980).
25. Gordon, M. R., "Joint Chiefs are said to resist Pentagon move that would ease obligations under the ABM treaty," *New York Times*, 4 July 15 (1988); Velikhov, Y., R. Sagdev, A. Kokoshin, *Weapons in Space: The Dilemma of Security*, (MIR, Moscow, 1986).

15

A Nonlinear Dynamical Systems Approach to International Security

Gottfried Mayer-Kress

Abstract

Nonlinear dynamical systems with the possibility of deterministic chaos, bifurcations, and sensitivity to external perturbations and noise show many similarities to the behavior of social and political systems. Simple computer models can be used to study general trends and possible counterintuitive consequences of decisions that appear to be good solutions at the time they are made.

We discuss general properties of nonlinear dynamical systems and apply them to a simple arms-race model of three nations. In our model, alliances are formed as soon as one of the nations becomes dominant.

The policies of the three nations are represented by a set of 12 parameters, and we typically test several million different modifications thereof in order to gain information about its robustness.

We simulate the outcome of the arms race over several decades and identify critical political configurations in which external perturbations of a very low level can induce major changes in the political configurations. In a neo-realistic context, we interpret these phenomena as the appearance of sensitivities similar to those observed in chaotic dynamical systems. We test the stability of different scenarios numerically against external perturbations. We also briefly discuss the application of Chapman-Kolmogorov equations for a direct simulation of the dynamical systems under influence of noise.

Introduction

In an increasingly complex and interconnected world, it becomes more and more difficult to estimate all possible consequences of international policy decisions. Situations in the past in which shortsighted decisions led to unexpected developments are numerous.

As an example, we can consider the United States' (U.S.) decision to introduce multiple warhead missiles (MIRV). This decision was made based on an analysis of the given political situation, and it brought a short-term advantage to the U.S. The response of the Soviet Union (S.U.) – to build their own MIRV systems – changed the situation rapidly due to the larger launch capacities of Soviet missiles. A more careful simulation of possible Soviet responses might have prevented the U.S. from taking steps that, in the long run, would have turned out not to be in their best interest. The same conclusion certainly could have been achieved by traditional analysis; in fact, such a consistency check should be performed with every computational result when we want to "understand" the implications [for example, see *(27, 28, 45)*].

Another decision with potentially immense political implications is the choice to develop and introduce strategic defense systems (SDI). We have simulated possible implications of such a decision on the arms race in a previous, simple model *(44)*.

A third example with potentially global impact is the strategic arms reduction talks (START). The potential impact of START on the stability of nuclear balance has been discussed on a technical level by Feiveson and v. Hippel *(17)*, for example.

A significant reduction in the number of

strategic weapons controlled by the two superpowers will diminish the difference between the nuclear capabilities of the two major forces and smaller nuclear powers like China. The dominant and special role of the two superpowers might no longer exist. Experts in international political theory see the possibility of an increased risk factor due to more complex political configurations and the chance of a politics of changing alliances *(14, 16, 51, 54)*.

The theory of Ken Waltz (*31*) would qualitatively predict a destabilizing effect of strategic arms reduction when the two superpowers reduce to levels at which a third nation might be able to compete. A likely candidate for the position of third superpower could be China *(31)*. Through the coupling of military strength to economic factors, Japan might also be put into the role of a third superpower, especially in cases that include an increasing U.S. trade deficit. In a world with three superpowers and changed political systems (*glasnost*, etc.) the question of large alliances and coalitions might become very important. A specific analysis of such a situation in the context of START was also given by Garwin *(20)*.

Our model is based on the classical arms-race models, first introduced by L. F. Richardson *(39)* to simulate the competition between two nations. In a previous study of a two-nation, arms-race model *(22)*, we found that the predictions from these models are typically stable arms-control situations or unlimited arms race. We also found rare situations with chaotic and oscillatory behavior [see also *(43)*].

A first goal of our simulations will be to study, under a large variety of configurations, the possible emergent scenarios for future interactions. In the model we observe stable scenarios, which lead to stable three-party configurations. On the other hand, we can also identify model parameters under which such a transition would lead to destabilization, which can be identified as unpredictable states of crisis.

In the current model, we observe conditions under which alliances will change and where the political configuration would become very sensitive to small external perturbations. Such a situation can be interpreted as crisis. It seems, however, that in our simulations most of the mathematical surfaces that describe these transitions are well-behaved manifolds without the complex and fractal structure indicative of chaos.

In this chapter, we mainly want to discuss questions of how modern mathematical developments in nonlinear dynamics and chaos theory could possibly help us better understand the possible implications of political decision. As an example, we focus on a hypothetical scenario in which three nations, X, Y, and Z, try to reach their political goals, respond to threats, and are restricted by finite economic resources. We assume that nation X will then decide about its policies, depending on the current realizations of these (and/or other) factors. As a simple example, for which it is easy to transform the problem into a computational framework, let us consider decisions about military expenditure. Here the decisions are made according to such questions as, what are the military goals? How large are the military threats? What are the economic constraints?

Whatever decision X makes will then influence the future decisions of an adversary Y and ally Z. The subsequent decisions of Y and Z then, in turn, will change the political reality for nation X. Some of the political conditions might have the consequence of X or Z deciding to form a new alliance. The dynamical process induced through this decision cycle can lead to behavior that is not directly evident from a direct analysis of the original political situation.

Being in an alliance can bring enough advantages to one of the weaker nations that it eventually can become dominant. In that situation, we assume in our model that the weakest player switches over to form an alliance with the previous opponent, who is now number two. We are aware that this type of behavior is very idealized and represents only one simple possibility. For simulational purposes, however, it gives us a basis on which we can study the possible modes of interaction. It is very easy to modify the model and allow other types of interactions.

A typical phenomenon based on the static and linear paradigm of a traditional approach is the implicit assumption of rigid constraints: the main current argument against talks between the U.S. and the S.U. on a reduction in short-range nuclear forces (SNF) is the possible consequence of political pressure, which could lead to a removal of all nuclear weapons from central Europe. This argument rests on the assumption that the presence of nuclear forces in central Europe is a rigid constraint. On the other hand, it is clear that nuclear weapons are deployed to achieve a certain strategic or political goal. That means their presence is a tool to achieve a goal and not a necessity by itself. Thus, there might be political configurations (for example, strongly reduced conventional forces of the S.U.) under which a nuclear-free Europe might be acceptable.

We are not in a position to make any recommendation for or against SNF. The only point we want to make here is that it is important to keep goals and means to achieve goals conceptually separated. In a modeling context, this concept is implemented very naturally. We do not expect that it will be possible to use computer models to accurately predict or control the behavior of nations. Computers are not even able to predict the evolution of a game of chess. Even the best chess programs will lose against a grand master. But, on the other hand, there are many programs on the market that will be better than an average player, and some of them will even beat good human chess players. One could imagine that a "team" of a human player and a chess computer together should be better than either one alone.

In a world where political power is less dependent on military strength and political decisions are more influenced by environmental, economic, and population factors, the complexity of the global system can be expected to significantly increase. We believe that it will be possible to develop "computer-aided politics" models that will match and eventually surpass political analysis methods based on traditional approaches; numerical simulations might be able to illustrate possible and even likely implications of policy decisions. This can then be used to find optimal policies with the fewest unwanted side effects. Some of these aspects can already be found in simulational computer games like "Balance of Power" *(11)* and "SimEarth" *(53)*.

This situation in simulations of political systems is similar to the one in weather prediction. The famous model by Lorenz *(32)* is overly simplistic in its design, but it is helpful to understand universal features and sensitivities of weather-related phenomena. It also gives us new types of insight on the limitations of long-term weather prediction.

Although our approach analyzes a political system through the behavior of its components, the method is not reductionistic: one of the insights from nonlinear dynamics is that we cannot derive the outcome of an arms race in which several nations are involved from the simple decision rules for each of the nations *(6)*.

A very important aspect of a useful simulational model is a powerful user interface. If we are interested in the dynamical behavior of any complex system, then we want to have a direct interaction with the simulation and as little contact with raw numerical values as possible. Especially output from simulations should be presented in visual form (except in special cases where specific evaluations have to be made) for a direct identification of domains of interesting behavior.

Besides graphical output from extensive numerical simulations, it is often desirable to simulate specific histories of special interest. For this purpose, we have been working on interactive models in which the outcome of a simulation is presented through the coloring of political, military, economic, and other types of maps. Parameters can be changed interactively by pointing and clicking a mouse at various countries, for example, and adjusting parameters through mouse-operated sliders. As the model evolves according to the simulation, the coloring of the map changes and makes it possible to directly observe changes in the international configuration. This method will be especially useful to communicate with social scientists who lack solid mathematical background. Furthermore, it will make it possible in the future to use actual political and economic data from the wire services to update the model parameter as in a political "weather map" *(1)*. This method comes closer to the traditional tools of political and military decision makers who use maps and sand box-type simulational games.

In more detailed models, we can also take into account internal processes among hawks and doves within each society, which then will lead to more complicated mixed strategies of a nation, depending on which political orientation controls each government *(1, 2)*.

Socio-Political Models

If we believe that we can learn from history, then we believe that there must exist some general rules that govern socio-political behavior. In more or less explicit form, some of these laws have been discovered and applied in the historical, political, and social sciences. We also know that these laws are typically much more complicated than laws in the natural sciences. Simple causal relationships often do not apply, many apparently identical causes can have widely different effects, and many different causes can lead to the same situation. For any forecast, many conditions have to be taken into account, and there are many exceptions to any rule. Probably one reason why many social scientists reject any mathematical approach is that socio-political problems apparently cannot be treated successfully in a reductionistic framework.

In a classical physical approach, one criterion for a successful model would be to accurately reproduce any epoch in history. Such a criterion is not especially useful in the context of chaotic nonlinear systems. It is known that there exist very simple chaotic systems that are equivalent to a Bernoulli shift, which means they can reproduce any sequence of events, even those generated by a random coin toss.

Since historical events are not reproducible, we can assume that the course of history could be translated into a stochastic process. Individual sequences of events are meaningless within a theory or model; only statistical properties are relevant. Thus, we think that any socio-political theory has to contain some stochastic elements. In our case these are represented by random perturbations acting on the system's parameters. A simple consequence of this fact could be that more detailed and complicated models do not necessarily improve the forecasting accuracy significantly but make the interpretation more difficult and take more time for numerical simulation.[1]

In the theory of nonlinear and complex dynamical systems, features of "multicausality" and "sensitivity to initial conditions" are abundant: indistinguishably small differences in causes might lead to completely different effects [see *(12, 23, 34, 35)*, for example]. This is true in spite of the fact that deterministic mathematical equations are at work.

This sensitivity, however, is not restricted to chaotic states: a similar unpredictable behavior of a nonlinear system can be observed, for instance, close to "bifurcations." In the arms-race models that we have simulated so far, the typical behavior seems to be nonchaotic [see, however, the cases discussed in *(22, 43)*]: for most of the observed policy choices (realized through a specific numerical parameter combination), the models tend to develop towards an equilibrium point. Quite frequently, however, there seems to be a bifurcation, i.e., a situation in which small changes in parameters (corresponding to slight modifications in political goals) lead to quite different developments. For example, a small change in the way one nation responds to a perceived threat from the other nation can lead to a transition from an arms-control policy to an unbounded arms race. As such a critical situation is approached, we observe similar sensitivity and unpredictable responses to small perturbations as in the case of chaos.

These and other surprising features of

1 This phenomenon has also been observed in large-scale climate models *(6)*.

complex nonlinear systems give us confidence that socio-political behavior might be very complex but still could be (at least approximately) described and simulated in a modern mathematical and computational framework.

A different perspective of mathematical modeling is that it forces one to be quite explicit about one's assumptions and policy posture. Thus, even in the case where it turns out that a given situation cannot be accurately simulated by a mathematical model, we think that the process of constructing a model can sharpen our perspective on a difficult socio-political problem. We can expect that this approach might also make any hidden agenda more explicit and apparent.

For the simple case of a conflict between only two parties, several mathematical models have been developed to study the general features of arms races and conflict resolution. The first such mathematical model was introduced by L. F. Richardson *(39, 40)*, who simulated the arms race between two nations. In a recent generalization of the Richardson model *(22)*, we discuss possible applications of some modern nonlinear mathematical developments and the connections between arms-race dynamics and economic constraints. We found a fairly regular behavior of the system with a transition to an unbounded arms-race if certain parameters in the model such as "mutual defense coefficients" increase beyond a critical threshold value.

In the context of a more detailed model, A. Saperstein and I have studied the impact of a new weapons system (SDI) on the arms race between the U.S. and the S.U. *(44)*. We showed that a very fast buildup of strategic weapons can lead to unstable and unpredictable outcomes, whereas a modest buildup with a reduced number of warheads per missile can lead to augmented stability and predictability, even to a reduction of offensive weapons.[2]

Very little has been done so far with respect to three and more party interactions. The fact that constellations of three or more dominant nations can become much more complex and unstable than corresponding interactions between only two nations has been discussed in the political science literature *(29, 50–52)*.

We find some preliminary indication that our conjectures are correct and that conditions exist under which instabilities and changes of alliances occur. There is so far, however, no indication of deterministic chaos in any of the chosen parameter configurations. On the other hand, we have found a very astonishing sensitivity of the model's behavior to small perturbations (less than 1%) in certain, well-localized domains of parameters. We are inclined to interpret these domains as conditions under which slight provocations might launch major political consequences.

Similar approaches to understanding political decision-making processes, especially with respect to the evolution of cooperation, have been studied in the context of game theory, especially modifications of prisoner's dilemma [for example, see *(4, 5, 47, 52)*].

A main difference between our simulations and game theoretical approaches lies in the fact that we assume that decisions are made based on certain strategic objectives such as gaining a certain military strength or responding to external military threat. In a game theoretical context, the main driving force is a payoff or reward as a result of any move or decision. In many situations it would be very difficult to estimate a direct payoff from a certain decision. In a mathematical context, this means that for typical dynamical systems, a generalized potential or Lyapunov functional does not exist, i.e., the payoff concept is not applicable. In that sense our approach is more general.

The synthesis of game theoretic and

2 In that paper we followed the arguments in the public discussion and made the unrealistic assumption of a mutually assured destruction (MAD) strategy. We also restricted ourselves to the discussion of land-based ICBMs.

dynamical models might lead to considerable improvements, including both explicit and implicit decision-making processes. A very exciting combination of both approaches would be to introduce a hierarchical structure in which each move would consist of a full set of simulations of a dynamical model with a given policy. The payoff then depends on the predicted outcome of the chosen policy. For example, we could try to minimize the total arms expenditures under the constraints of sufficient deterrence.[3] In a way then, each choice of a policy (set of dynamical parameters) would correspond to a single move in a policy game and the payoff would be determined by how well this policy fulfilled the political goals. Models using machine learning and genetic algorithms might then simulate the evolution of different policies toward desired future goals *(18)*. On the other hand, one can consider iterated games as discrete time dynamical systems with a set of possible choices as variables *(52)*.

A very important approach to modeling complex systems is known as "systems dynamics" and is based on work by J. W. Forrester *(19)*. Systems dynamics also has been applied to the study of international politics and conflict [see *(8–10, 49)* for an extensive discussion]. We think that the main strength in the systems dynamics approach lies in the systematic analysis of the systems elements and their connections and interactions. Once the analysis is done, there are several ways of translating the system into a mathematical framework. The traditional approach is based on partial or ordinary differential equations — Richardson's original equations *(39)* are of that sort — types of equations with which scientists have been familiar since the times of Leibniz and Newton and before the advent of computers.

For several reasons we do not believe that partial or ordinary differential equations are the appropriate tools for socio-political modeling. Differential equations assume that the system undergoes continuous changes in time. This is a good approximation for a falling apple but not necessarily for the dynamics of arms procurement or any other decision process. In basically all such processes, one can identify a characteristic or shortest reasonable time unit: humans certainly cannot make more than a few decisions every second. The second argument is related to the form in which the simulations are performed in computers: modern computers are digital and, therefore, intrinsically discrete. Any simulation of a differential equation will be transformed into a difference equation. This is done by numerical integration routines. The degree of accuracy of these routines is directly related to their speed. Highly accurate routines take orders of magnitude longer to compute than regular discrete models.

One of the most popular software packages of systems dynamics uses simple Euler integration routines for the simulations in order to stay within realistic computation times. For the Euler method it can be rigorously shown, however, that even in simple examples, the numerical error can become so large that the outcome is not even qualitatively close to the correct result. Thus it appears that, in a socio-political context, differential equations are a useless and inappropriate detour. Instead of approximating an intrinsically discrete system into a continuous equation and then translating it back into a discrete algorithm, we can make a shortcut directly to discrete models.

For digital simulations, the ideal mathematical structure would correspond to cellular automata. There all states of the system as well as space and time variables are discrete. If we compare the coarseness of the different discretizations on a computer, then we see that typical spatial or temporal resolutions are of the order of 10^{-4}, whereas numerical accuracy, i.e., the discretization of states is many

3 Sufficient deterrence could be defined in terms of retaliatory capability. Another criterion might be the distance from a perfect "balance of power" in which the arms expenditures of the two opposing sides are exactly equal. In the search for such solutions, genetic algorithms prove to be extremely useful in producing many counterintuitive scenarios to an arms control situation with perfect balance of power *(18, 21)*.

orders of magnitude better, e.g., 10^{-15}. If we study bifurcations or other responses of the system to subtle changes, then we need the quasi-continuous nature of the state variables and parameters. This is not the case, however, when we allow for parametric noise of realistic amplitude.

Since it is very inconvenient to construct Richardson-type models as cellular automata, we think that discrete maps have very strong advantages over the two other methods. We hope that through sophisticated models we might reach a point in which decisions with the promise of a short-term advantage will be discarded in favor of those that bring us closer to a long-term sustained evolution.

Structure and Properties of Nonlinear Dynamical Systems

In physics we know that the dynamics of two coupled systems (e.g., planets, modes of fluid motion) can, in most cases, be solved analytically and are stable and predictable. If the number of participating subsystems is increased to three or more, however, we have a qualitatively new situation: the emergent behavior can become complex with interspersed regions of regular and irregular dynamics. Since for most of these nonlinear, chaotic systems it is impossible to find analytical (global) solutions in closed form, it was possible only with the availability of fast computers to study the behavior of these systems. Computational and "experimental" mathematics based on extensive numerical simulations has become a new field of nonlinear science that provides powerful tools for gaining insight into complex chaotic dynamics.

The basic types of nonlinear dynamical systems in our study consist of coupled nonlinear maps. That means we introduce an intrinsic time scale, which we assume to be given by the decision-making processes. Our model thus approximates models in game theory, where each move defines a discrete time step. Game-theoretic models can be seen as special types of discrete-time dynamical systems for which a specific payoff function exists and where the number of possible moves is finite. For example, in a dynamical systems approach, player A would respond to a threat by player B with an increase in the arms expenditure. In a game-theoretic context, the response of A could be "hawkish," "doveish," or a mixture of both; each type of response is assigned a certain probability. A "flexible response" is not as naturally contained in game theory as in dynamical systems.

In socio-political models, we are not primarily interested in the asymptotic dynamics but probably have to assume that the interesting dynamics in our models will be transient. This assumption appears to be realistic, since even in the case of relatively simple, nonlinear dynamical systems, the transients can become extremely long. Thus, we try to describe different hierarchies of time scales exhibited by the system, i.e., the fast variables (for example, the behavior of individuals) collapse down to a smaller dimensional (possibly fractal) invariant manifold containing the attractor. An attractor could be a common mode of behavior based on a tradition, fashion, or political climate, which typically changes slowly compared to individual changes in behavior. At a later stage, we want to study this hierarchical reduction of dimensionality with many interacting parties. This reduction can reoccur several times on increasingly longer time scales.

For the same reason of limited observation time, we cannot directly apply statistical concepts from the theory of nonlinear dynamical systems such as Lyapunov exponents and Kolmogorov *(36, 46)* entropies. We have developed a method of analyzing local divergence rates of solutions that asymptotically converge to Lyapunov exponents *(13)*. This allows us to derive several independent methods for predicting stability properties of the model at any given instance. Thus, we shall be able to identify sensitive regions in which small perturbations will cause major changes in the observed behavior. In the same way, we should be able to predict when the system is close to a bifurcation, i.e., major state changes.

Sensitivity Analysis of the Simulations

Once we have formulated all the dynamical interactions of the model, we can ask for equilibrium solutions, i.e., configurations whose statistics do not vary in time. In traditional approaches the analysis often was confined to the search for fixed-point solutions and their local stability analysis. From complex nonlinear dynamics we know that, first, the equilibrium solutions are not usually unique (many coexisting, equivalent scenarios achieve the same goal) and, second, traditional, local-stability analysis does not provide us with convincing evidence that a certain solution will be stable in a realistic context when finite amplitude perturbations are present. We have demonstrated the strong variability in noise sensitivity of different attractors in a general context *(34, 35)*. We have shown that small amounts of external noise can drive stable, nonlinear dynamical systems into chaotic states, e.g., noise-induced intermittency *(34)*. But we have also observed the opposite effect: small external perturbations (smaller than 10^{-4}) applied to certain chaotic systems can drive them into ordered fixed-point states *(35)* ("noise-induced order"). These new methods should be especially applicable to the study of socio-political problems.

We shall use a method of sensitivity analysis that we have developed for our SDI model *(44)*, which simulates the intrinsic uncertainties in socio-political processes. Since the parameters that describe the political environment as well as factors that influence political decision making are never very accurate, classical deterministic approaches to modeling have been extremely limited and unreliable. With the help of high-speed supercomputers it is possible to simulate political uncertainties and their possible consequences for the degree of accuracy of predictions from the model. In the simulation we determine a set of roughly plausible parameter values that attempt to define and describe the socio-political situation. In this part we have to rely on input from experts in the social and political sciences. Second, we need an estimate of the accuracy with which these parameters can be determined and weighted. Then we evaluate the dynamics of the model for a specific time, relevant for the problem under consideration.

In the model discussed here, we take decision units as time steps, e.g., years in the case of national budgets. Next, we repeat the same calculations for the model but with slightly different sets of values for the model parameters, chosen at random within the uncertainty range. Then we repeat the simulation for a whole ensemble of these perturbed configurations of times (on the order of 10^3 to 10^7). In previous simulations, these statistical methods have not been used for basically two reasons: (i) the simulations were mainly continuous in time, which makes the computations very slow, and (ii) most of the simulations were performed under a linear paradigm, i.e., under the assumption that deviations from the correct model would, in their effects, add up in a linear fashion. This implies that it is sufficient to study the impact of perturbations in the individual factors separately. It is well known, however, in the context of nonlinear dynamical systems that this assumption is generally wrong: parameter regions for which we have a given type of behavior can have a fractal boundary [see, for example, *(33, 37)*], i.e., even in the case that small perturbations in individual parameters do not cause significant changes in the system's behavior, small perturbations to a combination of the parameters might cause a synergistic amplification of the effect and result in a transition to a new, qualitatively different behavior.

The mathematical, theoretical, and numerical tools are becoming available now to make such a modeling effort valuable for advisory purposes in politics. Some of the most important new concepts are (i) the theory of synergetic processes, which describes how in complex systems unspecific changes of external stress parameters can lead to self-organized, coherent global behavior *(23)*; (ii) the theory of chaotic dynamical systems, which shows how regular and ordered behavior of a complex system can undergo transitions to unpredictable chaos and vice versa *(46)*. This theory allows us to analyze the erratic dynamics of complex systems and optimum means of controlling them; and (iii) the evolution of computer algorithms that can "learn" and "adapt" to new situations that the original programmer never could have envisioned at the time the model was developed *(18, 21)*.

On the hardware side, we see the emergence of highly parallel computers which will greatly exceed the computational capacity of even the largest supercomputers today. The problem of interacting individuals, groups, or nations appears to be extremely well suited for the new technologies; therefore, we think it is very important to start pioneering work today to develop the theoretical foundations for global security models. Incidentally, many of the technologies and software developments funded for the generation of a strategic defense initiative will become very well suited for the type of modeling described above.

Nonlinear Models and the Linkage Problem in Negotiations

The classical scientific method, which entails abstraction in the sense of isolating a subproblem from its context, also has become a major paradigm in diplomacy. In negotiations among nations, it has become a rule that no "linkages" with other negotiations or treaties are admitted. Otherwise, the risk of arriving at a stalemate is too high when the topics are too complex. The justification for such a restriction seems to be obvious. The human brain is able to store and process a limited amount of information, especially if the elements are interconnected in a complex manner.

On the other hand, it is clear that treaties are not formed in a vacuum. Tactical nuclear weapons are deployed with the justification of a conventional imbalance. Strategic forces are coupled to conventional forces in a doctrine of flexible response. Also, negotiations on confidence-building measures will have an influence on the perceived threat from the adversary and vice versa.

Terms like "synergy" or "synergetic effect" are used to describe these "nonlinkage" interactions. This gives us evidence that diplomats have discovered nonlinearity. It is clear that, taking into account these nonlinear connections, it becomes virtually impossible for negotiators to keep track of all possible implications of each individual proposal or suggested treaty plan.

A brute force simulation of a large-scale battle on a "microscopic" level (considering individual units with their detailed technical specifications) apparently is of limited use; the run of a large model often takes too much time and is too unreliable in its predictions, since it is very difficult to test its robustness.

We propose using a nonreductionistic approach to investigate nonlinear dynamical models (possibly taking into account some parameters obtained from the outcomes of specific war games simulations) that describe the synergetic system of factors that constitute international relations. This approach is also reflected in the nonformal way of a "new thinking," which emphasizes mutual security, common problems, international security, and sustainable development.

We are convinced that the scientific basis for this new thinking is given in nonlinear dynamics and chaos theory. We propose to extend war games based on national security interests to a synergetic approach to international security (peace games). We also expect that we will be able to find a common language and basis for a new type of diplomacy in which computer-literate negotiators study a variety of scenarios to learn which serves best for the national and common interest of all parties.

Numerical Simulation of a Simple Three-Nation Generalized Richardson Model

Numerical models of arms-race dynamics based on Richardson-type equations have not been very successful in predicting real political situations *(41)*. Probably this is true for any theory or general method that attempts to predict political behavior. In general, for every example in history in which a certain theoretical model applies, there is a counterexample that contradicts the theory. Therefore, we do not try to reproduce or predict individual, specific histories. (In a sense, we take here a position similar to the "many world" interpretation of quantum mechanics.)

We consider here the dynamics of three nations, X, Y, Z, which are competing according to a generalized version of Richardson's equations *(22, 39, 40)*. The expenditures of each country for strategic weapons during year n is denoted by x_n, y_n, z_n measured as a fraction of the GNP, i.e., all values x_n, y_n, z_n lie between 0 and 1. The decision of each country to modify their arms expenditure depends on three different factors: first, the self-armament level (quantities with index s: x_s, y_s, z_s), which is the level of armament each country wants to achieve independent of the policies of other hostile or friendly nations. These goals are approached at a rate given by parameters k_{11}, k_{22}, k_{33}. The second factor determining the arms expenditure of a country is the perceived threat from other countries. The perceived threat not only increases when hostile nations spend more money on arms but also when allies spend less. So we take the overall perceived threat to be the difference between the expenditures of the enemy minus expenditures of the allies. The rate at which a country responds to external threats is given by coupling or response parameters k_{23} (rate for country X in response to the levels of countries Y and Z), k_{13}, k_{12}, similarly. Finally we know that economic factors have to be taken into account which, in turn, are influenced by military expenditures.

In order to account for this coupling in the simplest way possible, we include the influence of the economy as a limiting factor: if the economical resources x_m, y_m, z_m (measured in percent GNP) are much larger than the amount spent on armaments (in year n this is x_n, y_n, z_n), then the constraints are small. If all or most of the economic resources are spent in the military budget, then the flexibility in the response to changing threats is significantly reduced or goes to zero.

Up to this point we have assumed perfectly rational decision makers acting according to these simplified criteria for arms procurement. There are, however, many factors involved in the procurement decisions that are not accounted for in our simplistic model (or any arbitrarily sophisticated model for that matter). We incorporate these unknown influences most efficiently if we make the assumption of maximal ignorance: if we cannot recognize any rule or law, then we treat these factors as unpredictable random perturbations ξ_n of magnitude σ. The specific distribution of the noise fluctuations seems not to be very important for the response of the system *(34)*. Thus, we choose for the simulations identically distributed, uncorrelated, random numbers with standard deviation σ. These random fluctuations can either be applied directly to the total amount spent on armaments or, perhaps more realistically, to the policy decision parameters in the models, i.e., to the quantities k_{ij}, x_s, x_m, etc. Typically these Monte Carlo-type simulations are very computer-time intensive. Instead of simulating individual histories and then taking the statistics, it is also possible to write down an explicit and deterministic integral equation for the time evolution of the probability distribution *(24, 30)*. This could be especially useful in analyzing possible amplifications of perturbations in extremal directions over a short time span. This method is, however, computationally limited to moderately high numbers of noisy parameters.

Without loss of generality, we can assume that $x_n > y_n > z_n$. The rule for alliance formation is chosen such that the two weaker nations, Y and Z, team up against the strongest power, X. This seems to be the most realistic and interesting case. If the strongest nation has an ally, this alliance dominates the action completely, and we have basically a two-nation model. This means that X perceives the joint armament of Y and Z as threat; whereas

Fig. 1. A two-dimensional slice through the 12-dimensional parameter space of the model through the standard parameter set. The slice consist of 64 × 64 = 4096 time histories over T = 100 time steps (years). The x-y coordinates correspond to the range of parameter values of the model. (In this example, the parameters k_{11} is displayed on the x-axis and the parameter z_s on the y-axis. The z-coordinate represents the arms expenditures x_t of country X after a certain number of time steps (ten, for example) starting from an initial configuration of zero expenditure for all countries. The gray coding corresponds to the relative strength of different countries: bright = X is strongest (i.e., $x_t > y_t$ and $x_t > z_t$), medium = Y is strongest, dark = Z is strongest. The boundary between regions of different shades of gray corresponds to a change in the alliance configuration, a case that we associate with a major crisis.

the threat for Y is reduced by the arms expenditure of Z and vice versa.[4] For this situation, the discrete Richardson-type equations with economic constraints are given by

$$x_{n+1} = x_n + [k_{11}(x_s - x_n) + k_{23}(y_n + z_n)](x_m - x_n)$$

$$y_{n+1} = y_n + [k_{22}(y_s - y_n) + k_{13}(x_n - z_n)](y_m - y_n)$$

$$z_{n+1} = z_n + [k_{33}(z_s - z_n) + k_{12}(x_n - y_n)](z_m - z_n)$$

The model parameters,

$$p_n \in \{k_{ij}, x_s, y_s, z_s, x_m, y_m, z_m, \text{ for } i,j = 1, 2, 3; i \leq j\},$$

have stochastic time dependence, i.e., we have $p_n = p_0(1 + \xi_n)$. Let us assume the generic case that the expenditures of each country are different. The alliance (and therefore the model) changes, according to the rule mentioned above, every time the leading nation is replaced by one of the previous alliance partners.[5] For a sample simulation, we choose a set of standard parameters at *(22)*.

$$(k_{11}, k_{22}, k_{33}, k_{23}, k_{13}, k_{12}) = (1, 1, 1, 0.7, 0.7, 0.7)$$

$$(x_s, y_s, z_s, x_m, y_m, z_m) = (0.1, 0.1, 0.1, 1, 1, 1).$$

From this set of parameters, we explore a neighborhood of policies in basically two different ways: by making two- and three-dimensional slices through a cube around our standard parameter set and by exploring a noise cloud around the core by perturbing all relevant parameters simultaneously by bounded random influences as described before.

In our model, four policy parameters must be specified for each of the three nations such that we have a total of 12 parameters for each simulation. In our parameter search, we have performed two-dimensional cuts through the 12-dimensional parameter space on a 64 × 64 grid, i.e., for each slice we simulated 4096 parameter scenarios. This parameter search was performed by systematically varying the parameters by 20% around a given set of standard parameters. We obtain surfaces that represent the arms expenditure of one of the three nations. A color coding scheme indicates the rank of the given nation with respect to the two other competitors (Fig. 1).

We observe large areas of robust and stable behavior — e.g., when the self-arma-

4 The actual (wall clock) time period associated with our time index n depends on how fast each of the nations can respond to a change in the perceived threat from the opponent. For simplicity reasons, we assumed that n is measured in "years."

5 In the spirit of L. F. Richardson we use the arms expenditures in the current year as criterion for the intensity of the arms race, not the accumulated amount of armaments. Should the accumulated amount of armaments be considered, the equations would become more complicated since factors like aging of weapons systems and maintenance would have to be treated explicitly.

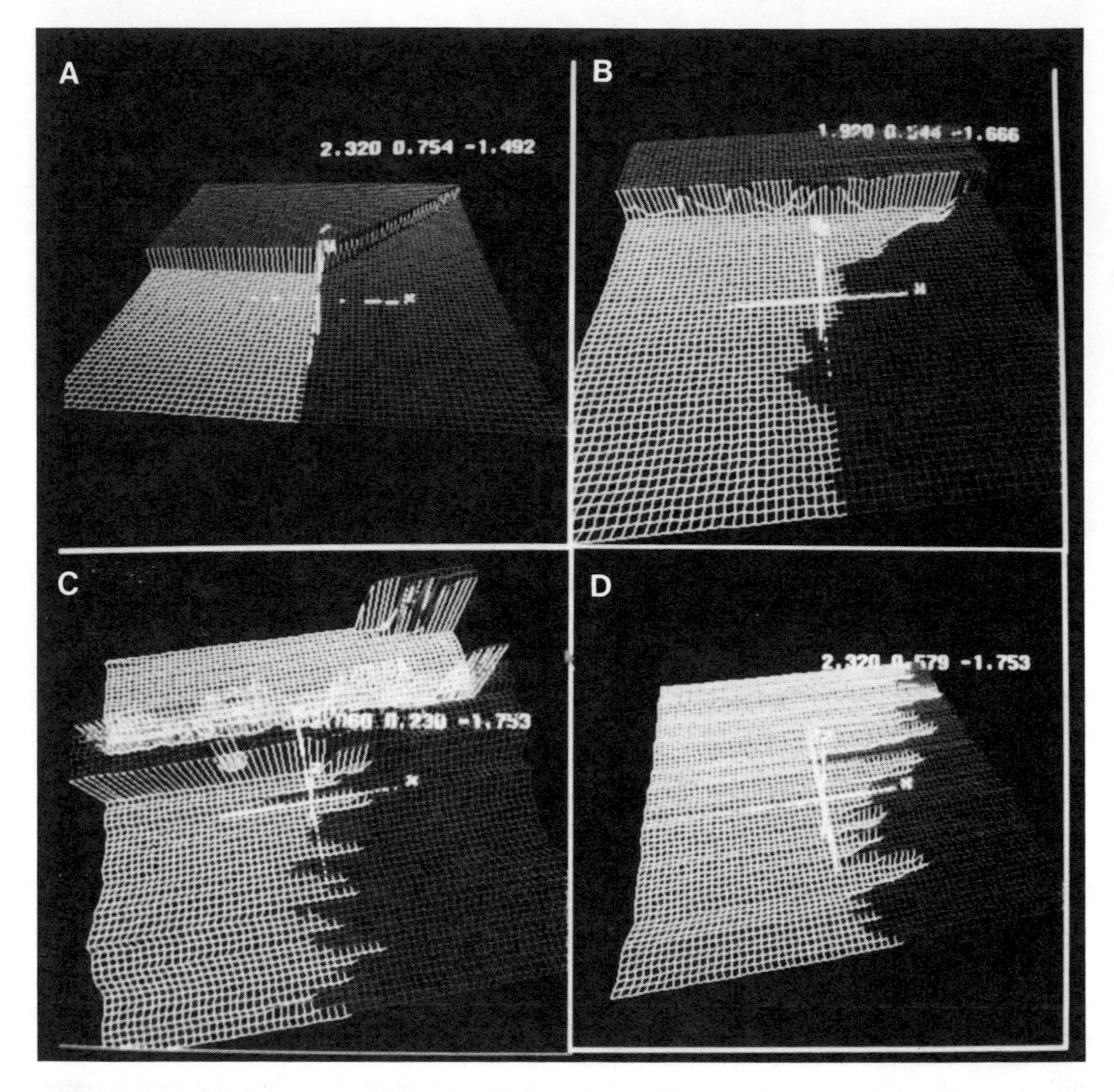

Fig. 2. Effect of low-level perturbations in the system parameters on a slice in the k_{11}, z_S plane. (**A**) no noise, $\sigma = 0$ (this corresponds precisely to the surface of Fig. 1). (**B**) noise, $\sigma = 0.001$ averaged over N = 1000 time histories, (**C**) $\sigma = 0.003$, N = 1000 (**D**) $\sigma = 0.003$, N = 5000

ment level of a country is low — when it only spends a relatively small amount for armament in the absence of external threats, for example.

The interpretation of the graphical representation of the transition surfaces in a three-dimensional cut through parameter space becomes much clearer in an interactive visualization on the computer screen or on a video when the surfaces can be viewed from different directions. With the help of interactive simulation/visualization tools, it is possible to interactively move through the multidimensional parameter space, exploring boundaries of stability regions.

Associated with these transitions is an increase in sensitivity to small external fluctuations in the parameters that we associate with uncertainties in policy decisions, as manifested in the uncertain outcome of voting behavior. In Fig. 2 we demonstrate the effect of low-level noise ($\leq 1\%$) on the structure of the transition surfaces. We observe that the clear-cut transition regions become more and more

Fig. 3. Location of the set of all initial points within the unit cube ($0 \leq x, y, z \leq 1$) that come to within $\varepsilon < 0.01$ - distance of a fixed point located at $(x^*, y^*, z^*) = (0.431, 0.236, 0.236)$ after $n = 20$ iterations with standard parameters.

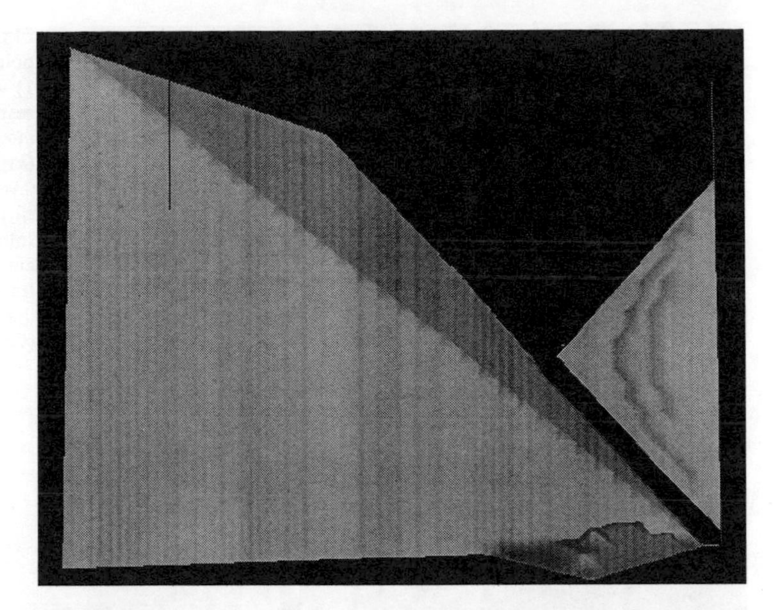

fuzzy or "fractal." This is a very typical phenomenon for nonlinear dynamical systems. At the same time, we can still identify large regions (away from the original transition surfaces) of robust behavior.

For the examples that we have studied, it seems that some configurations disappear after averaging over several thousand similar scenarios, which we would interpret as lack of robustness. Where we saw transitions from one alliance to an other, we now see that under the influence of perturbations only one type of alliance is strongly favored. Transition to deterministic chaos could only be found in nontypical small regions for this model.

Whereas we keep the initial conditions, x_0, y_0, z_0, at a fixed value for the simulations in Figs. 1 and 2, it is also very common that the final outcome of the simulation depends sensitively on the starting values. That means that for nondegenerate parameter values there exist multiple attractors in the unit cube. These are separated by two-dimensional basin boundaries that can have fractal structures. We can approximate these manifolds by locating all points that come within a given distance of the fixed point after a given number of time steps (Fig. 3). These iso-surfaces undergo rapid changes close to basin boundaries *(7)*.

The structure of these basin boundaries varies dramatically, depending on the position in parameter space, i.e., the political configuration. Smooth surfaces with only small transition regions would correspond to very robust configurations. The extreme example of such a configuration was found with the help of a genetic algorithm, in which the parameter space is searched for configurations that satisfy specific goals like perfect balance of power between the opponents [see *(18)* for a detailed discussion]. There the solutions all converge to a single fixed point, independent of the starting value within the unit cube. As we modify parameters such that the arms-control equilibrium with perfect balance of power is destroyed, we observe small-scale chaotic attractors and a very fragmented, possibly fractal basin boundary, i.e., a surface with a dimension between 2 and 3 (see Figs. 4 and 5).

Future Perspectives

For models to have direct impact upon daily political decisions, fast and accurate access to global parameters relevant to global security is required. With the coming availability of electronic news services and further increases in computer memory and processing speed, we think the concept of a global political "weather map," where information about critical political situations will be immediately avail-

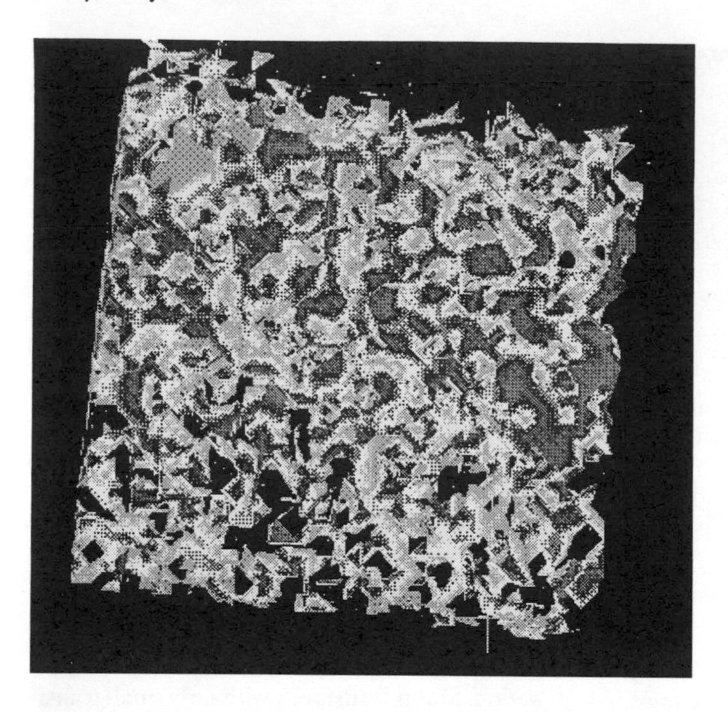

Fig. 4. Location of the set of all initial points within the unit cube ($0 \leq x, y, z \leq 1$) which converge to an attractor corresponding to nation Z being dominant after $n = 20$ iterations with parameters: $(k_{11}, k_{22}, k_{33}, k_{23}, k_{13}, k_{12}) = (0.44, 0.59, 0.96, 0.50, 0.97, 0.77)$ $(x_s, y_s, z_s, x_m, y_m, z_m) = (0.44, 0.98, 0.29, 0.68, 0.50, 0.18)$. Intertwined with these basin boundaries are surfaces which correspond to different types of solutions (X dominant, Y dominant, or unbounded arms race). [Courtesy of T. Ohsumi, see *(48)*.]

Fig. 5. Same as in Fig. 4 but with parameters: $(k_{11}, k_{22}, k_{33}, k_{23}, k_{13}, k_{12}) = (0.44, 0.59, 0.96, 0.50, 0.97, 0.77)$ $(x_s, y_s, z_s, x_m, y_m, z_m) = (0.44, 0.98, 0.21, 0.68, 0.50, 0.18)$. [Courtesy of T. Ohsumi, see *(48)*.]

able to the public, will become feasible. Especially with these new information technologies, it will become realistic to have fast and increasingly accurate updates of estimates of model parameters, which allows a steady improvement and evolution of models. Specifically, we can extract trends about the relative strengths and influences of individual countries to estimate which three or more nation configuration will be the dominant factor for global security. We see our efforts as paralleling current developments in new information exchanges between the two superpowers, which, according to the *New York Times* (27

July 1988), are "an effort to prevent the kinds of miscalculations that have historically been the prime cause of war."

Because of the universality of these mathematical methods, one could also apply the formalism to studies varying from the stability of family structures to those of worldwide corporate competition. (It is interesting to note that the only family structures that have proved stable are either monogamous relationships or those in which there exists one dominant parent with several nondominant adult family members.) It might also be interesting to compare the behavior of our model simulations to the dynamics of political party coalitions in pluralistic societies with different numbers of political parties *(3, 25, 42)*.

This project will contribute to the understanding of the general mechanisms underlying the dynamics of striving for dominance, as is epitomized by the arms race between superpowers. As part of the study, new tools will be developed for identifying, analyzing, and predicting critical configurations in which the probability of the outbreak of violent aggression or war are drastically increased.

Acknowledgments

I would like to thank R. Abraham, B. Arthur, B. Ault, A. Beyerchen, D. Campbell, G. Cowan, N. Choucri, S. Doughton-Evans, F. Dyson, P. Ford, S. Forrest, M. Koscielniak, R. Rice, A. Saperstein, and M. Simmons for many stimulating discussions and helpful suggestions. I also would like to thank A. Keith and J. Milton for allowing me to use their scientific visualization codes and Y. Kevrekidis and M. Jolly for the use of scigma to find the fixed points of the model and for their hospitality during my stay at Princeton University where part of this work was finished. I also would like to thank T. Ohsumi for providing Figs. 4 and 5 from his computational mathematics project *(48)*. I am also grateful to P. Ford, R. Mainieri, H. Peck, S. Pope, and N. Rapagnani for their help in computer matters. Most of the simulations were performed on the Cray supercomputers of the Los Alamos National Laboratory and the Air Force Weapons Laboratory in Albuquerque.

References

1. Abraham, R., *IS Journal* **2**, no. 2, 27 (1987).
2. Abraham, R., *IS Journal,* in press.
3. Abraham, R., A. Keith, M. Koebbe, G. Mayer-Kress, "Double Cusp Models, Public Opinion, and International Security," *J. Bifurcations and Chaos*, in press.
4. Axelrod, R., *The Evolution of Cooperation* (Basic Books, New York, 1984).
5. Brams, S. J., *Superpower Games* (Yale University Press, New Haven, CT, 1985).
6. Campbell, D. K., and G. Mayer-Kress, *Nonlinearity and Chaos: The Challenges and Limitations of Mathematical Modeling of Environmental and Socio-Political Issues,* Proc. Conf. on Technology-Based Confidence Building: Energy and Environment. *CNSS Papers* no. 22 (July 1989).
7. Challinger, J., "Interactive Graphical Exploration of the Three-Nation Richardson Model," Project paper, Math-206 (University of California, Santa Cruz, 1990).
8. Choucri, N., *International Interactions* **1**, 63 (1974).
9. Choucri, N., and R. C. North, *Nations in Conflict* (Freeman and Co., San Francisco, 1975).
10. Choucri, N., and T. W. Robinson, Eds., *Forecasting in International Relations* (Freeman and Co., San Francisco, 1978).
11. Crawford, C., *Balance of Power* (Mindscape Inc., Northbrook, IL, 1990).
12. Crutchfield, J., J. D. Farmer, N. H. Packard, R. S. Shaw, *Sci. Am.* **254** no. 12, 46 (1986).
13. Dressler, U., G. Mayer-Kress, G. Mitagunta, "Local Divergence Rates in Nonlinear Dynamical Systems," in preparation.
14. Dittmer, L., *World Politics* **33,** 485 (1981).
15. Eisenhammer, Th., A. Hübler, G. Mayer-Kress, P. Milonni, "Resonant Stimulation of a conservative nonlinear oscillator: classical and quantum-mechanical calculations, Los Alamos preprint, LA-UR-90-1911, Submitted to *Phys. Rev. A.*
16. Fader, P. S., and J. R. Hauser, *J. Confl. Resolut.* **32,** 553 (1988).
17. Feiveson, H.A., and F. v. Hippel, *PU/CEES Research Report 234* (Princeton University, Princeton, NJ, March 1989).
18. Forrest, S., and G. Mayer-Kress, in *The Genetic Algorithms Handbook* (Van Nostrand Reinhold, New York, in press).
19. Forrester, J. A., *World Dynamics* (Wright-Allen

Press, Cambridge, MA, 1971).
20. Garwin, R., *Bull. Atomic Scientists* **10** (March 1988).
21. Goldberg, A., *Genetic Algorithms* (Addison-Wesley Publishing Co., Reading, MA, 1989).
22. Grossmann, S., and G. Mayer-Kress, *Nature* **337,** 701 (1989).
23. Haken, H., *Advanced Synergetics* (Springer, Berlin, 1983).
24. Haken, H., and G. Mayer-Kress, *Z. Phys. B-Cond. Matter* **43,** 183 (1981).
25. Holsti, O. R., and J. N. Rosenau, *J. Confl. Resolut.* **32,** 248 (1988).
26. Hübler, A., and E. Lüscher, *Naturwissenschaften* **76,** 67 (1989).
27. Intriligator, M. D., and D. L. Brito in *Arms Races: Technological and Political Dynamics,* N. P. Gleditsch and O. Njolstad, Eds. (Sage Publishers, London, 1990).
28. Intriligator, M. D., and D. L. Brito in *Dynamic Policy Games in Economics,* F. van der Ploeg and A. J. de Zeeuw, Eds. (Elsevier, North Holland, 1989).
29. Keohane, R. O., *International Security* **13,** 169 (1988).
30. Koebbe, M., J. Krause, G. Mayer-Kress, to be published
31. Lin, C.-P., "China: Nuclear Wild Card," *New York Times,* 29 July 1988.
32. Lorenz, E., *J. Atmos. Sci.* **20,** 130 (1963).
33. Mandelbrot, B., *The Fractal Geometry of Nature* (W. H. Freeman, New York, 1982).
34. Mayer-Kress, G., and H. Haken, *J. Stat. Phys.* **26,** 149 (1981).
35. Mayer-Kress, G., and H. Haken, *Physica* **10D,** 329 (1984).
36. Mayer-Kress, G., Ed., *Dimensions and Entropies in Chaotic Systems,* Springer Series in Synergetics, vol. 32 (Springer-Verlag Berlin, Heidelberg, 1986).
37. Mayer-Kress, G., *Phys. Bull.* **39,** 357 (1986).
38. E. Ott, C. Grebogi, J. Yorke, *Phys. Rev. Lett.* (1990).
39. Richardson, L. F., *Statistics of Deadly Quarrels* (Boxwood, Pittsburgh, 1960).
40. Richardson, L. F., *Arms and Insecurity* (Boxwood, Pittsburgh, 1960).
41. Roten, C. D., and J. M. Orient, *Nature* **312,** 107 (1984).
42. Rosen, L., "The Role of National Consensus in the Enhancement of National Security," *CNSS Briefs* (Los Alamos, 11 June 1987).
43. Saperstein, A. M., *Nature* **309,** 303 (1984).
44. Saperstein, A., and G. Mayer-Kress, *J. Confl. Resolut.* **32,** 636 (1988).
45. Saaty, T., *Mathematical Models of Arms Control and Disarmament* (John Wiley & Sons, New York, 1968).
46. Schuster, H. G., *Deterministic Chaos, An Introduction* (Physik-Verlag, Weinheim, 1984).
47. Smale, S., *Econometrica* **7,** 1617 (1980).
48. Ohsumi, T. K., Project paper, Math 207 (University of California, Santa Cruz, 1990).
49. Väyrynen, R., Ed., *The Quest for Peace* (Sage Publications, London, 1987).
50. Walt, S. M., *The Origin of Alliances* (Cornell University Press, Ithaca, NY, 1987).
51. Waltz, K., *Theory of International Politics* (Addison-Wesley Publishing Co., Reading, MA, 1979).
52. Wendroff, B., "On Cooperation in a Three-Way Arms Race," Los Alamos Preprint LA-UR-90-931 (1990).
53. Wright, W., and J. Kalin, "SimEarth: The Gaia Concept," demonstration at Santa Fe Institute.
54. Yorke, H., Talk given at the Center for National Security Studies, Los Alamos National Laboratory (July 1988).

16

The Dimensions of Cosmic Fractals

Reuben Thieberger, Edward A. Spiegel, Leonard A. Smith

Abstract

Chaotic behavior is often described by *flows*, which are vector fields in appropriate spaces. Vector fields, such as the magnetic field in the sun, are important in astrophysics, and they are typically chaotic. Thus, when these fields are solenoidal, they display a mixture of regular and chaotic portions that qualitatively rationalize the spotty appearance of the solar magnetic field. A less tangible example is the flow that can be imagined to sweep galaxies forward in cosmic time. The evidence suggests that this flow forms a fractal object whose lacunae or voids are analogues of the stellar spots.

The textures of these objects will help the cosmologist in understanding some large-scale dynamics. We describe and assess the cosmologists' schemes to quantify the galactic distribution by determining its dimensions. Theoretical sets and a set of galaxy positions measured at the Observatory of Nice provide the illustrations. We find that the data suggest a value close to 2, which is rather larger than the dimension of the distribution of galaxies than has been generally accepted. The methods described for finding such results may be helpful in other astrophysical studies involving complex sets.

Cosmic Chaos

Reliable identification of chaotic behavior in astrophysical systems is usually much more difficult than for controlled laboratory systems. In astrophysics as in any *observational* science, it is hard to develop the extensive data sets needed to detect chaos. It is also difficult to isolate the phenomena of central interest without contamination from other processes in observational sciences.

Consider pulsars, for example. The flashes seen from pulsars are believed to emanate from polar caps of rotating neutron stars. The rotation rates of these stars decrease slowly with time, but the rates of decrease normally fluctuate slightly. To decide whether these fluctuations are deterministic, we need a good series of measures of the rotation rate. Those might be obtained from the blinking light from the stars' polar caps, which are thought to arise like flashes from light houses. But the neutron stars are intrinsically inconstant, like most stars. They may well emit their light chaotically; hence it is hard to say from the observations whether the rotational braking, or the emission process, or both are chaotic.

Even though it is difficult to establish the existence of deterministic temporal chaos in astrophysics, the basic ideas of chaos theory are helpful in astronomical thinking. They can rationalize apparently aperiodic behavior that we encounter in the universe as deterministic processes. Astromathematicians no longer need to invoke the *deus ex machina* called noise to understand aperiodicity. In chapters 17 and 18 by Buchler and Regev in this volume, one sees examples of how astrophysicists are incorporating the ideas of temporal chaos into their work.

Beyond this, the ideas of chaos theory can influence our vision of the mathematical objects with which we describe the contents and structure of the universe. Chaos theory helps us to cope with irregular temporal behavior; it provides us with ways to think about complexity in spatial structure; it provides tools for quantifying complicated behavior. This chapter is about some of the issues of such quantification. We shall describe methods for determining the dimensions of fractal sets and illustrate their use with theoretical sets and an astrophysical example.

Chaotic Vector Fields

Erratic temporal behavior is but a superficial manifestation of chaos. Astrophysicists can avail themselves of other, far-reaching aspects of chaos theory. For example, the property of being chaotic can apply to mathematical structures such as vector fields, not only to temporal behavior itself. This is a useful notion, and it is easy to grasp.

The trajectory of a particle is described by giving its position, x, as a function of time, t. Such information often emerges in the solution of a differential equation of the form

$$\dot{x} = v\,(x, t). \tag{1}$$

Here the vector field v is a specified *flow* that carries the particle about. Equation 1 is called a dynamical system by mathematicians. Fluid dynamicists might think of the vector field v as the velocity field for the flow of a real fluid. For them though, the content of Eq. 1 is kinematic, and it describes how a fluid particle moves. A trajectory of Eq. 1 would be called a *streak line* in fluid dynamics.

Suppose that we are working in a three-dimensional (3D) space, as for ordinary physical flows. A way to visualize the nature of a trajectory $x(t)$ is to cut the space with a plane into portions called left and right. Whenever the trajectory crosses this plane, going from left to right, for example, we mark the place. If a particle moves for a long time, its path will typically pierce the surface of section quite a few times, given a certain amount of judgment in choosing this *Poincaré surface*. If you put the plane where the orbits are plentiful, you will see something like what you see on cutting open a coaxial cable.

In learning to use the Poincaré section, it is well to practice on simple examples. For a periodic orbit, the surface of section may consist of just one point because the particle keeps coming back to the same place. Indeed, so long as the set of marked points is finite, it must always be the trace of a periodic orbit. Other simple situations may arise. For example, suppose that the orbit is wrapped on the surface of a torus. The points in the surface of section will then lie on a closed curve. Orbits with such simple spoors are often called regular.

When the series of points in the surface of section does not repeat, go off to infinity, nor remain on a simple closed curve, we must have a case of highly complicated motion. The points often do not fill the surface of section nor any piece of it, so they may be said to form an object of fractional dimension, as we shall explain presently. The motion in such cases may be called chaotic. Since the motion is dictated by v, we may consider that v itself is the chaotic object. The notion that chaos is basically the property of a vector field is of importance in many disciplines. For astrophysics, a telling example is the solar magnetic field.

In a certain sense, what we see of the solar magnetic field is a surface of section. Where the field is strong, it inhibits the convective motions. Near the solar surface, such motions are responsible for the outward flux of heat. Hence, magnetic inhibition of this flow makes for relatively dark regions called sunspots. So the places where strong ordered fields protrude from the solar surface are distinguished by being darker than their surroundings.

To understand this as a surface of section, think of any snapshot of the solar magnetic field $\mathbf{B}(x, t)$ as a flow as in Eq. 1, with a fictitious particle moving along some trajectory. The parameter that tells us how far along in a trajectory of the field the particle has gone is, of course, not the real time, for that is fixed. Rather, we introduce a fictitious time, s, increasing along the trajectory of a particle in the flow $\mathbf{B}(x, t)$. Then the position of the particle, for fixed t, traces out a path governed by the equation

$$\frac{dx}{ds} = \mathbf{B}\,(x(s), t). \tag{2}$$

Such trajectories of $\mathbf{B}$, for fixed t, are called the *streamlines* of $\mathbf{B}$.

We can make a surface of section for the streamlines of $\mathbf{B}$ and get some idea of its topology in this way. And there is no reason we should not use a spherical surface to make our surface of section. Then we can hope to get something that looks like the solar surface for a reasonable choice of **B**. The regular parts of the field correspond to flux tubes that would appear as sunspots, while the chaotic portions resemble the general field over the solar surface.

Another way to think theoretically in terms of a surface of section is to use a return map. That is, instead of following a trajectory round and round waiting for it to cross the Poincaré surface, we can derive or invent a rule that tells us where the system will next cross the surface, given the location of its previous crossing. Such a rule saves us the effort of having to solve a differential equation like Eq. 2 to understand the structure of the field.

Although we do not know how to make a solar return map (as yet), we can anticipate the real thing by simply looking at typical return maps that have been constructed in the study of Hamiltonian chaos. (We specify Hamiltonian chaos because magnetic fields are divergence-free, so we need to use the so-called area-preserving maps of Hamiltonian dynamics.) Such maps generally produce distributions of magnetic field that look qualitatively quite right. From such maps, we typically get intense concentrations into spots of very regular field surrounded by a background chaotic field. The spots in this theoretical image correspond to flux tubes of the basic vector field, at least in this imagery *(29)*. In Fig. 1 we show a section of a simple area-preserving return map invented by M. Hénon

In studying such maps, without going into complicated physical processes, we can see the possible qualitative structure of solar fields. That is, chaos theory, with no appeal to any physical processes, predicts that spots ought to exist for any generic magnetic field other than the highly special cases that are constructed for classrooms and certain carefully designed devices. Without this realization, we might have been tempted to give special credence to models that predict a field whose surface of section is spotty. But since this aspect of vector fields is generic, we realize that any sensible theory ought to predict sunspots. The theory of chaotic vector fields suggests, moreover, that such surfaces of section ought to have spots of smaller and smaller sizes. What we need to look for are the quantitative aspects of such fields as a test of our theories, as indeed some people have been doing *(30)*.

Similar general conclusions may change our picture of other cosmic arrangements such as the spatial distribution of galaxies. It is now believed that the luminous portions of galaxies contain only a very small fraction of the total mass in the universe. Therefore, a galaxy is like the test particle whose motion is described in Eq. 1. The flow, $\mathbf{v}$, in that equation is presumably determined by the invisible mass of the universe, but the way in which this happens may well surprise us if we can unravel it. Naturally, to be a bit more precise, we should write the four-dimensional version (at least) of Eq. 1 for galaxies. Then what we would be observing is a surface of section made with a space-like hypersurface, which for present purposes is a surface of (nearly) constant time.

Observational limitations lead us to think differently about the construction of such surfaces of section in cosmology than in more tractable examples. Instead of a galaxy going round and round on a single trajectory, we adopt an approach like that used to visualize flows by fluid dynamicists. We start out with a uniform distribution of particles and let them all run according to Eq. 1. Then we mark the points where they cross a particular space-like hypersurface. That is, we take a photograph at some moment. If the cosmic flow is generic, we naturally would not be surprised to see a fractal distribution of the test particles or galaxies. This is, indeed, what the observations indicate. Again, we appreciate that a theory has to do more than just predict this outcome if it is to command our respect.

Cosmic Cascades

When a swarm of points is swept along by a flow, we shall see the points rearrange themselves as we look at successive time-like cuts of

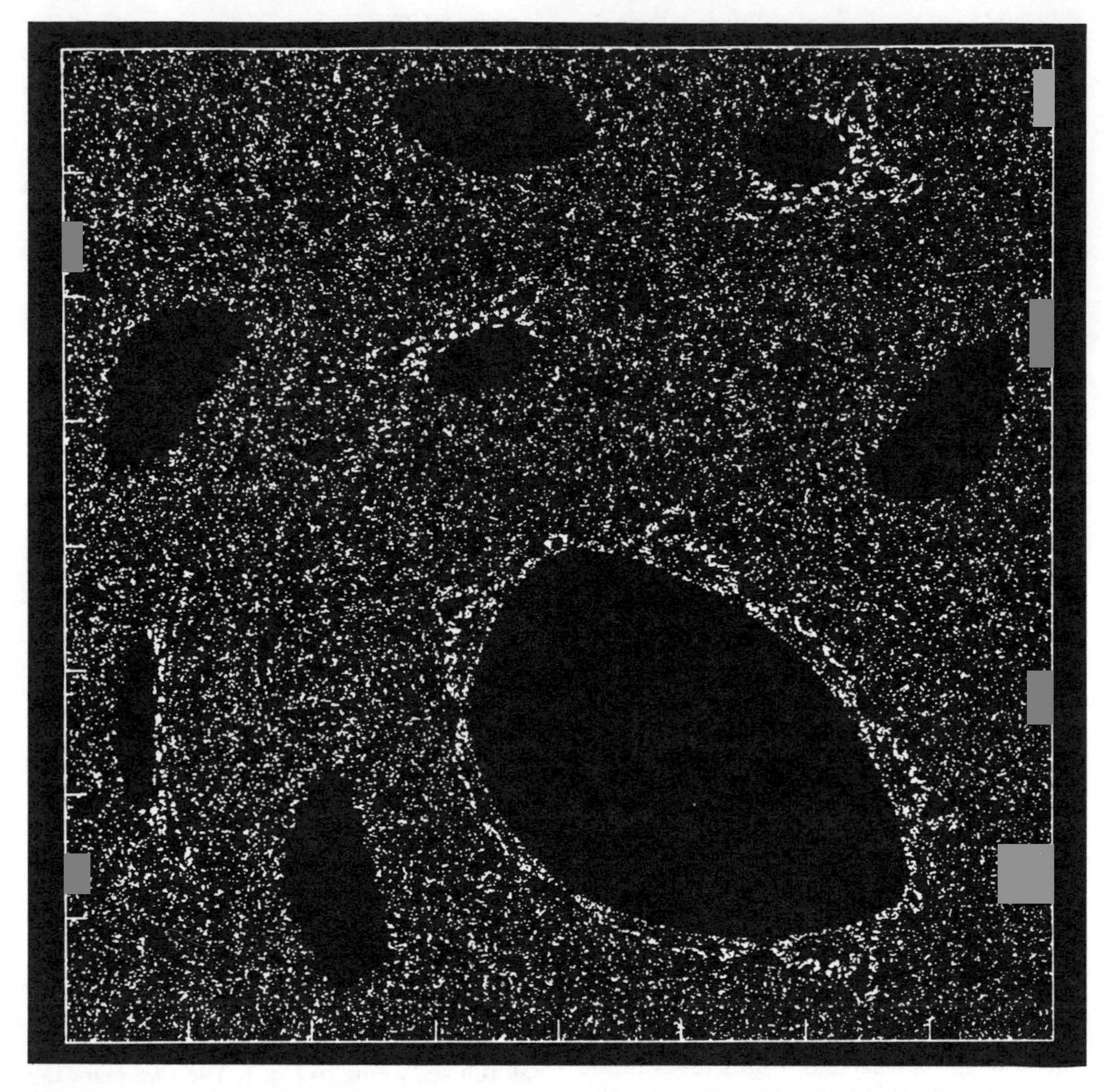

Fig. 1. A chaotic trajectory from the area preserving *Hénon map* of chaos theory [e.g., *(29)*]. The dark areas are the integrable or regular regions of the field, which we liken to flux tubes.

their world lines. If the flow has chaotic parts, we can see the emergence of a hierarchical structure from an initially rather uniform configuration. This begins with a slight redistribution of points leading to the first traces of clustering on a particular scale. In the simplest cases, these first clusters occur either on the largest or the smallest scale allowed by physical or geometrical constraints of the real world.

In a flow where the first inhomogeneities appear on the largest scales, we first see large clusters appear. Within these, the distribution of points is initially rather uniform. These first clusters of points will behave like the initial set, and clusters will form within them. In the simplest possible cases, a self-similar hierarchy of clusters will emerge from this geometrical cascade. On the other hand, if the initial clusters are formed at the smallest scale allowed by the physics, we may see them coming together to form larger structures that cluster together in their turn. The continuation of this latter process is called an inverse cascade. If the initial scale is determined by some forcing extrinsic to the process of interest, we may get cascading in both senses at once.

This way of thinking about how a flow moves points about is often much more convenient than trying to deal with the flow directly. Mandelbrot *(17)* considers the way that Cantor first made his hierarchical sets as a form of cascade. Students of turbulence have used such thinking for years, beginning with the work of Novikov and Stewart *(20)*. Unno *(31)* sees cascades throughout natural history, referring to Hoyle's *(11)* discussions of the hierarchical formation of cosmic bodies in the astronomical case. Although the mathematical rationalization for this way of thought is still in its infancy, Kida *(12)* has used a model equation for the cascade of fluid turbulence related to the so-called Kolmogorov-Ford equation of statistics. The same kind of modeling may be helpful in the cosmological cascade as well.

Consider a great cosmic ball of matter subject to the Hubble expansion of the universe. As cosmologists often do, let us think of the motion of the galaxies in comoving coordinates, in which the uniform Hubble expansion has been subtracted out of the flow. With the appearance of the first large-scale inhomogeneity in the cosmic fluid, we get our first generation of objects (clumps, aggregates, groupings, or what you will). These cosmic lumps are objects with characteristic scale l_1. In the ball there are N_1 such entities.

In the following stage, the N_1 objects create the next generation. Suppose that the first generation spawns a total of N_2 second generation objects of size l_2. If we let this process run its course through a great many generations, we shall have a basis for writing an equation for N_n, the number of objects in the nth generation. Let the creation of objects of the nth generation take, on average, a time τ_n. Then, we can write a so-called master equation of this form:

$$\frac{dN_n}{dt} = -\frac{N_n}{\tau_n} + \zeta_n \frac{N_{n-1}}{\tau_{n-1}}, \qquad (3)$$

where ζ_n is the mean number of offspring produced by the objects in the $(n-1)^{th}$ generation.

The meaning of the first term on the right in this equation is that the members of the nth generation are destroyed with a characteristic lifetime τ_n. If that were the only term, we would find that $N_n \propto \exp(-t/\tau_n)$. But the dying members of each generation may be used to make the next generation. That is the meaning of the second term on the right of Eq. 3. To use this equation, we have to provide appropriate formulae for ζ_n and for τ_n. In the simplest cases,

$$\zeta_n = \left(\frac{l_{n-1}}{l_n}\right)^3. \qquad (4)$$

The exponent 3 in this formula indicates that the objects are embedded in a three-dimensional space. Kida sought greater generality by replacing the 3 in Eq. 4 with a parameter, s. It appears that he wanted the freedom to choose s, not an integer, in modeling some features of turbulence. As we see next, the outcome of Sreenivasan's measurements on turbulent flows suggest that 3 is a good choice (see the appendix to this volume for references).

Suppose that we have statistically steady turbulence. Then, if we apply the steady state version of Eq. 3, we have

$$\frac{N_n}{\tau_n} = \zeta_n \frac{N_{n-1}}{\tau_{n-1}}. \qquad (5)$$

For the example of homogenous turbulence, Kolmogorov *(13)* has provided a similarity theory that gives us $\tau_n \propto l_n^{2/3}$. With this result, we can find a *particular* solution of Eq. 5 with $N_n \propto l_n^{-7/3}$.

To extract the meaning of this result, we observe that the total mass M_n in the nth generation is (to within a constant factor) $\rho_n l_n^3 N_n$, where ρ_n is the mass density of the typical object in its generation. In laboratory turbulence, the motion is effectively incompressible, so ρ_n is a constant. If the mass involved in the cascade were conserved, M_n would be constant and independent of n. This would require that $N_n \propto l_n^{-3}$. In that case, the objects in any generation would fill a finite volume. Instead, we have $l_n^3 N_n \propto l_n^{2/3}$, so the volume occupied by each generation is tending to zero with decreasing l_n. This means that less and less mass is participating in the cascade as we go down through the generations. There must be *lacunae* where the cascade has been arrested.

We could even improve on the exponent of 7/3 in the solution. Kolmogorov's formula for the average lifetime of an eddy is implicitly based on the assumption that the eddies are space-filling. Now that we have the first estimate of the departure from this condition, we can correct the expression for τ_n and get an improved estimate to replace the 7/3. However, that estimate is close enough for our present purposes, so we shall not pause to do this here.

The fact that the total volume occupied by the smallest of the objects goes to zero has an interesting interpretation. If, in ordinary space, a set of points fills a piece of that space and so has a finite volume, we would say in ordinary language that this object is three-dimensional. When the volume of the object vanishes, we are tempted to quantify this by assigning a dimension to the object that is lower than 3. We shall be going further into this in "Dimension," pages 206–208. For now, let us call the exponent in the power law for N_n the dimension of the set of objects. As we shall see, there are many ways to define the dimension of a set of points, so this is really just one particular dimension; perhaps we could call it the cluster dimension. It resembles a dimension discussed by Unno *(31)*.

In the case of turbulence, the dimension of the set of points on which the dissipation is appreciable in the experiments is quite close to 7/3 as Sreenivasan has found. This encourages us to think that Eq. 5 may provide some useful insights into other cascades. We are therefore tempted to try it out on the cosmic cascade. But before doing that, let us remark that there is a more general solution to Eq. 5 in a geometrical cascade, when the ratio of successive sizes is a constant, that is, ζ_n does not depend on n. In that case, we may reasonably suppose that

$$\tau_n = const.\, l_n^{\kappa}, \tag{6}$$

where the parameter κ is characteristic for the particular cascade. Then we may readily verify the existence of the following solution to Eq. 5:

$$N_n = (l_n/L)^{-d}\Psi\,[\log\,(l_n/L)], \tag{7}$$

where

$$d = 3-\kappa, \tag{8}$$

and

$$\Psi\,(x) = \Psi\,(x + \log\beta), \tag{9}$$

where $\beta = \zeta_n^{1/3}$ and L is a unit of length.

The function Ψ measures the *lacunarity* of fractals. It generalizes Mandelbrot's definition of the term as a constant prefactor in the statistical description of fractals. Instead, we may call Ψ a prefunction. So far, it remains an arbitrary function, found only by observing real fractals. The ripple described by Ψ can also be found in inhomogenous fractals or multi-fractals as some call them. This ripple has long been noticed both in mathematical *(18)* and physical *(23)* theories involving hierarchies and cascades. In connection with his early model of the turbulent cascade, Novikov *(19)* pointed out that the ripple would occur in turbulence in principle, but would not be detectable. However, it may have been seen in turbulent flows *(28)*. It will take many more data than we have now, or a new way to analyze them, to see the ripple in statistical analysis of the distribution of galaxies.

To formulate a cascade model for a cosmic process like the creation of the hierarchical structure of the galaxy distribution *(32)*, we need to look harder at Eq. 3. First, we have to depart from the usual imagery of fluid turbulence, in which eddies of a given size, l_n, are normally considered to be effaced after a typical time, τ_n, when they give rise to smaller eddies. In the usual situation of fluid turbulence, we need an energy source to obtain a statistically steady state. In the cosmic example, we need to allow for the possibility that the eddies persist for much longer than this decay time. That is, they may have produced the next generation without being destroyed themselves. If the memory of the early generations can linger on in the cosmic structures, we can expect to observe the coexistence of objects on all scales. We want to suggest a way of doing this that seems suited to some versions of cosmic clustering theory.

If only the first term on the right of Eq. 3 existed, with no feeding from previous generations, we would find that N_n decays exponentially. And for turbulence, this makes good sense. The linear parts of the turbulent processes with constant coefficients do give an exponential behavior like this. For any process where such exponential behavior is expected, Eq. 3 is likely to be helpful. A measurement of d will give a value of κ, which will give clues to the basic mechanism causing the cascade. But the universe is expanding, hence it is time dependent. Therefore, describing the cosmic cascade calls for a modification of this feature of Eq. 3.

Some theoretical pictures of the cosmic cascade suggest that the decay of individual structures in the Hubble flow may be algebraic rather than exponential. For example, a density anomaly in a self-gravitating medium grows as a power of t, as in Lifshitz's well-known work *(21)*. Then we ought to try a master equation, which in the absence of cascading, gives $N_n \propto t^{-\alpha}$. Such behavior comes from an equation like

$$\tau_n \frac{dN_n}{dt} = -N_n^{\mu},$$

where $\mu = (\alpha + 1)/\alpha$ and τ_n is the time constant for the nth generation. This version mollifies the sudden-death, exponential aspect of Eq. 3. When we also include a cascade term, we get this equation:

$$\tau_n \frac{dN_n}{dt} = -N_n^{\mu} + \zeta_n N_{n-1}^{\mu}, \qquad (10)$$

if we assume that the exponent in the decay rate is the same for each generation.

Suppose that galaxies were formed in a burst of activity some billions of years ago. They are then approximately frozen into the Hubble flow. Though clustering is still going on, we shall be cavalier and look at the equivalent of Eq. 5 for this modified situation. That is, we look at the steady-state version of Eq. 10, $N_n^{\mu} = \zeta_n N_{n-1}^{\mu}$. In the case of a geometrical cascade, we get the solution Eq. 7 again, with the condition Eq. 9. But this time, for the dimension derived from the size distribution, we get $d = 3\alpha/(\alpha + 1)$ instead of Eq. 8.

According as β is greater or less than unity, we have a cascade or an inverse cascade. That is, either big objects spawn smaller ones, or small ones gather together to make bigger ones. Both possibilities have been considered for galaxy formation, but the appearance of β in the size distribution $N_n(l_n)$ may help settle this matter by observational means. Unfortunately, the ripple in the sizes implied by the periodic nature of Ψ is hard to detect without abundant and accurate data, and that is why it is so little discussed. But de Vaucouleurs *(32)* has suggested that a size ripple indeed exists in the clustering of galaxies. His interpretation of the observed clustering of galaxies is interesting, and it suggests a value of β in excess of unity. If that is correct, it implies that a direct cascade is responsible for the cosmic clustering. This naturally needs confirmation, but already we see how observations together with the new developments in mathematical thinking may lead us to some significant astrophysical conclusions, if we can solve the calibration problems. In perceiving evidence of a size ripple in the galaxy distribution, de Vaucouleurs has, in effect, descried an indication of the existence of the cosmic voids *(14)*.

Even more can be learned from comparing Eq. 7 to observations. If the hierarchical arrangement of galaxies is gravitational in origin, then we can say something about the possible values of α. Those values are model dependent and may be different according as the universe is closed or open. In particular,

$$\alpha = \frac{d}{3 - d}. \qquad (11)$$

Hence the clustering dimension, d, will give us rather direct information on global cosmic properties.

Although this is the simplest possible kind of theory, it serves our purpose well, for it brings out nicely how parameters of the distribution of galaxies in space actually can emerge explicitly from theoretical calculations. Next, we want to explain how one goes about getting such parameters, especially the dimension. We begin that task by reviewing some background on statistical physics.

Statistical Tools and Methods

We have spoken of sets of points that can be observed in astronomy such as the points in which the solar magnetic field pierces the solar surface or the way galaxies lie in space at fixed time. We need ways to characterize such sets of points, both to describe the flow fields that lie behind their arrangements and to test theories that we hope will bear on such questions. If we know too little about such details, we can fall back on a less detailed kind of discussion, as in "Cosmic Cascades," pages 200–204, and use some statistical property of the sets such as the distribution of cluster sizes. In simple cases, a statistical distribution of points may be characterized by one or two parameters, including the dimension. In fact, there are many statistical quantities that have been used to extract dimensions from cosmic fractals. In this chapter, we introduce some of the main statistical descriptors of fractal sets that may be distilled down into the dimensions of the sets, along with one or two other parameters such as the period of Ψ.

To describe some of the statistical methods that may have use in cosmic applications, we shall suppose that the basic notion of a space of dimension D is already intuitively plain when D is an integer. We shall consider a set of points such as idealized galaxies, embedded in such a space of integral dimension D. When the set is generated numerically or observed, the total number of points in the set, J, is finite; ideally, J is very large. For convenience, we can label the points by an index, j = 1, 2,..., J. The data of the problem are the coordinates, $\mathbf{r}_j$, in the space to which the J points belong. We shall assume that the embedding space is Euclidean and the coordinates are Cartesian.

When dealing with a set of points such as one pictures for the atoms in a gas or liquid, one of the first quantities that one tries to determine is the density. This may at first seem a simple thing to work with, but it becomes elusive in lacunar sets such as we are now discussing. In these cases we have a problem underlined in cosmology by de Vaucouleurs *(32)* almost twenty years ago: "What precisely do we mean by average density? What is the evidence to support the notion that a mean density can be defined?" In short, we shall have to watch out. We shall simply use physical intuition to avoid the pitfalls of this topic, for we cannot here enter into the mathematics called *measure theory* that is appropriate to the general discussion. [For an intermediate approach, see *(16)*.]

Thus forewarned, we draw a spherical surface of radius λ around any point in the set. This spherical surface is called S^{D-1}, in our space of integral dimension D. (This space is called R^D.) Following current usage, we shall refer to the interior of the sphere as a ball. It is useful to make this distinction between the spherical surface and its interior, though it is not always done in ordinary scientific discussion when the meaning is clear from the context. We let V (λ) be the volume of the ball of radius λ whose surface is S^{D-1}.

Now, to define the number density function $n^{(1)}(\mathbf{r})$ we count the number of points in the ball centered on the point at $\mathbf{r}$. Designate that number N (λ). For a continuous medium, we would define the density $n^{(1)}(\mathbf{r})$ as the small-λ limit of N (λ)/V(λ). The awkwardness in this is like the one we face in defining the density of a real fluid composed of atoms. In that case, as here, we dare not take the limit of λ going all the way to zero, when N(λ) is not large. For the fractal sets that we are dealing with in the cosmological case, the inhomogeneity on all scales is so great that $n^{(1)}(\mathbf{r})$ is an even more awkward quantity to deal with. Because of the voids in the galaxy distribution *(14)*, this definition is not only sensitive to λ, the radius of the ball; $n^{(1)}$ also depends on $\mathbf{r}$, even for a statistically homogenous system. This is what led Pietronero *(22)* to suggest the introduction of other statistical quantities into the description of the galaxy distribution.

Just as $n^{(1)}(\mathbf{r})$ gives the density of points at $\mathbf{r}$, there is a function $n^{(2)}(\mathbf{r}'', \mathbf{r}')$, which gives the density of *pairs* of points separated by Δ =

$\mathbf{r}' - \mathbf{r}''$. To get $n^{(2)}(\mathbf{r}, \mathbf{r} + \Delta)$, suppose that there is already a point in the set at $\mathbf{r}$. Then, $n^{(2)}(\mathbf{r}, \mathbf{r} + \Delta)$ is the number of points whose displacement vector from $\mathbf{r}$ lies inside a sphere of (small) radius δ, centered on $\mathbf{r} + \Delta$, divided by the volume within the sphere of radius δ, and multiplied by the density at $\mathbf{r}$.

As in statistical physics, we define the *pair distribution function* $\tilde{g}$ through the relation

$$n^{(2)}(\mathbf{r}', \mathbf{r}'') = n^{(1)}(\mathbf{r}')\, n^{(1)}(\mathbf{r}'')\, \tilde{g}(\mathbf{r}', \mathbf{r}'' - \mathbf{r}'). \quad (12)$$

If no point in the set is special, we may assume that the pair distribution function has this form:

$$\tilde{g}(\mathbf{r}', \mathbf{r}'' - \mathbf{r}') = g(\mathbf{r}'' - \mathbf{r}'). \quad (13)$$

The density at $\mathbf{r}''$, given that a point is already at $\mathbf{r}'$, is therefore $n^{(1)}(\mathbf{r}'')g(\mathbf{r}'' - \mathbf{r}')$.

We have already said that the idea that no points are special is risky for complicated distributions of points. A point at the edge of a large void may be different than other points in some ways. But now, we have pushed our use of the ideas of homogeneity back one level to the relations between pairs of points, and we shall see where this gets us. It is reasonable to adopt the convenience of choosing a particular origin, at $\mathbf{r}'$, for example. Then the conditional density becomes $n^{(1)}(\mathbf{r}'')\, g(\mathbf{r}'')$ and Eq. 12 becomes

$$n^{(1)}(\mathbf{r}'')\, g(\mathbf{r}'') = \frac{n^{(2)}(\mathbf{0}, \mathbf{r}'')}{n^{(1)}(\mathbf{0})}. \quad (14)$$

In the same spirit, let us consider the sets in which there are, on average, no preferred directions. We may then assume that the dependencies in this expression do not involve angles. Hence there is a function $\Gamma(r)$ such that

$$n^{(1)}(\mathbf{r})\, g(\mathbf{r}) = \Gamma(r), \quad (15)$$

where r is the magnitude of $\mathbf{r}$. The function Γ is equal to that introduced by Pietronero *(22)*, albeit defined slightly differently. Γ may be used to define a dimension of a set of points, as we shall see in "Dimension," page 206. But other, more popular, statistical moments have been widely used.

The correlation is a very frequently studied object in all sorts of statistical investigations. This is a standard quantity defined, for example, by Peebles *(21)* for cosmological purposes. In the study of fractals, the related correlation integral *C(r)* is used by Grassberger and Procaccia *(8)*, defined as the

$$\lim_{J \to \infty} J^{-2} \cdot [\text{number of pairs of points } (i,j)],$$

such that the separations between the members of each pair, $|\mathbf{r}_i - \mathbf{r}_j|$, is less than r. They call C an integral because it is the integral over the ball of radius r of the correlation itself. We mention this as an intuitive way to think about the correlation, whose usual definition may seem less vivid. In statistical physics, this definition is

$$\xi(r) = \frac{n^{(2)}(\mathbf{r}', \mathbf{r}'')}{n^{(1)}(\mathbf{r}')\, n^{(1)}(\mathbf{r}'')} - 1, \quad (16)$$

where $r = |\mathbf{r}' - \mathbf{r}''|$. But this brings out the relation to Γ, whose use in the study of fractals has been urged by Pietronero.

Finally, we conclude this small selection of statistical tools for analyzing sets of points by mentioning the ways that cosmologists have confronted the problem of limited data samples in calculating statistical quantities like ξ. A good example is provided by a scheme described in Peebles' book *(21)* on cosmology; this scheme is used in many discussions of galaxy distributions. A brief discussion may help in assessing the results under discussion.

To illustrate the use of this amelioration scheme, we describe it for the pair correlation determination. The first step, of course, is to derive the distances between the J^2 pairs of points in the set. Let $N_r^{(2)}$ be the number of such pairs with separation in the interval $[r - dr, r + dr]$. Points too close to the edge of the set may be excluded *(6)*. Next generate a random set with the same boundaries and the same number of points as in the data set. If we combine this set with the data, we can calculate the distance between a randomly selected set of J^2 pairs in the combined set. In doing this, we always choose the first point from the data and the second from the random set. Let $M_r^{(2)}$ be the number of such pairs, with separation in the interval $[r - dr, r + dr]$. We then calculate

$$\Xi(r) = \frac{N_r^{(2)}}{M_r^{(2)}} - 1 . \tag{17}$$

This quantity is the ameliorated pair correlation function.

We may use similar amelioration methods to determine a pair distribution function. A fiducial random background set may be used as described. We are then able to calculate

$$\Gamma(r) = n_0 [1 + \Xi(r)], \tag{18}$$

where n_0 is the integral of Γ over the ball of radius r.

Another way to reduce the problem caused by finite data sets has been to use coarse-graining. For example, some workers use a coarse-grained Γ defined by *(6)*

$$\Gamma^*(r) = V^{-1} \int_V \Gamma(r')\, dr', \tag{19}$$

where V is the volume within a sphere of radius r. But problems remain, whether we use $\Gamma(r)$ or $\Gamma^*(r)$; the finite sets that arise from observations present practical difficulties.

Perhaps, methods like these have been used because the occurrence of significant fluctuations in plots of various statistical moments vs. r were thought to be mainly a result of insufficient data samples. The amelioration methods tend to wipe out such fluctuations. Although small samples may contain spurious fluctuations, *real fluctuations* are also inherent to the statistical moments of fractal sets. As we saw in "Cosmic Cascades," the ripples in such moments contain valuable information. Hence, we should stress that in describing these amelioration methods we are not advocating them. Indeed, we hope that soon there may be sufficient cosmological data to permit the dropping of amelioration. If we sometimes will use it here, it is to facilitate comparison with current work in cosmology.

Dimension

In the early stages of development of a subject, one prefers to paint with a broad brush. So it is useful to try to characterize the statistical distributions of points by a few parameters such as their dimensions. There is more than one kind of dimension that can be associated with a set of points, but for the simplest sets, the various dimensions that are conventionally defined are normally not very different in value. The evidence is that the cosmic fractals are not too far from this simplest state, so we need not go into the finer points of dimension determination. Nevertheless, it is best to begin with the most direct method, though it may not be the easiest one to carry out.

To describe some of the dimensions that may have use in cosmic applications, we continue to suppose that our set is embedded in a space of integral dimension D. We shall reserve the symbol D to denote the dimension of the set of points under study such as the points in a surface of section.

An operational definition of the dimension of the set of J points, assuming the J is as large as is required, is provided by the following simple procedure. We recall that $N(\lambda)$ is the number of points contained in S^{D-1}, the surface of the sphere in D dimensions, and introduce the quantity $\langle N \rangle(\lambda)$, which is the average of N over the points in the set. If we measure this quantity over a good range of values of λ, we can make a reasonable determination of the dependence of $\langle N \rangle$ on λ. For these results, we make a plot of log $\langle N \rangle$ vs. log λ. We shall call the slope of the line that best fits this plot (allowing for the intrinsic ripples) the fractal dimension of the set.

This prescription makes the notion of fractal dimension operational, but it may not provide a very convenient way to determine the dimension. First of all, there are pitfalls when we deal with the limitations of real data, as detailed in the various books on the subject. In particular, the plot is limited in the available scales of λ, above by the size of the system and below by the limit of resolution. Furthermore, the best fit is normally not a line, but a line with superposed wiggles, as we can anticipate from the discussion of "Cosmic Cascades," pages 200–204. Inevitably, other definitions of dimension have been proposed, and these are based on asymptotic properties of various statistical moments of sets of points.

After all, the quantity ⟨N⟩ is a statistical moment on the set, and there is no reason why dimensions based on any other moments, such as those described in "Statistical Tools and Methods," pages 204–206, could not be defined. Renyi actually did just that and Halsey *et al.* *(10)* have described the usefulness of some of these dimensions and their interrelations. In recent years, much attention has been devoted to the distinction among these generalized dimensions for inhomogeneous fractal sets, or multi-fractals.

We may also base the definition of dimension on the conditional density function, Γ. This quantity has the units of number density. So the quantity $\int_{V(r)} \Gamma(r) d^D r$ is analogous to the number of points in the ball. For small r, we may expect it to grow like r to a power; that power could be used to define a dimension. However, an advantage in using a moment like Γ to define a dimension is that we avoid doing the integral. For power laws, the integration simply raises the exponent by D. So it is natural to define a dimension based upon Γ in this way:

$$D_\Gamma = D - \lim_{r \to 0} \frac{\log \Gamma(r)}{\log r}. \tag{20}$$

In evaluating D_Γ, we have to make allowances for the lack of resolution, owing to the finiteness of the data sample. In practical terms, we should think of Γ as having an expansion like

$$\log \Gamma = -(D - D) \log r + \cdots \tag{21}$$

for small r. The dots stand for higher terms in the expansion. We expect the first of such higher terms to describe the ripple associated with hierarchical processes *(27, 28)* or fluctuations associated with the finite data sample. For self-similar structures we can readily see why, in the following manner.

Self-similarity of a structure implies that statistical moments are scale invariant. If $C(r)$ is any such moment, we take self-similarity to mean that there are constants A and B such that

$$C(r) = AC(Br). \tag{22}$$

This functional relation has the solution

$$C(r) = \Psi(\log r)\, r^{-D_C}, \tag{23}$$

where $D_C = \log A/\log B$ and Ψ, like its counterpart in Eq. 7, is an arbitrary function satisfying $\Psi(\log r) = \Psi(\log r + \log B)$. In this case, there is only one higher term in the expansion Eq. 21, and that is the ripple we have already discussed *(28)*. However, for real data sets, there will be, in effect, corrections because of sampling errors, system boundaries, or systematic inhomogenities, and these may obscure the lacunarity of the set. In smoothing the data, we tend to obliterate such sources of error, but we eliminate the lacunarity ripple as well.

We may also define dimensions based on the various correlations, and the pair correlation is the one most commonly employed. It may appear from Eq. 18 that a dimension determined from the slope of a log Ξ-log r plot would not be different from D_Γ, in the limit of small r. However, we find that, in practice, the power law dependence of Ξ may differ significantly from that of Γ. To make the comparison, we expand the correlation function for small r in terms of log r. For the ameliorated version, this gives and expression like

$$\log \Xi = -\gamma \log r + \cdots, \tag{24}$$

where γ is a constant that is characteristic of the set. We then define the correlation dimension as

$$D_\Xi = D - \gamma. \tag{25}$$

This is the quantity ν defined by Grassberger and Procaccia *(8)*.

We shall be comparing the values of these two dimensions for the cosmic clustering. Of course, it is always uncomfortable to compare just two methods when there is a chance that they may disagree. To help in deciding between them, we first try them on sets for which we know the fractal dimensions on theoretical grounds. And, in the belt and suspenders mode, we shall also use a third functional on fractal sets described by Badii and Politi *(1)*. Following them, we select a subset from the data. From each point in this subset, we find the distances P to K of the remaining points. Let the minimum among these distances be called P_{min}. Badii and Politi show that the quantity D_P in

$$P_{min}(K) = const. \times K^{-\frac{1}{D_P}} \tag{26}$$

gives a good approximation to D. By varying K, one can therefore get an estimate of D.

We now turn to the determination of these three dimensions in a few specific cases. We find that these are more straightforward to obtain than d of "Cosmic Cascades," Eq. 8. But, in the cosmic context, we shall mention some possible values for d also.

Theoretical Examples

To illustrate and compare the dimensions, we find them for two theoretical fractal sets in this section. In each case, the fractal dimension is known on theoretical grounds, and this is useful in assessing the methods. We shall, however, not rederive the theoretical results for the fractal dimensions since they are explained in books *(2)*.

(i) We start with an interval of length unity and mark the end points and the midpoint. Then we make two copies of this interval and scale them to a new length, $\sigma < 1$. We attach these two copies onto the original interval by attaching the midpoint of each to one of the endpoints of the original unit interval. Next we pivot each secondary with respect to the primary interval to some selected value θ. Then we make four scaled copies of the second generation, each with length σ^2. We attach these by their midpoints to the four free endpoints and set the angles at the contact points all equal to θ. We continue this to the nth generation, where n is very large. The 2^{n-1} sticks of length σ^{n-1} will have 2^n free endpoints. These points are the model fractal. [A realization in a three dimensional space is described in *(27)*.]

The only differences between the present construction *à* la Barnsley *(2)* and that of Groth *et al.* *(9)* are that the latter chose the angles at the attachment at random and worked in three dimensions. In Fig. 2, we show such a set for (constant) $\theta = 22.5°$ and $\sigma = 0.7$. The density of points in the set is qualitatively indicated as if the points were individually bright. A bright region in the picture is very dense and a dark one is void. Two qualitative features of this set emerge on inspection: the presence of voids (or lacunae) and the filamentary distribution of the highest density of points. In these respects, the set resembles the observed distribution of galaxies in producing both voids and dense filaments. Such features may be caused mechanistically as in some theories, but they are a natural feature of a hierarchical universe *(32)*.

Coming back to this particular fractal set, we recall that, on theoretical grounds, its fractal dimension is

$$D = \ln 2/\ln \frac{1}{\sigma}, \tag{27}$$

provided that the set is totally disconnected *(2)*. Qualitatively, we may say that the set is disconnected when the different parts produced by the generating algorithm do not overlap. It is not hard to see that Eq. 27 follows on applying the *ansatz* Eq. 22 to N(r), on which the definition of fractal dimension is based. When the parts of the fractal set begin to overlap, we expect Eq. 22 to fail, and so too will Eq. 27. The formal definition of a connected fractal, based on such an idea of overlap, can be found in Barnsley's book *(2)*. A mathematical discussion of these points is given in *(4)*.

Notice that although D does not explicitly depend on θ, the largest σ for which Eq. 27 holds does depend on θ. It is not hard to work out that for $\theta = 0, \sigma_{max} = 1/2$ and for $\theta = \frac{\pi}{2}, \sigma_{max} = \frac{1}{\sqrt{2}}$. We illustrate the transition from a disconnected to a connected region in Fig. 3, where we fix $\theta = 0.40$ and vary σ. Clearly, as σ increases, we go from unconnected to totally connected cases.

When we are within the totally disconnected region of a fractal, and for some connected cases as well, then different values of θ will correspond to different lacunarities, as indicated by the appearance of voids in these sets. We can see this in Fig. 4, where we draw just a portion of the relevant (x, y) plane and use a specific $\sigma = 0.693$ with two different values of θ. We see that sets with the same dimension give quite distinct maps for the two

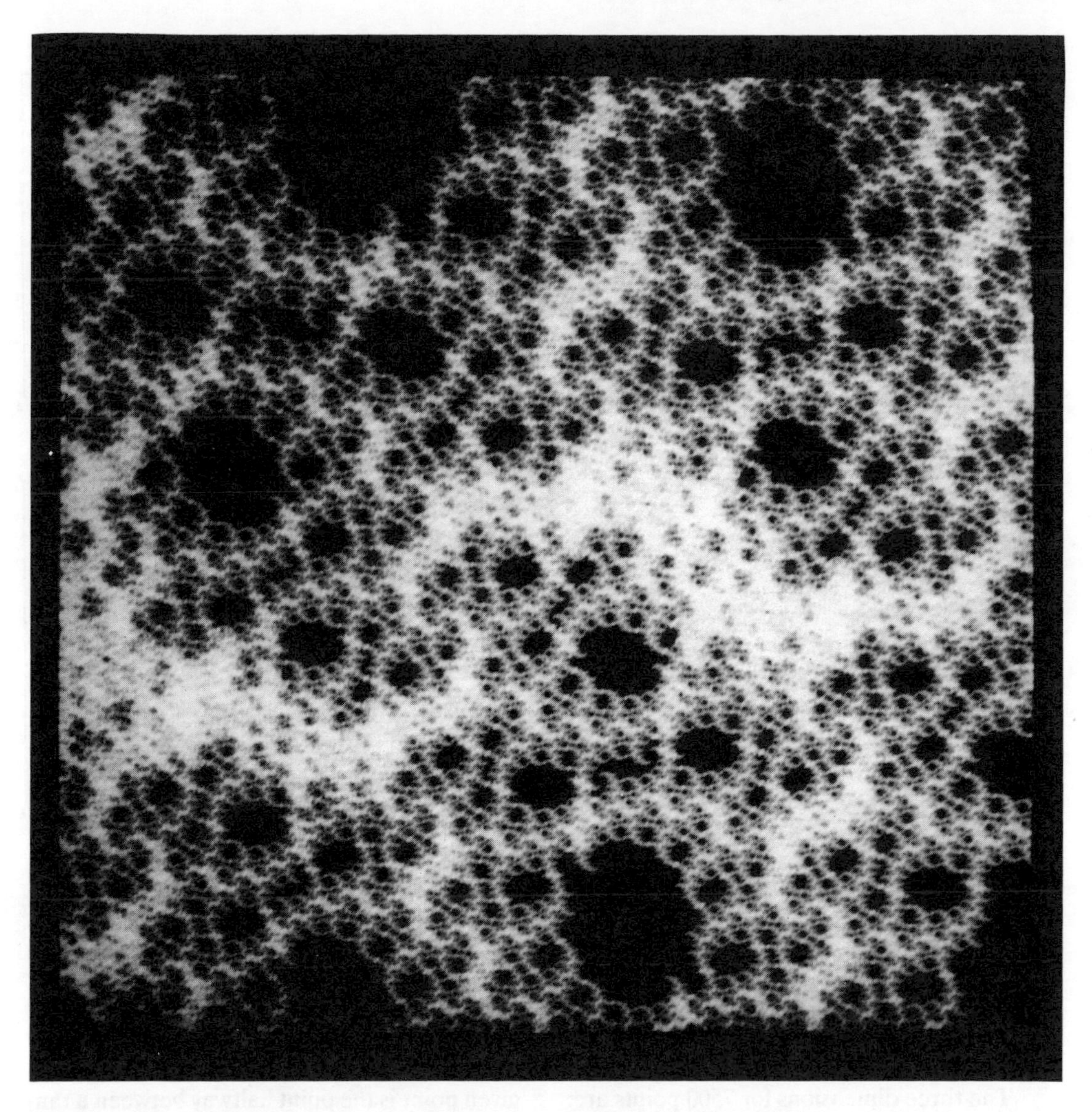

Fig. 2. The set of points generated by the map described in the text for $\theta = 22.5°$. The black spots are voids where the map generates no points unless one is introduced initially, in which case, a regular orbit arises.

different lacunarities.

We give the results of dimension calculations in this case for $\sigma = 0.636$, and $\theta = 0.40$ rad, which have been used in Fig. 3(1). As can be seen in the figure, this case is totally disconnected and Eq. 27 gives $D = 1.532$. We took 7990 points and obtained these values:

$$D_{\Xi} = 1.51 \pm 0.01$$

$$D_{\Gamma} = 1.56 \pm 0.01$$

$$D_{\mathrm{P}} = 1.59 \pm 0.03$$

(ii) Another well-studied set for which we know the fractal dimension is an equilateral triangular Sierpinski gasket. To construct it, we pick a point inside the triangle. Then we select one vertex at random and mark the point midway between the chosen point and the selected vertex. This midpoint becomes the new chosen point. Again, we select a vertex at random and find the point halfway between it and the second interior point. This process is continued, and the points generated

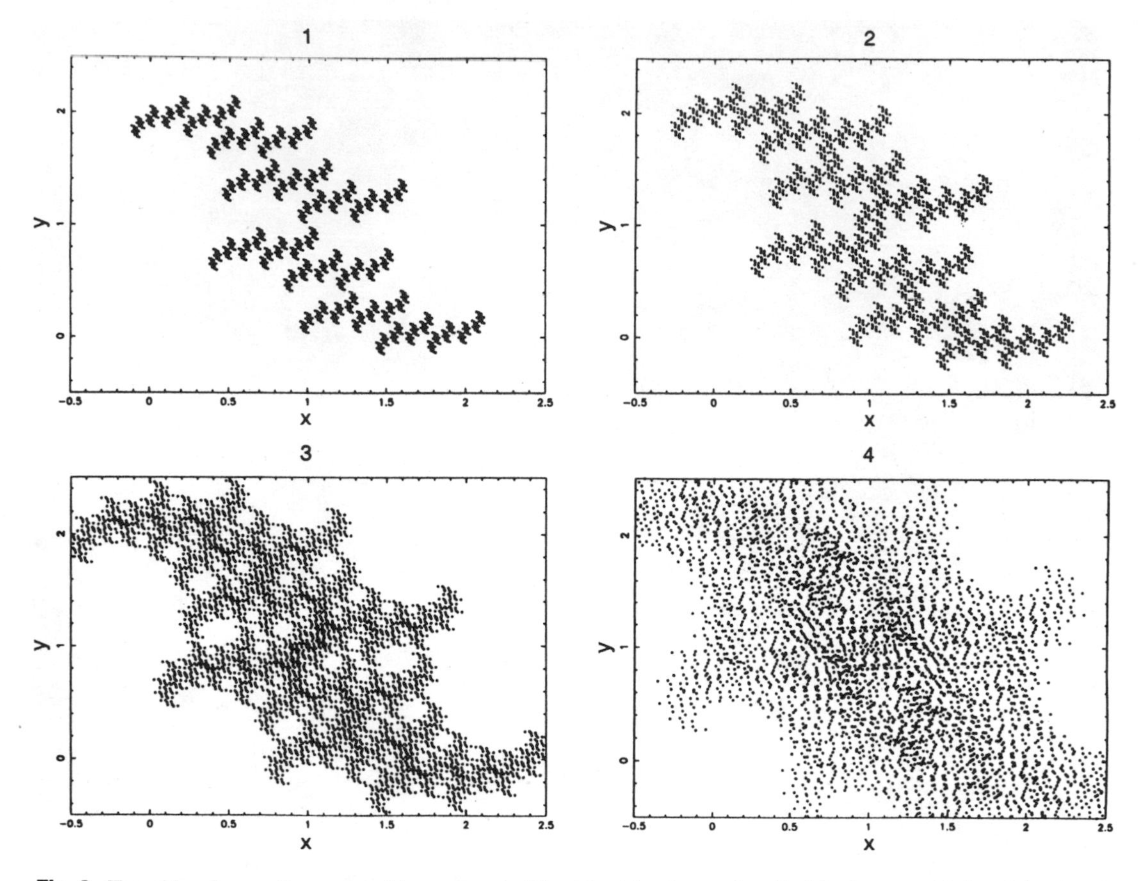

Fig. 3. Transition from a disconnected to a connected fractal set for the set described in the text with $\theta = 0.4$ rad and for (1) $\sigma = 0.636$, (2) $\sigma = 0.665$, (3) $\sigma = 0.700$, (4) $\sigma = 0.750$.

in this way form the set shown in Fig. 5A, where we have discarded the first dozen points. The theoretical dimension of this object is $D = \ln 3/\ln 2$, close to 1.58 *(17)*.

The three dimensions for 7500 points are:

$$D_\Xi = 1.37 \pm 0.02$$

$$D_\Gamma = 1.62 \pm 0.02$$

$$D_P = 1.60 \pm 0.03$$

We see that, in this case, D_Ξ is noticeably different from D_Γ, which is close to the known fractal dimension. This is consistent with our experience in such tests: D_Γ is typically close to D for self-similar sets, with D_P almost as close, but D_Ξ is sometimes well off D. Next, we turn to the case of the galaxy distribution, where these effects are more pronounced. But in passing, let us mention that the Sierpinski construction need not be restricted to triangles.

The distribution of Fig. 5B is a generalization of the Sierpinski gasket to the case where it is generated by the vertices of a regular pentagon. As in the triangular case, the image of a given point is the point halfway between a randomly chosen one of the five vertices. The mapping is continued, and a fractal set is generated (actually, it is a so-called fat fractal). This object shows ripples in its statistics.

Galaxies on the Sky

For many years, the positions of the galaxies on the celestial sphere have been measured by astronomers. This has never been an easy task. Because of our own position within a spiral galaxy, the distribution of galaxies on the celestial sphere is given an apparent large-scale nonuniformity. But if we restrict our at-

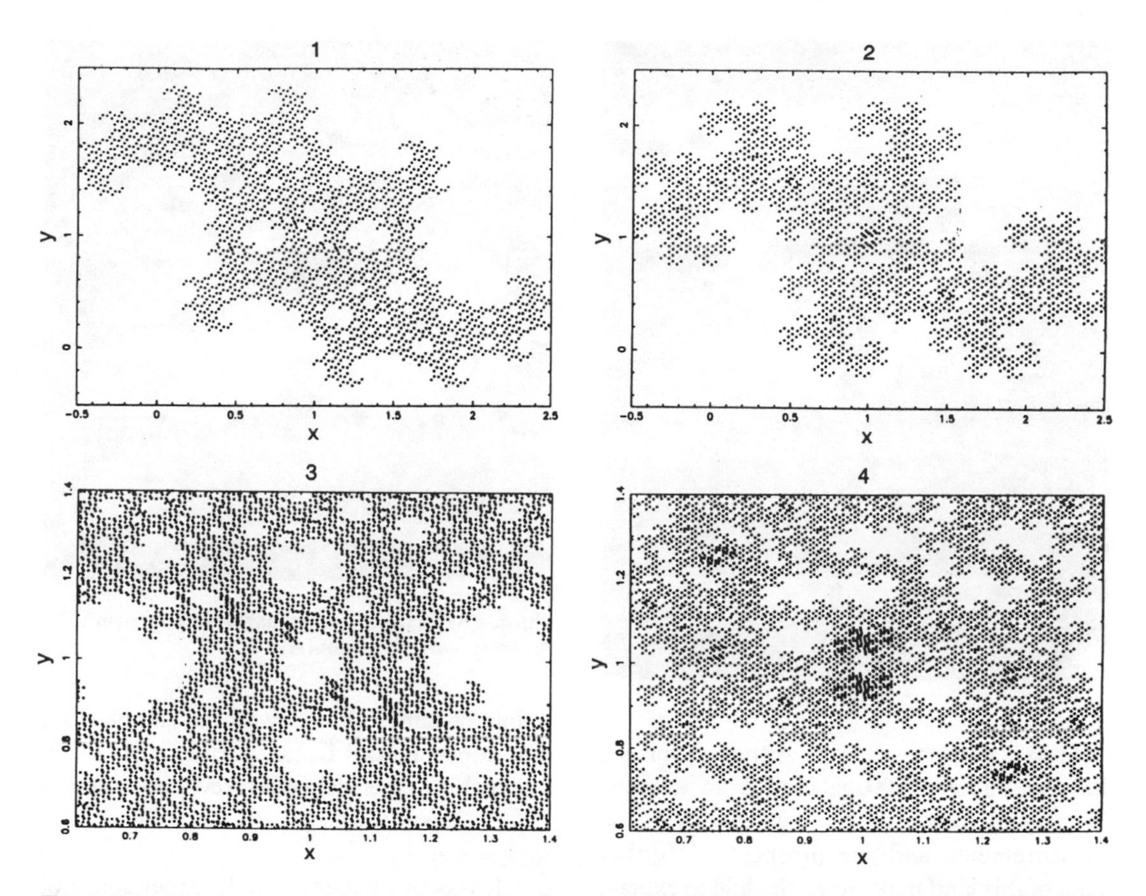

Fig. 4. The map for $\sigma = 0.693$ with two different lacunarities: (1) $\theta = 0.523$, with the whole range shown; (2) $\theta = 0.785$ with the whole range shown; (3) $\theta = 0.523$, with restricted range; (4) $\theta = 0.785$ with restricted range.

tention to selected regions of the sky, information bearing on the true current distribution of these markers in the cosmic flow may be derived. The story is well described in the books by Peebles who, with various collaborators, has been studying the statistics of the measured galaxy coordinates on the celestial sphere. These coordinates are simply two angles on S^2, in the language we have been using. So when Peebles and co-workers derive correlations of the second and higher order, the separation variable is no longer linear distance, but an angle on the sky called ϑ. To keep this in mind, we use the cosmologist's designation of ω for the pair correlation. (However, we do not use a special notation for Γ when using angular coordinates.)

Determinations of the plot of log ω vs. log ϑ have revealed a nonintegral slope. This has led Mandelbrot *(16)* to suggest that the galaxies, regarded as points, form a fractal set. This observation nicely focuses our thinking about hierarchical universes and the cascade that lies behind the galaxy distribution. The implied task is to derive the basic properties of the cosmic fractal. Fortunately, Peebles has long since begun this task, so we can follow his lead. Here we undertake the limited goal of describing how to get dimensions for cosmic fractals. In doing this, we treat the coordinates of the galaxies on the sky as given columns of numbers and, with regret, omit the story of how difficult they are to obtain and how much they have improved in recent decades.

There are several sets of measurements that are relevant. To illustrate the determination of the dimensions of the cosmic fractal, we use a set of coordinates compiled by a group at the Nice Observatory led by A. Bijaoui *(25)*, who kindly provided us with their

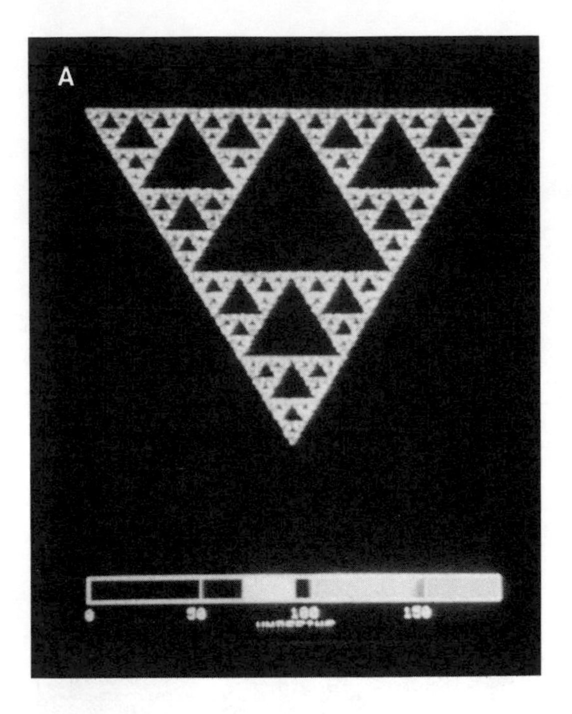

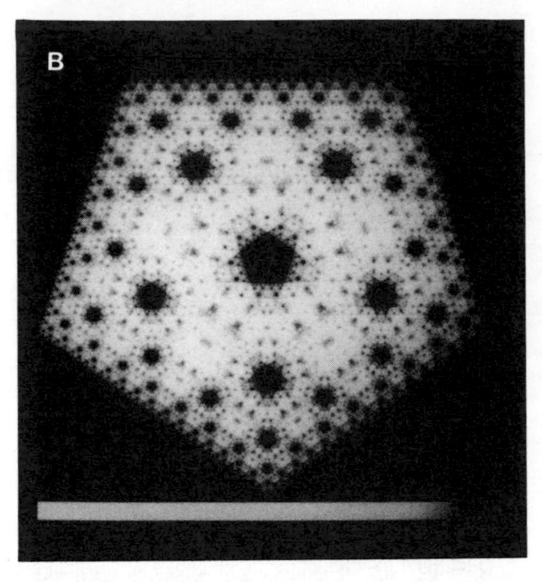

Fig. 5. (A) The Sierepinski triangle with 6000 points. (B) A pentagonal analogue of (A).

data in a convenient form. Although this set has only about 7500 galaxies from a single Palomar plate as yet, the high quality of the measurements and the prospect of further data of this kind make it worthwhile to expend some effort on them.

The most extensive statistical studies use the pair correlation $\omega(\vartheta)$. It was early known that the log-log plots of this function showed a straight line portion for small ϑ and a break to a less simple behavior. The break is known as the "knee" in this subject. Figure 6 is a plot of log ω vs. log ϑ for the Nice data showing the knee. The slope of the straight line portion of the plot is called γ and from these data we find $\gamma = 0.88 \pm 0.03$. For the fit, the values of $\omega(\vartheta)$, between $\vartheta_1 = 0.0001$ rad and $\vartheta_f = 0.006$ rad were used. For larger ϑ we obtain the usual picture of a breakdown of the fit due to a more rapid decrease in $\omega(\vartheta)$ *(9)*.

The problem with the knee is that its location changes with the size of the sample. This has been recently studied with data from a new position catalogue *(7)*. We show some values of ϑ_f, the position of the knee, for various data samples in Table 1; the first two values are from the quoted work and the third comes from Fig. 6. Not only do the break points or knees change with the sample size, the slopes in the linear portion also differ noticeably. Thus the measured dimension is $D_\Xi = 2 - \gamma$, which clearly depends on the sample size.

It was this uncomfortable dependence of ϑ_f on sample size that led Pietronero to suggest that Γ, the conditional density function, might provide a more robust statistical object to characterize galaxy positions. While the work performed by him and his collaborators seems to bear this out, the close connection between Γ and the pair correlation in Eq. 17 seemed to suggest that the dimensions D_Γ and D_Ξ would not differ by much. Indeed, Pietronero called the slope of the correlation function and "intrinsic quantity," and assumed that it is not much affected by sample size. On the other hand, it has been suspected *(32)* that galaxy "clusters occasionally overlap," so, given the lesson of "Theoretical Examples," pages 208–210, we may expect that this is not so.

The dimension that is normally accepted for galaxies on the sky is obtained from the correlations measured by Peebles and collaborators *(9)* from data in the Shane-Wirtanen catalogue. Following the suggestion of Mandelbrot, from their value of γ, one gets a value

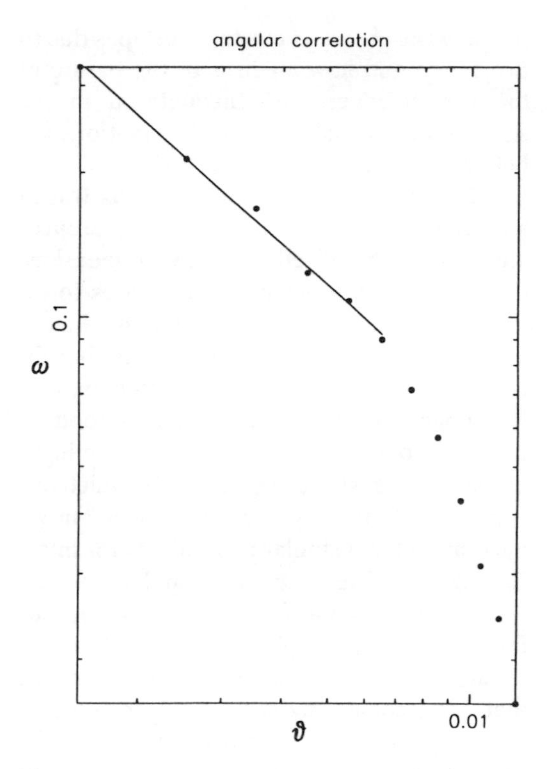

Fig. 6. Angular correlation ω, using the catalogue of *(26)*.

of about 1.2. The value we get from the Nice data is

$$D_\Xi = 1.12 \pm 0.03. \tag{28}$$

(We might have called this D_Ω, but the notation would have become too cumbersome.)

In Fig. 7 we show a plot of $\Gamma(\vartheta)$ vs. ϑ. From these results, we get $D_\Gamma = 1.88 \pm 0.01$. The values of $\Gamma(\vartheta)$ used for this fit were from the domain from $\vartheta_i = 0.0001$ rad to $\vartheta_f = 0.015$ rad. The wider range of usable ϑ supports Pietronero's *(22)* proposal to use Γ rather than Ξ. However, we should mention that changing ϑ_f, or using $\Gamma^*(\vartheta)$ instead of $\Gamma(\vartheta)$, results in small differences that are larger than the statistical error. Therefore, we prefer a more conservative error estimate and write

$$D_\Gamma + 1.88 \pm 0.03. \tag{29}$$

We may mention that, for this data set, the calculations done without amelioration give essentially the same dimensions. Finally, we add that, for the Nice data, we also get

$$D_P = 1.89. \tag{30}$$

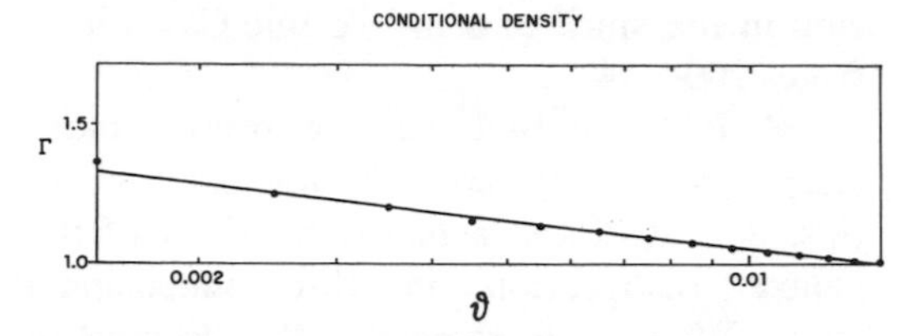

Fig. 7. Conditional density, Γ, using the catalogue of *(16)*.

The dimension of the set in Fig. 2 also has this value.

We reach the evident conclusion that D_Γ and D_Ξ are disparate. Table 1 indicates that D_Ξ increases with increasing sample size. It is not guaranteed that D_Γ will not change when the sample size is increased, but the studies done so far suggest that such a change will not be substantial unless the region of the sky used is unusual in some way. Completion of the Nice study will provide the data needed to be firm on this point. But the relation between ω (or ξ) and Γ make it unlikely that the discrepancy could be due to physical causes like the inhomogeneity of the cosmic fractal.

Wiedenmann and Atmanspacher *(32)*, whose recent study points to a discrepancy such as we have found, attempt to reconcile the two dimension values by associating them to two different domains of ϑ. This view is reminiscent of the interpretation that has been placed on observations of the solar granulation *(24)*. In any case, in the appropriate domain of ϑ, they get $D_\Gamma = 1.89$, consistent with our results, with data from the Zwicky catalog.

Another recent cosmic dimension determination, for the Virgo cluster of galaxies *(5)*, leads to the high value of 1.98 ± 0.23. Of course, this is a special region of the sky and only 200 galaxies were used. However, the clustering method employed gives a dimen-

Table 1. Values of ϑ_f, the position of the knee, for selected data samples.

γ	ϑ_f	No. of plates	Calculation
0.63	0.030	24	*(7)*
0.71	0.015	6	*(7)*
0.88	0.006	1	Present work

sion in the spirit of *d* in "Cosmic Cascades," pages 200–204.

A question that must be probed more deeply before we can make too much of all this, is that of the actual dimensions of the galaxy distribution in three-dimensional space. We need to remember that, in most of the statistical studies of the cosmic fractal, the embedding dimension D is 2. To determine what the effect of the projection onto the sky is on the true dimension of the galactic set is a difficult matter, since galaxies have a wide range of intrinsic luminosities. The correction will be positive, but we do not know its magnitude. We can only surmise that the true dimension in three-dimensional space is greater than the value measured on the sky. (We ignore cosmological corrections that are likely to be quite small for existing data.)

Some hints about the three-dimensional effects may be contained in the study of de Vaucouleurs *(32)* who made reasonable assumptions about the three-dimensional aspects of the objects under study in assessing the data. His plot of the log of the mean density of a ball of radius *r* vs. log *r*, has the slope -1.7. This implies a dimension $3-1.7 = 1.3$. Since de Vaucouleurs used a single particle density, we infer a dimension compatible with D_{Ξ}, conceivably marginally in excess of it because of the differences in D. On the same page, he also gave a plot of what corresponds to a plot of log N_n vs. log l_n, in the notation of "Cosmic Cascades." The negative slope of that curve is slightly in excess of 2—around 2.1. So there is also a result in reasonable agreement with D_{Γ}, slightly in excess of our 1.9. (There is even a third graph on the page suggesting a lacunarity ripple.)

Obviously, without considerable further measurement and discussion, no definite conclusions will be possible. But we do have enough information to propose that the difference between D_{Ξ} and D_{Γ} is significant and is a result of the cosmic fractal being "connected" *(2)* (de Vaucouleurs speaks of "interlocking" clusters). This effect is no doubt enhanced by projection effects. It would be valuable if we could ultimately discern how much of the observed cluster overlap is due to projection and how much is of the nature of the connectedness we encountered in the study of theoretical fractals. Observations are helping to unravel such effects *(14)*.

The structure of inhomogeneous fractal sets can be probed by studying differences among the generalized (or Renyi) dimensions *(10)*. It may be possible in some cases to remove significant differences among dimensions by retreating to higher embedding dimensions, D. Thus, one might conjecture that inhomogeneous sets may be understood as homogeneous sets projected from some higher dimensional space. The extent to which this is possible seems at present unknown, but we may suspect that similar possibilities for interpreting the difference between D_{Γ} and D_{Ξ} exist for the cosmic fractal. That is, the difference is in a sense physical and could result in large measure from projection effects. In that case, comparing D_{Ξ} and D_{Γ} may be of some use in ongoing attempts to deduce statistical information on three-dimensional distribution of galaxies. For now, however, we do suggest that the fractal dimension associated to the clustering of galaxies on the sky is rather closer to 1.8 than to 1.2 and that it is the former value that ought to be used in evaluating theoretical predictions such as those of "Cosmic Cascades."

Conclusion

In its modern incarnation, chaos theory began with the study of erratic, which is to say, aperiodic and unpredictable behavior. We now know a lot about how this sort of thing can come out of equations of various kinds. In many fields, including astrophysics, practitioners are trying to decide how to use the new understanding of these equations to unravel some of their own problems. The aim is to take observations of complex temporal variability and to try to understand them in terms of equations that are known to produce aperiodicity. That part of the subject is covered by other contributors to this volume and in many standard works on chaos. In brief, the idea is to describe the variability of a system by letting it move along a trajectory in a space whose coordinates characterize the state of the system. In certain cases, where the data cover a number of cycles of the system adequately, it has been possible to extract such trajectories by ingenious methods. Then, one can hope to get some idea of the kind of mathematical model that might generate a trajectory that resembles the one reconstructed from the observations. At the very least, one may hope to discover how many state variables are required to permit the prediction of the future of the system in principle.

When the number of variables, call it F, needed for complete predictability turns out to be finite, we have a deterministic system. In fact, for a given F, the prediction is possible for only a limited time in practice, and that time decreases with increasing F, tending to be effectively zero at some upper, critical value of F. The critical value, F_c, will depend on existing computational means. When it is quite large, predictability becomes a hopeless prospect, as people have realized for centuries.

Once F comes near to that critical value, and sometimes even sooner, we give up trying to predict in detail, use statistical methods instead, and call the system stochastic. In this uncomfortable situation we study systems with conveniently small F, replacing the missing variables by something called noise. This process has worked well in a number of problems, such as continuum physics. The transition between the two situations is discussed in a number of standard works and by others in this symposium.

When the effective F is quite small, we can make detailed studies and develop a deep understanding of the trajectories of the system, even to the point of grasping the topology of the orbits. In these circumstances, the issue of predictability in practice can now be probed. In the ideal situation, we can come to something as clear as the flow of Eq. 1. But sometimes, without making us pass through this rite of unraveling an erratic time signal, nature presents us with flows or trajectories that are revealing and fundamental. In the two cases we have discussed most here, we are led directly to some kinds of Poincaré maps or surfaces of section. Of course there may be other examples that are so simple that we hardly think in these terms. But when there is a wealth of structure, our interest perks up. Even then, we might not think in terms of Eq. 1 in a case like the galaxy distribution were it not for the sensitive awareness engendered by the excitement over chaos theory.

We have written much more about the galaxy distribution than about the other more evident example, the sunspot topology, because we know much less about the galaxy distribution. Naturally, in that case, it is easier to be expansive. de Vaucouleurs *(32)* has cautioned us with this quote from Otto Struve: "...the observer knows too many facts to be satisfied with any theory." Cosmic data are, however, beginning to accumulate, and our outlook will surely be changing rapidly.

As we have seen, it is not at all obvious what to do with the time slice of the cosmic flow that we do have at our disposal. We have attempted to describe some of the available tools and the issues involved in their use. We have concluded, with some surprise, that even the currently accepted value of the fractal dimension of the distribution of galaxies has to be reexamined; indeed, we think it has to be seriously revised. But the truth is, that this was not the reason for our delving into the statistics of galaxy distributions.

We set out in this study to look for the lacunarity ripple in the statistics that we have repeatedly mentioned. It is not surprising that we have not detected it with the low-order statistics of quantities like Γ or ξ. It is quite clear that to find the scaling factor β we need to go to higher moments of the galaxy distribution *(27)*. To do that we need more data. There can be no doubt that they are forthcoming, and we can then hope to proceed along the lines hinted at in "Cosmic Cascades."

In these circumstances, it is not appropriate for us to start discussing the theoretical consequences of lower order statistics, until some more is known of the higher terms in the developments. In "Cosmic Cascades," we merely gave a sample of the most primitive of possible theories, with no mechanistic aspects, just to give an idea of how the lore of the fractal may begin to help us understand the flow that propels our part of the universe. We have not, and could not, give any indication of the diversity of opinion on the theoretical side. We did, however, emphasize that many of the results that theories may claim to produce are generic properties of any sensible models and not to be fussed over unduly. We hope that no one will overlook this message or take it amiss.

But what of the sun, which is reaching a maximum of activity as we finish this report? We have introduced the notion *(29)* that when you cut open any vector field and see what looks like the innards of a coaxial cable, you are looking at the surface of section of the equivalent flow. This vision leads to the understanding that spots and a fractal structure are normal for most magnetic fields. In this case, the observations are abundant, and the sort of issue that hampers the galaxian discussion ought not to pose a real problem. The real issue is how to think about such problems theoretically. Normally, one tries to attack this by leading the magnetic field through the full dance of the equations of magnetofluid-dynamics. But there may be simpler steps to take as hinted in "Cosmic Cascades."

In oceanography and planetary sciences, the subject of vortex merger *(15)* has intrigued all who encountered it. It is not understood in detail (and it has been detailed only in two dimensions), but it does seem to be an inverse cascade process *par excellence*. Can we make a Kolmogorov-Ford-Kida type of model for this process? Can we do something similar for the spots on the sun that are mathematical analogues of vortices? These are the issues that we think are of interest in understanding the sunspot morphology. They loom large now that our minds have been opened to these prospects by thinking about the implications of chaos for astrophysics. Indeed, the main impact of that subject is the tremendous increase in our mental arsenal for formulating questions that previously seemed too complex to express. That perhaps is the main message of this volume.

Acknowledgments

In preparing this chapter we have had help of diverse kinds from many people to whom we express our gratitude. These include Jim Anderson, Albert Bijaoui, Jacqueline van Gorkom, Jim Peebles, Itamar Procaccia, Oded Regev, Ed Shaya, Charles Tresser and Bill Young. We are especially grateful to Wang Quingda for his kind collaboration in the use of AIPS for making figures.

Support for this work came from the National Science Foundation under grant PHY87-04250 the Air Force under grant AFOSR89-0012 and the NASA Cooperative Agreement NCC5-29. We thank the Geophysical Fluid Dynamics Program at Woods Hole Oceanographic Institution for its hospitality in the summer of 1989.

References

1. Badii, R., and A. Politii, *Phys. Rev. Lett.* **52,** 1661 (1984).
2. Barnsley, M., *Fractals Everywhere* (Academic Press, San Diego, CA, 1988).
3. Barrow, J. D., S. P. Bhavsar, D. H. Sonoda, *M.N.R.A.S.* **210,** 19p. (1984).
4. Bedford, T. J., *J. Lon. Math. Soc.* **2,** 89 (1986).
5. Chikawa, M., W. Unno and M. Yuasa, *Sci. Technol.* **1,** 35 (1989).
6. Coleman, P. H., L. Pietronero, R. H. Sanders, *Astron. Astrophys.* **200,** L32 (1988).
7. Collins, C. A., N. H. Heydon-Dumbleton, H. T. MacGillivray, *M.N.R.A.S.* **236,** 7p. (1989).
8. Grassberger, P., and I. Procaccia, *Phys. Rev. Lett.* **50,** 346 (1983).
9. Groth, E. J., P. J. E. Peebles, M. Seldner, R. M. Soneir, *Sci. Am.* **237,** 76 (1977).
10. Halsey, T. C., M. H. Jensen, L. P. Kadanoff, I. Procaccia, B.I. Shraiman, *Phys. Rev.* **A33,** 1141 (1986).
11. Hoyle, F., *Ap. J.* **118,** 513 (1953).
12. Kida, S., *Prog. Theor. Phys.* **67,** 1630 (1982).
13. Kolmogorov, A. N., *C. R.(Doklady) Acad. Sci. U.S.S.R.* **30,** 301 (1941).

14. de Lapparent, V., M. J. Geller, J. P. Huchra, *Ap. J.* **302,** L1 (1986).
15. McWilliams, J. M., *J. Fluid Mech.* **146,** 21 (1984).
16. Mandelbrot, B. B., *Fractals: Form, Chance and Dimension,* (W. H. Freeman and Co., New York, 1977).
17. Mandelbrot, B. B., *The Fractal Geometry of Nature* (W.H. Freeman and Co., New York, 1982).
18. Mulhaupt, A. P., *Phys. Lett.* **122A,** 403 (1987); Mulhaupt, A. P., *Phys. Lett.* **204A,** 151 (1987).
19. Novikov, E. A., *Sov. Phys. (Doklady)* **1,** 497 (1966).
20. Novikov, E. A., and R.W. Stewart, *Isv. Akad. Nauk. S.S.S.R. Seria Geofiz.* **3,** 408 (1964).
21. Peebles, P. J. E., *The Large Scale Structure of the Universe* (Princeton University Press, Princeton, NJ, 1980).
22. Pietronero, L., *Physica* **144A,** 257 (1987).
23. Pietronero, L., and E. Tosatti, Eds., *Fractals in Physics* (North-Holland, Amsterdam, 1986).
24. Roudier, T., and R. Muller, *Solar Phys.* **107,** 11 (1986).
25. Slezsak, E., G. Mars, A. Bijaoui, C. Balkowski, P. Fontanelli, *Astron. Astrophys. Suppl. Ser.* **74,** 83 (1988).
26. Smith, L. A., *Phys. Lett.* **133A,** 283 (1988).
27. Smith, L. A., Thesis, Columbia University (1987).
28. Smith, L. A., J.-D. Fournier, E.A. Spiegel, *Phys. Lett.* **114A,** 465 (1986).
29. Smith, L. A., and E.A. Spiegel, *Chaotic Phenomena in Astrophysics*, J. R. Buchler and H. Eichhorn, Eds., Ann. N.Y. Acad. Sci. **97,** 61 (1987).
30. Tarbell, T., S. Ferguson, A. Title, G. Scharmer, P. Brandt, private communication.
31. Unno, W., *Sci. Technol.* **1,** 41 (1989).
32. De Vaucouleurs, G., *Science* **167,** 1203 (1970).
33. Wiedenmann, G., and H. Atmanspacher, Preprint (Max Planck Institute, Garching, FRG, 1989).

17

Chaos in Stellar Variability

J. Robert Buchler and Oded Regev

Abstract

This chapter reviews studies of an astrophysical system characterized by chaotic behavior. The system consists of *pulsating stars* that show time variability in their light curves, sometimes resembling typical deterministically chaotic signals. The observational assessment of the underlying low-dimensional dynamics is not easy because of the insufficient quality and poor time coverage of the data. There are theoretical models with varying degrees of sophistication, however, which permit a systematic study of the mechanisms behind the chaotic behavior as well as its occurrence in the relevant parameter space. Other variable stellar sources are also mentioned in this chapter; however, work concerning these stars is not discussed in detail.

Introduction

Composed of fluids and containing a multitude of nonlinear physical processes, astrophysical systems are natural candidates for exhibiting chaotic behavior. Indeed, investigations of chaos in celestial systems of various scales have been made and are in progress. Studies of chaotic phenomena in systems ranging from planetary rings to the large-scale structure of the universe (see chapters 16 and 18 in this volume) are being pursued.

This chapter focuses on a specific type of astrophysical fluid and concentrates on the exploration of *dissipative* chaotic behavior within it. The choice is biased, of course, by the authors' interests and work. We consider stars whose observable luminosity (together with other properties) varies in time. Prominent among these are the *intrinsic variable stars*, whose variability results from *internal instabilities*. These nonlinear oscillators are naturally best suited for chaotic behavior [see *(1)* and *(2)* for general reference]. Despite obvious observational difficulties, which stem from insufficient accuracy and incomplete time coverage of the signals, significant progress has been made in elucidating the nature of the nonlinearities and chaos in stellar pulsators with the help of theoretical modeling. In recent years, progress has been made from the first simplistic models, through the discovery of bifurcations and chaos in sequences of realistic hydrodynamic numerical models, to first attempts at a systematic exploration of the relevant parameter space. These developments are reviewed in the following section, "Chaotic Stellar Pulsations."

In another class of stellar sources, light variability arises for different reasons. These are the *accreting variables*, in which matter (usually originating from a companion star) is accreted (usually on a compact star) and releases its gravitational potential energy, additionally giving rise to thermonuclear runaways and thus emitting radiation (usually in the form of UV or X-ray photons). Temporal variability can be caused in these systems by a variety of mechanisms whose nonlinear nature is obvious. Again, the observational determination of the existence of low-dimensional attractors in the signal has not yet become conclusive. Also, the theoretical situation is much less developed than in the case of pulsating stars. We will not deal here with the accreting variables. The interested reader is referred to several recent attempts to detect low-dimensional chaos in the electromagnetic signals from accreting variables [*(3, 4)*; however, see *(5–9)* for a general approach to the analysis of chaotic time series in astronomical data]. Theoretical approaches and models in this context include the early attempts to apply the study of autoregressive schemes and Markov chains to irregular variability [see *(10)* and more recent nonlinear dynamical approaches discussed in *(11–13)*].

We shall conclude by giving a brief summary of this review together with a list of some prospects for future work.

Chaotic Stellar Pulsations

There are at least superficial observational indications of period doublings, chaos, and intermittency in the pulsations of variable stars [see *(14)* for a nice review of the observational data]. For example, among the classical variable stars, period doubling seems to manifest itself as so-called RV Tauri behavior, which is the alternation of deep and shallow minima in the light curves *(15)*. Analysis of the long-term visual observations of the RV Tau-type star, R scuti *(16)*, is in agreement with such an interpretation. Similarly, the analysis of observations of the short-period ZZ Ceti stars seem to indicate the presence of subharmonics *(17)*. However, more accurate and suitable data gathering is necessary to establish incontrovertible evidence that the observations indeed indicate an underlying low-dimensional attractor.

On the theoretical side, a thorough study of nonlinear pulsations is only possible with numerical hydrodynamic techniques. Up to the present time, for computational reasons, these studies have been limited to the radial pulsations of the most regular stars, viz., the classical variable stars *(18)*. Even so, the physical complexity of these "natural heat engines" is great because of the presence of partial ionization regions, complicated radiation transport, and the presence of shock waves, to mention but a few. Since the pioneering work of Christy *(19)*, numerical hydrodynamic studies have given a good understanding of the gross features of the regular pulsations of the classical variable stars (Cepheids, RR Lyrae), but, no doubt, a number of problems remain *(20)*.

In contrast, irregular intrinsic stellar variability has been the stepchild of stellar pulsation theory, perhaps for want of a theoretical framework within which one could understand such behavior. In fact, the belief seems to have prevailed that limit cycles, or at most double-mode pulsations, are the only attractors of the hydrodynamic equations. To explain irregularities, some ad hoc "mechanisms" have been proposed such as, for example, the interaction of the pulsation with convection *(21)* or with closely spaced nonradial modes *(22)*. None of these could really be accepted as an explanation of the ubiquitously observed irregularities. Irregular pulsations of stellar models have been sporadically reported in the literature *(19, 23–26)*. Generally these were isolated models, and no follow-up was performed to understand the underlying causes. It is only in the last couple of years that a systematic study has been undertaken. This work, to be reviewed below, has shown that low-dimensional chaotic attractors can arise in the dynamic of the stellar models, and what is more, that they occur along well-established routes to chaos, viz., period-doubling cascades and tangent bifurcations.

In the appendix we list the relevant stellar structure equations, which by finite differencing, can be converted into a dynamical system suitable for such studies. Model studies with a direct numerical approach to Eqs. A1–A4 are expensive, and the exploration of parameter space is necessarily limited. The thorough study of simple models and maps is therefore essential for a correct interpretation of the complicated numerical hydrodynamical investigations. Analytical attempts have been made to reduce the complexity of the problem either in the form of "one-zone" models *(27–29)*, through the use of the amplitude equation formalism *(30, 31)*, through the use of a Galerkin truncation *(32, 33)*, or through the derivation of algebraic maps *(34)*. Although oversimplifications, such approaches are often very useful for giving a better understanding of the physics of the pulsations.

A systematic study of one-parameter families ("sequences") of radiative Population II cepheid (W Virginis) models was undertaken in an effort to expose the nature of the bifurcation from regular to irregular pulsations *(31, 35–38)*. In order to mimic approximately the evolution of these stars, the mass, luminosity, and composition were fixed, and the effective temperature was chosen as the control parameter along each sequence. The following systematical behavior was found as the effective temperature was lowered along a sequence: (i) in the low-luminosity sequences, the fundamental limit cycle stayed stable to very low effective temperature; (ii) for intermediate-luminosity sequences, pulsations of increasing complexity arose, going through a period-doubling cascade to chaos. In the chaotic regime, a reverse cascade of period doublings, characteristic of the Feigenbaum map, was observed; (iii) for the more luminous model sequences, the fundamental limit cycles were interrupted by irregular bursts, which as the effective temperature was lowered, became increasingly frequent until the pulsations were fully chaotic. This last behavior has also been found by Aikawa *(39)* in more massive stellar models in which the mass was used as the control parameter. More recently, it has been found that for sequences between categories (i) and (ii), one can have one or several period doublings, which reverse themselves back to regular limit cycles without ever achieving chaos *(31, 39)*. This can be understood from the behavior of two very simple model oscillators *(31)*.

The observed two routes to chaos are strongly reminiscent of the Feigenbaum period-doubling cascade and the Pomeau-Manneville *(40)* tangent bifurcation, respectively. First return maps turn out to be almost one dimensional. In the case of the period-doubling cascades, they display a hump characteristic of the Feigenbaum map, whereas for the tangent bifurcations, they exhibit the typical near-tangency to the 45 degree line. In the chaotic regime, phase-space reconstructions in three dimensions reveal very simple structures reminiscent of the Rössler band *(37)*. In contrast, a Fourier spectral analysis reveals nothing particular aside from a great amount of noise.

It is remarkable that two of the universal routes to chaos, which had been found in rather simple mathematical models such as the Rössler or the Lorenz equations, should also appear in a complicated dynamical system such as our stellar models. We stress that these routes to chaos occur for different sequences, different zoning, and even different hydrodynamics codes, even though the exact bifurcation points are sensitive to such changes. Preliminary work with convective codes has also turned up period-doubling cascades *(43)*, but we repeat that convection is not the cause of the chaotic behavior. The occurrence of chaos is thus a very robust feature of Population II Cepheid models. In addition, it develops for stellar model parameters that put it approximately in the right place in the Hertzsprung-Russell diagram. This is very suggestive that the observed RV Tauri and Semiregular stars owe their existence to a low-dimensional, underlying chaotic attractor.

Intermittent pulsational behavior, as it is observed in many irregular variables, has not yet been modeled, although very erratic hydrodynamic behavior has been found *(41, 42)*. It has been conjectured that intermittency may be due to the proximity of dynamical instability (very long fundamental period) in some red giant envelopes [*(30)*; also compare with *(27)*]. The appropriate amplitude equations show very rich temporal behavior indeed, including intermittency. The physical reason is that the nonlinearities effect a resonant energy sharing between the first vibrational overtone and the fundamental mode. Excitation of the latter thus appears as a slow contraction or expansion, and the pulsations pick up again when the energy is transferred back to the overtone. A systematic numerical hydrodynamic study will be necessary to verify the occurrence of this scenario for intermittency.

Conclusion

Nonlinear stellar oscillators can be modeled as a dynamical system in which modal expansions lead to a system of ordinary differential equations in time. It is thus natural to investigate bifurcations, period doublings, intermittency strange attractors, and *temporal chaos* in these systems.

In the last ten years, a substantial body of work has accumulated on chaos in pulsating stars. These studies have come a long way from the first attempts of "one-zone" models to the discovery of period doublings, tangent bifurcations, and chaos in realistic pulsating star models. We have reviewed these developments briefly but comprehensively, referencing most relevant papers. Work on this subject is continuing, and more progress is expected. The hope is that detailed analysis of data, using new and sophisticated techniques, will soon shed new light on the behavior of irregular pulsators. On the theoretical side, the Hertzsprung-Russell diagram has to be systematically scanned by building detailed numerical stellar models and checking their behavior. Convection, an old enigma of stellar structure theory, itself connected to turbulence and chaos, continues to be a major obstacle. In addition to these relatively "brute force" approaches, one hopes for the advent of robust, relatively simple models which can extract the richness of the nonlinear behavior of pulsating stars. Amplitude equations are a good possibility here, and progress has already been made in their derivation and interpretation.

Acknowledgments

O.R. would like to thank Ed Spiegel and the Astronomy Department at Columbia University for their hospitality. Saul Krasner and the AAAS are also acknowledged for their assistance with the meeting and publication of papers from this symposium. This work was supported by the National Science Foundation.

References

1. Buchler, J. R., J. Perdang, E. A. Spiegel, Eds. *Chaos in Astrophysics* (D. Reidel, Dordrecht, 1985). General reference.
2. Buchler, J. R., and H. Eichhorn, Eds. *Chaotic Phenomena in Astrophysics. Annals New York Acad. Sci.* **497** (1987). General reference.
3. Spiegel, E. A., in *Chaos in Astrophysics*, J. R. Buchler, J. Perdang, E. A. Spiegel, Eds. (D. Reidel, Dordrecht, 1985), pp. 91–136.
4. Voges, W., H. Atmanspacher, H. Scheingraber, *Astrophys. J.* **320**, 794 (1987).
5. Norris, J. P., and T. A. Matilsky, *Astrophys. J.* 346, 912 (1989).
6. Lochner, J. C., H. H. Swank, A. E. Szymkowiak, *Astrophys. J.* **337**, 823 (1989).
7. Canizzo, J. K., and D. A. Goodings, *Astrophys. J. Lett.* **334**, L31 (1988).
8. Morfill, G. E., H. Atmanspacher, V. Demmel, H. Scheingraber, W. Voges, in *Timing Neutron Stars*, H. Ögelman and E. P. J. van del Heuvel, Eds. (Kluwer, Dordrecht, 1989).
9. Scargle, J. D., *Astrophys. J.* **343**, 874 (1989).
10. Lortet-Zuckermann, M. C., *Ann. Astrophys.* **29**, 205 (1966).
11. Celnikier, L. M., *Astron. Astrophys.* **60**, 421 (1977).
12. Livio, M., and Regev, O., *Astron. Astrophys.* **148**, 133 (1985).
13. Hanami, H., *Mon. Not. R. Astr. Soc.* **233**, 423 (1988).
14. Perdang, J., in *Chaos in Astrophysics*, J. R. Buchler, J. Perdang, E. A. Spiegel, Eds. (D. Reidel, Dordrecht, 1985), pp. 11–89.
15. Tsesevich, V. P., in *Pulsating Stars*, J. Kukarkin, Ed. (John Wiley and Sons, New York, 1975), p. 112.
16. Kollath, Z., *Mon. Not. R. Astr. Soc.*, in press.
17. Vauclair, G., M.-J. Goupil, A. Baglin, M. Auvergne, M. Chevreton, *Astron. Astrophys.* **215**, L17 (1989).
18. Cox, J. P., *Theory of Stellar Pulsation* (Princeton University Press, Princeton, NJ, 1980).
19. Christy, R. F., *Rev. Mod. Phys.* **36**, 555 (1966).
20. Buchler, J. R., in *The Numerical Modelling of Stellar Pulsations; Problems and Prospects*, J. R. Buchler, Ed. (Kluwer, Dordrecht, 1990), p. 1.
21. Deupree, R. G., and S. W. Hodson, *Astrophys. J.* **208**, 426 (1976).
22. Mantegazza, L., *Astron. Astrophys.* **151**, 270 (1985).
23. Stobie, R. S., *Mon. Not. R. Astr. Soc.* **144**, 485 (1969).
24. Davis, C. G., *Astrophys. J.* **172**, 419 (1972).
25. Bridger, A., in *Cepheids: Theory and Observations*, B. F. Madore, Ed. (Cambridge University Press, Cambridge, U.K., 1985), pp. 246–249.

26. Worrell, J. K., *Mon. Not. R. Astr. Soc.* **223**, 782 (1986).
27. Buchler, J. R., and O. Regev, *Astrophys. J.* **263**, 312 (1982).
28. Auvergne, M., and A. Baglin, *Astron. Astrophys.* **142**, 388 (1985).
29. Takeuti, M., in *The Numerical Modelling of Stellar Pulsations; Problems and Prospects*, J. R. Buchler, Ed. (Kluwer, Dordrecht, 1990), pp. 121 – 142.
30. Buchler, J. R., and M.-J. Goupil, *Astron. Astrophys.* **190**, 137 (1988).
31. Moskalik, P., and J. R. Buchler, *Astrophys. J.* **355**, 590 (1990).
32. Perdang, J., in *Lecture Notes (in English), 3e Cycle Interuniv. en Astron. et Astrophys.* (FNRS, Brussels, 1978).
33. Perdang, J., and S. Blacher, *Astron. Astrophys.* **112**, 35 (1982).
34. Perdang, J., in *The Numerical Modelling of Nonlinear Stellar Pulsation; Problems and Prospects*, J. R. Buchler, Ed. (Kluwer, Dordrecht, 1990), pp. 11 – 90.
35. Buchler, J. R., and G. Kovács, *Astrophys. J. Lett.* **320**, L57 (1987).
36. Buchler, J. R., M.-J. Goupil, G. Kovács, *Phys. Lett.* **126**, 177 (1987).
37. Kovács, G., and J. R. Buchler, *Astrophys. J.* **324**, 1026 (1988).
38. Buchler, J. R., in *Multimode Stellar Pulsations*, G. Kovács, L. Szabados, B. Szeidl, Eds. (Kultura, Budapest, 1988), pp. 71 – 86.
39. Aikawa, T., *Astrophys. Space Sci.* **139**, 281 (1987).
40. Pomeau, Y., and P. Manneville, *Comm. Math. Phys.* **74**, 189 (1980).
41. Tuchman, Y., N. Sack, Z. K. Barkat, *Astrophys. J.* **219**, 183 (1978).
42. Takeuti, M., M. Nakata, T. Aikawa, *Science Rep. Tohoku Univ. Eight Ser.* **5**, 180 (1985).
43. Glasner, A., and J. R. Buchler, in *The Numerical Modelling of Stellar Pulsations; Problems and Prospects*, J. R. Buchler, Ed. (Kluwer, Dordrecht, 1990), pp. 109 – 120.

Appendix: Stellar Pulsation Equations

In the case of radial pulsations, for which the star preserves spherical symmetry at all times, it is convenient to use a Lagrangean description where the mass, as measured from the center, is used in lieu of the radial coordinate. The equations of motion [*(18)*, for example] are given by

$$\frac{\partial^2 R}{\partial t^2} = -4\pi R^2 \frac{\partial p}{\partial m} - G\frac{m}{R^2}, \tag{A1}$$

$$T\frac{\partial s}{\partial t} = -\frac{\partial L}{\partial m} + \varepsilon, \tag{A2}$$

$$\frac{\partial R^3}{\partial m} = \frac{1}{4\pi\rho}, \tag{A3}$$

where $R(m\ t)$ denotes the distance from the stellar center of that spherical shell that encloses an amount of mass, m, and $s(m, t)$ denotes the specific entropy at m at time t. All the other symbols have their usual meanings.

The equation of state $p(\rho, s)$, which is used in state-of-the-art numerical hydrocodes, is a complicated function as it takes into account the variable ionization of hydrogen and helium; it has to be evaluated through an iterative numerical solution of the Saha ionization equation. Similarly $\varepsilon(\rho, s)$, the local energy- generation rate, is supposed to be a known function of the state variables, resulting from a nuclear physics calculation. The luminosity in Eq. A3 represents the angle integrated heat flow, both radiative and convective. For radiative envelopes, a conductive approximation to L_{rad} is usually used:

$$L = L_{rad} = (4\pi R^2)^2 \frac{ac}{3\kappa} \frac{\partial}{\partial m} T^4. \tag{A4}$$

Here a is the radiation constant, c is the speed of light, and $\kappa(\rho, s)$ is the opacity, again, like the equation of state, a complicated but known constitutive relation.

18

Complexity from Thermal Instability

Oded Regev

Abstract

A recent effort to investigate the nonlinear development of thermal instability is reviewed in detail. Thermal instability is thought to operate in diverse astrophysical settings from the solar corona up to cooling flows in clusters of galaxies. In particular it is thought to be one of the triggers of interstellar cloud formation. A very simplistic model of a thermally unstable medium that gives rise to the appearance of an inverse cascade of clouds of larger and larger sizes with increasingly longer lifetimes is discussed. Inclusion of essentially any driving is sufficient to cause the formation of complex, spatially chaotic cloud patterns. This basic mechanism may have relevance to the observed complexity of the interstellar medium.

Introduction

Astrophysical fluids are often subject to thermal instabilities when, for example, a temperature drop leads to even stronger cooling. Radiation-driven thermal instability is thought to play an important role in the formation of different thermal phases in the interstellar gas. Other astrophysical settings for thermally unstable media are also proposed (see following references). In this chapter we shall use a simplistic example of the interstellar medium (ISM) to introduce a technique for treating the nonlinear development of patterns in a thermally unstable medium.

The stellar objects discussed in chapter 17 of this volume are observed as point sources. The focus is, thus, on their *temporal* behavior. In contrast, the interstellar medium consists of vast regions between the stars whose *spatial* characteristics can be studied in addition to their temporal variability. Observations in different wavelength regimes of the electromagnetic spectrum (notably in the radio and infrared bands) reveal complex patterns of what can be described as relatively dense clouds with a rare intercloud medium.

Many different physical processes are thought to be responsible for shaping the interstellar medium. On the one hand, all the complexities of a compressible, often turbulent, fluid are present. In addition, self gravity and magnetic forces may be important on some relevant length scales. On the other hand, intricate heating and cooling mechanisms contribute to the system's behavior and instabilities. The modeling of such a complex system is not easy if one is not prepared to give up some of the physical ingredients in the purpose of elucidating the role of the others. Spiegel *(1)* and, more recently and explicitly, Scalo *(2)* have suggested the study of the possibility of chaos in nonlinear model equations in an attempt to better understand the ISM complexity ("interstellar turbulence" in Scalo's nomenclature). Since the discovery of instabilities in such simple ISM models *(3, 4)*, several specific attempts of this kind have been made. The possible occurrence of *temporal* chaos was investigated in simplified *local* model equations for the ISM phases and/or clouds and star formation. These works include Ikeuchi and Tomita *(5)*, Regev *(6)*, and recent comprehensive models by Scalo and Struck-Marcell *(7)* and Struck-Marcell and Scalo *(8)*.

We shall not deal with these contributions here. Rather, we would like to describe a first step in the investigation of the origin and nature of complex *spatial* patterns of the ISM. Thus, the main part of this chapter reviews an attempt to apply techniques of nonlinear pattern theory [see *(9)* for general reference] to just one aspect of the interstellar complexity, viz., nonlinear development of thermal instability. We conclude with a brief summary of this review and a list of some prospects for future work.

Thermal Instability and Interstellar Complexity

The extreme complexity of the ISM is apparent in almost any observation. Clouds on many scales, having various shapes and consisting of several phases of the interstellar matter, are observed. It has recently been argued that interstellar cloud structures are so complex that their boundaries are *fractal (10)*. On the theoretical side, the basic physical processes responsible for shaping the interstellar medium have been identified and investigated in considerable detail. [For excellent recent reviews on both observations and theoretical studies of ISM physics, see *(11)*.] Theoretical investigations have included studies of the thermal instability brought about by the special shape of the interstellar cooling function *(12–17)*, identification of the different phases of the ISM and physical processes operating in them [see *(18)* for a review], and formulation of cloud formation and destruction physics *(19)*. Work on these questions has followed three main lines: (i) linear theory; (ii) detailed numerical modeling [*(20, 21)*; see *(2)* for additional references]; and (iii) phenomenological approaches, in which judicious applications of physical arguments have led to general scaling laws for clouds [for example, *(22, 23)* and references therein].

All these approaches help to understand what is happening, but there is certainly room for yet another approach in which techniques used in nonlinear dynamics are brought to bear on the development of interstellar complexity. As mentioned in the introduction, Scalo *(2)* strongly advocated such an approach by proposing that the ISM be viewed as a nonlinear system and suggesting that advantage be taken of recent progress in turbulence theory. Here we review just one recent work of this type, which concentrates on formation of *spatial* complexity [due to Elphick, Regev, and Spiegel *(24)*, henceforth referred to as ERS]. That approach consists of performing a nonlinear analysis of thermal instability in a one-dimensional medium in which gas dynamics, gravity, magnetic fields, and other effects are ignored. Its purpose is to introduce a mathematical language for investigating interstellar complexity.

Taking advantage of their assumptions, ERS impose *hydrostatic equilibrium* with zero velocity and constant pressure (in both time and space) and concentrate only on *thermal* effects [as did Zel'dovich and Pikel'ner *(15)*; hereafter ZP]. Only one differential equation, the heat equation, describes the system in which spatial couplings are also allowed via a thermal diffusion term

$$\rho c_p \partial_t T = -\rho \mathcal{L}(T, p) + \partial_x(\kappa \partial_x T), \qquad (1)$$

where $\mathcal{L}$ is the heat-loss function per unit mass, per unit time, and p,T,ρ,c_p, and κ are the pressure (a parameter), temperature, density, specific heat, and thermal conductivity, respectively. This equation is essentially the one investigated by ZP.

The function $\mathcal{L}$ can be calculated using various degrees of sophistication in the treatment of the many different cooling and heating processes of the ISM *(25)*. As pointed out previously, ERS were interested in exploring the basic nonlinear behavior of the thermal instability, and this could be served well by approximating $\mathcal{L}$ by an analytic expression that represents $\mathcal{L}$ in a qualitatively faithful way. The S-shaped topology of the $\mathcal{L} = 0$ curve in the p-ρ plane, giving rise to the two linearly stable states and an unstable one between them [see Fig. 1 in *(15)* and Fig. 1 in this chapter], must be retained from the realistic case. It is very helpful to introduce natural units into Eq. 1. For example, the cooling time can serve as the time unit and the thermal diffusion length as the length unit (see the caption of Fig. 2 for details). Scaling all variables by their characteristic values and using the perfect gas law, Eq. 1 becomes

$$\partial_t Z = F(Z, p) + Z\partial_x^2 Z, \qquad (2)$$

where a power law dependence of the thermal diffusivity $\kappa \alpha T^{\alpha}$ is assumed, and a new variable $Z \equiv T^{1+\alpha}$ is defined. The new heat function is $F(Z, p) \equiv -(1+\alpha)T^{\alpha}\mathcal{L}(T, p)$. For details of the above and of what follows, we refer the reader to *(24)*, where all the derivations and discussions are given in considerable detail.

Thermal equilibrium is achieved when $\mathcal{L}(T, p) = 0$, and thus, $F(Z, p) = 0$. In Fig. 1, the locus of these thermal equilibria in the p-ρ plane is shown for the above-mentioned analytical approximation to the cooling function. The values of Z at the equilibria are the roots of $F(Z, p) = 0$ for some fixed p. As required for preserving realistic behavior, the middle equilibrium, Z_2, is linearly *unstable,* while Z_1 and Z_3 are *stable.* If the pressure is equal to the special value chosen as the unit (so that $p = 1$), the stable equilibria are symmetrically disposed about the central, unstable one, i.e., $Z_1 = Z_0 + \Delta$, $Z_2 = Z_0$, and $Z_3 = Z_0 - \Delta$, where Δ is a constant. This is a convenience of the above choice of the pressure unit; ERS show in detail how this value of p plays a role in the phase rule derived by ZP as a condition for the existence of a stationary front between the two stable phases (see below).

ERS wish to investigate the nonlinear development and motion of fronts between different phases of the ISM. Thus, it is necessary to go beyond the equilibrium solutions $Z = Z_i$ and analyze the *traveling wave* solutions. These are solutions of Eq. 2 which can be expressed as functions of the single variable $\chi \equiv x - ct$, where c is the constant wave speed. Upon transformation to this new independent variable, the following ordinary differential equation (associated ODE) obtains

$$Z Z'' + cZ' + F(Z) = 0, \qquad (3)$$

where the prime denotes differentiation with respect to χ and, for simplicity, the dependence on the parameter p is dropped.

The solutions with constant $Z = Z_i$ are fixed points of this associated ODE. However, their linear stability, as such, depends on the value of c and is not expected to be identical to that of the corresponding solutions of the original partial differential equation (PDE). It turns out that each of the outer fixed points is always a saddle point (i.e., has one unstable and one stable root). On the other hand, the stability for the central equilibrium depends on c. This defines a critical speed $c_0 = 2\Delta Z_0$. For $0 < c^2 < c_0^2$, $Z = Z_2$ is a spiral point and is unstable (stable) for c negative (positive). For $c^2 > c_0^2$, $Z = Z_2$ is a node, stable or unstable, according to the sign of c.

In the phase plane of Eq. 3, with coordinates Z and Z′, trajectories spiral into (out from) Z_2 for $c > 0$ ($c < 0$) and around it when $c = 0$. Such trajectories translate into fixed spatial structures of the original problem in the frame moving at speed c *(26)*. Among the possible trajectories, certain ones are singled out as being useful in the description of the global dynamics. These are the trajectories that connect the equilibria to each other or to themselves, *heteroclinic* and *homoclinic* orbits, respectively *(27)*. The interest in these special orbits is that they have infinite periods and so correspond to solitary *localized* structures in space. Heteroclinic orbits translate into fronts, sometimes called kinks. In space,

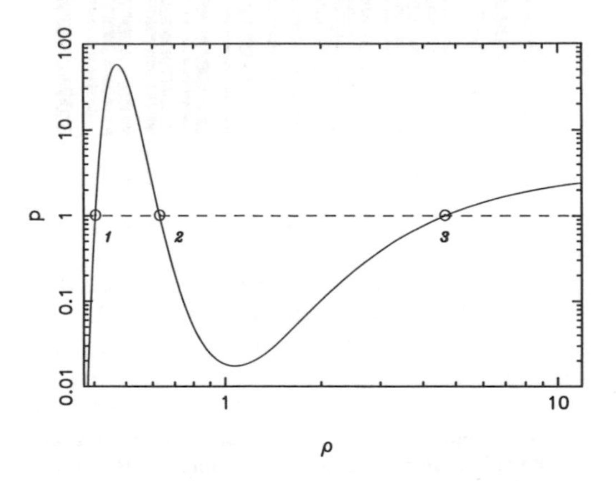

Fig. 1. The cooling curve in the p-ρ plane. p and ρ are in our nondimensional units. The solid line is the locus of points for which $\mathcal{L}(p, \rho) = 0$. The dashed line is the line $p = 1$. The intersection points, marked 1, 2, and 3, correspond to the constant solutions of the PDE and fixed points of the associated ODE. The plot is for the parameter values specified in the text.

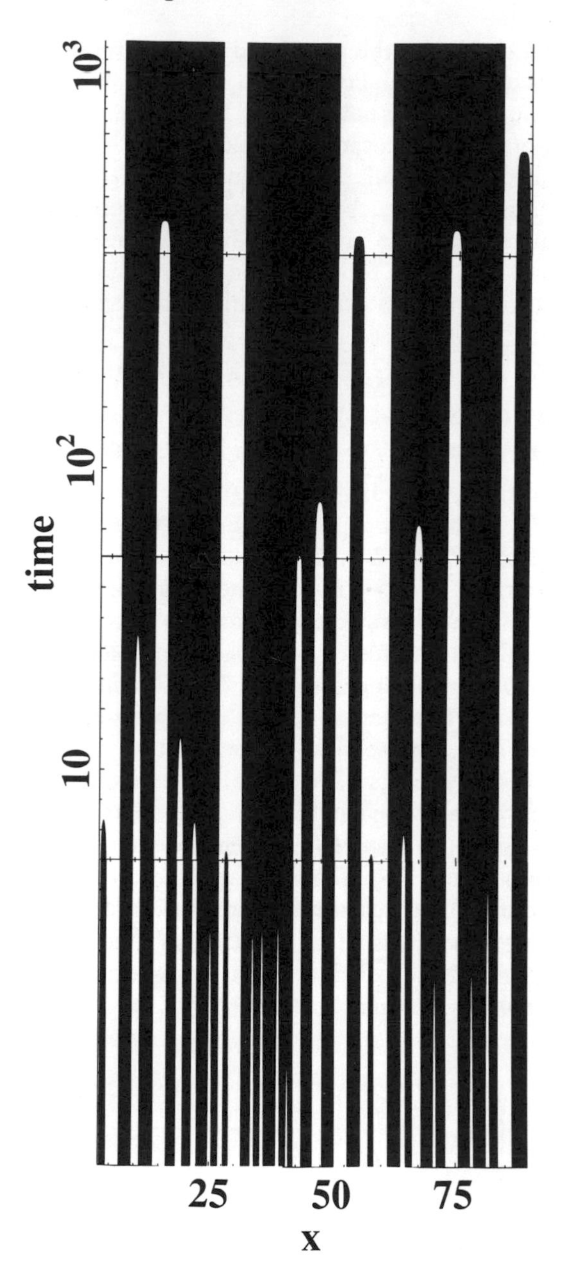

Fig. 2. Time evolution of a 50-front system. Phase Z_3 is represented by the shaded regions. x is in units of $\tilde{L} = 1.7\, 10^{16} (\tilde{\kappa}/\tilde{\kappa}_{ZP})^{1/2}$ cm. With $\tilde{T} = 2500°K$ and $\tilde{p} = 250\, k_B$ (k_B is Boltzman's constant) and (for $p = \tilde{p}$) the ZP thermal conductivity ($\tilde{\kappa}_{ZP}$) yields $\tilde{L}_{ZP} = 1.7\, 10^{16}$ cm as the thermal diffusion length. For other thermal conductivities, it scales as above. The inverse cascade mentioned in the text is apparent.

fronts join up regions with two different values of Z and so describe the motion of the boundaries between different phases of the ISM. The width of such a localized front is approximately the thermal diffusion length, which we have selected to be our length unit. In the same way, the homoclinic orbits describe spatial pulses because the same value of Z is achieved at $\chi \rightarrow \pm \infty$.

ERS find that several types of heteroclinic orbits for the ODE (front solutions of the PDE) exist. The ones involving a front between a stable and an unstable phase of the medium, of course, move in such a way as to diminish the extent of the unstable phase and replace it by the stable one. Another set of heteroclinic orbits joins the saddle points Z_1 and Z_3. These correspond to fronts between the stable cloud and intercloud phases. Among these, ERS chose to focus on the case $c = 0$, when Eq. 3 is easily integrated to give a stationary front, $K(x)$, or its conjugate, $\overline{K}(x)$, where the notation reflects that such fronts are sometimes called kinks and antikinks. Since $c = 0, \chi = x$, one can find

$$K(x) = Z_0 + \Delta \tanh\left[\frac{\Delta}{\sqrt{2}}(x - x_0)\right]$$

$$\overline{K}(x) = Z_0 - \Delta \tanh\left[\frac{\Delta}{\sqrt{2}}(x - x_0)\right], \tag{4}$$

where the integration constant x_0, locating the front, arises because Eq. 1 has *translational invariance.* These stationary fronts are symmetric around Z_0, and their widths are of order 1. They are representative of a family of related *slow fronts* with $c \sim 0$. The significance of the special value of p chosen by ZP is associated with stationary fronts connecting Z_1 to Z_3. Only for $c = 0$ are such fronts steady, and this happens for $p = 1$. As mentioned above, ERS were able to show, by introducing a *Lyapunov functional* of the problem, that such stationary fronts play a privileged role in the general stability theory of Eq. 1. The expression of that reality is closely connected with the equal area phase rule of ZP.

To construct a simple, one-dimensional description of a thermally unstable medium, ERS adopt the view that the medium consists of regions of constant properties joined by the fronts given by Eq. 4. The description of a complex medium by long chains of only such kinks and antikinks in alternation is an extreme oversimplification. However, it seems to be a reasonable first step in an exploration of this kind, and the choice of the stationary fronts of Eq. 4 also seems natural. Among the fronts connecting the two stable phases, the stationary ones last longest. Thus, qualitatively, they should be dominant. By the same token, the fronts that join the unstable phase to the other two phases are not expected to last long. They propagate in the direction of the nearest unstable phase and quickly disappear through front merging (kink-antikink annihilation).

ERS's strategy has been to use a superposition of fronts as the first term in an expansion of the solution for the state of the thermally unstable medium described by Eq. 2. However, the calculation of higher order corrections by *regular* perturbation theory cannot work well. Hence they use the method of *singular* perturbation theory [for example, see *(28)*], in which the parameters of the leading order term are allowed to vary so as to improve the description by perturbation theory. The parameters at one's disposal here are the positions of the fronts, which are the values of the symmetry group parameter implied by the translational invariance of Eq. 2. These parameters are the first to adjust to disturbances like frontal interactions. Such an adjustment serves to determine the equations that govern the kink and antikink positions, $\{x_j\}$.

The description of a configuration of the medium as a superposition of elementary structures is inspired by multisoliton theory *(29)* where, however, the elementary solutions may be superposed even when they are very close. The use of similar superpositions for nonintegrable systems is also expected to work well in some circumstances *(30)*. In such cases, it has to be assumed that the fronts are separated by distances significantly greater than their half-widths, w (here ~ 1). Thus, the natural small parameter for the approximation is

$$\varepsilon = e^{-(d/w)}, \tag{5}$$

where d is the characteristic separation of fronts.

The underlying assumption, which justifies the superposition idea, is that for small ε the fronts retain their basic structure (approaching the homogenous state exponentially on either side). They also may be expected to move slowly in response to their distant interactions so that the x_j's are slowly varying functions of time, that is $x_j = x_j(\varepsilon t)$. For a general configuration with a front at $x = x_i$, ERS thus use an *Ansatz* of superposition to fronts of the above type, together with a small correction term because of the nonlinearity. The detailed form of the Ansatz is given in the Appendix (see pages 231–232). Although that Ansatz is developed as an approximate solution for Eq. 2, under the assumption of small ε, it is possible to show that its behavior when the fronts come close together is also very reasonable. It correctly describes the process of a kink-antikink pair annihilation when two mirror-image fronts meet after coming from opposite directions. On the basis of the stability theory referred to above, it can be anticipated that the final state of the medium will be a single kink (or antikink) in the whole space when the initial number of them, N, is odd or that it will be one of the homogeneous phases when N is even. However, the lifetime of the chain is typically very long, as can be seen in the Appendix.

Carrying out the algebra of the perturbative calculation, in the course of which a nearest neighbor approximation is added, ERS obtain a set of ODEs for the front positions, x_i. Again, the details of the derivation are given in *(24)*. The final ODEs as well as some important implications thereof are given in the Appendix. Here we just describe the main findings. For any initial distribution of many sizes of clouds and their spacings, the smallest objects will disappear first as the system tries to become homogeneous. The time scale of this transient behavior becomes progressively longer during the process. The lifetime of the whole chain is on the order of $\exp(\sqrt{2}\Delta D_m)$, where D_m is the system's largest cloud diameter (see Eq. A5). Thus, the system time is of order ε^{-N}, where N is the initial number of pairs of kinks and antikinks. In retrospect, it seems reasonable to have left the faster moving fronts out of this model. Since the speed of the retained slow fronts is $O(\varepsilon)$, the former ones would have annihilated quickly on average anyway.

These developments can be further illustrated by numerical simulations on Eq. A2. In Fig. 2 the temporal evolution of a kink-antikink chain is shown. The initial condition consists of N = 50 fronts at random separations D_i, with $0.53 < D_i < 3.16$. As expected, the sizes of the homogenous phases grow because of merging of fronts and kinks and antikinks as larger and larger separations arise. Consequently the evolution time scale grows exponentially. In Fig. 2, the fronts are the boundaries between the white and dark regions, and the vertical axis is t, normalized by a constant γ (see Appendix). The dark regions are clouds, domains of $Z = Z_3$, and thus of high density and low temperature. The calculation is interrupted when only three big clouds (six fronts) are left, since the time scale for further evolution $\sim \exp(\sqrt{2}\Delta d)$, becomes extremely long (on the order of 10^{20} in the units of the figure). In any case, the outcome is clear: the final mergers give rise to one huge cloud, whose two bounding fronts attract and ultimately annihilate.

This situation, as illustrated in Fig. 2, is a rundown process, in which the initial state of excitation decays monotonically away. It is analogous to decaying turbulence. In turbulence theory, in order to study statistically steady processes theoretically, one often adds a simple, time-independent, excitation term at a given length scale to avoid commitment to a specific process. Then, when the turbulence is studied at scales disparate from this input scale, one may hope for some measure of universality in the results. In this spirit, ERS add a driving term to the basic Eq. 1, which amounts to one spatial Fourier component of a general forcing term. For the ISM, such small-scale driving would presumably come from various stellar mass injection processes. It follows then that the basic ODEs for the front positions are likewise modified by the inclusion of a spatially sinusoidal driving term (see Appendix). Equations A7 admit steady solutions, in which the positions of the successive stationary fronts are given by a two-dimensional pattern map (Eqs. A10).

The map (Eqs. A10) is an area-preserving map, and it has similar properties to the well-studied *standard map* associated with Hamiltonian systems [see for example, *(31)* and *(32)*, p. 211]. In particular, the map exhibits periodic, quasiperiodic, and *chaotic* orbits *(33, 34)*. In this context, a chaotic orbit is a steady solution of Eqs. A10 with large N, in which the positions of the fronts are aperiodic. Thus, a complex steady pattern of clouds of any size may result. This can be illustrated by formally taking the limit of large separations so as to make the periodic forcing term dominant (this requires $e^{-\sqrt{2}\Delta d} << a$, which is a minor violation of the conditions of the derivation of Eqs. A7. The positions are given by $x_i = 2\pi n_i/q$, where the n_i are arbitrary integers and may be chosen at random.

This behavior can also be verified numerically. Starting with the same initial conditions as in the previous simulations but now including the forcing term (with $a = 0.1, q = 3.8$; see Appendix), ERS obtain the time evolution depicted in Fig. 3. Initially, as long as the separation between two fronts is small compared to the wavelength of the perturbation, the behavior is as in the unforced case. That is, kinks annihilate antikinks and the domains grow. When the separations become comparable to the perturbation wavelength, the front positions lock onto the external forcing, producing a steady complex pattern in accord with the pattern map. In this way, a steady, complicated array of clouds and anticlouds on any scale is obtained up to a cutoff scale, depending on the detailed initial conditions.

To conclude this section, we summarize the basic findings of ERS. The simplest outcome of that study is that isolated clouds grow in size (as do the states between clouds) and that the time for disappearance of a small cloud increases exponentially with size. On the other hand, a regular array of fronts is a steady solution, albeit an unstable one. As the medium passes through various configurations in state space, on the way to complete phase separation, it may tend to linger near such spatially periodic solutions, and that will very much influence the nature of the observed ISM. However, the excitation is clearly moving to larger and larger scales, so we have what is called an *inverse* cascade. Like the direct cascade of turbulence, it also may be expected to lead to a hierarchical range of scales. One may conjecture that this phenomenon is related to observed fractal features of the ISM, but it would be premature to say more on the basis of a one-dimensional study (see the conclusion to this chapter for discussion of possible extensions of this).

This simple conclusion – that an initial inhomogeneity dies away continually, leaving only very few homogeneous domains, which last essentially "forever" – does not seem to be sufficiently appreciated in the literature. Perhaps that is why one encounters so much reliance on the standard scenario of cloud-cloud collisions and coalescence as a source of gravitationally bound structures. It seems that thermal instability provides a direct way to form gravitationally unstable configurations through the creation of larger and larger clouds, persisting for increasingly longer times. Because the life times of clouds increase exponentially with cloud size, the prospects of long-lived large clouds is natural. Nevertheless, it does seem more fruitful to think of interstellar complexity as a result of the combined effects of continual forcing at some small scale together with the transfer of the excitation through the spectrum of ever larger scales by the process studied here. This again is a picture inspired by studies of statistically steady turbulence.

To get some impression of the nature of a statistically steady cloud hierarchy, a spatially periodic, weak forcing may be introduced into the problem to schematize the disturbances generated locally by star formation and mass injection. The problem then has steady solutions, which are chaotic in space, for appropriate values of the parameters. The resulting pattern map (Eqs. A10) that describes this spatial chaos, and others like it, have been explored in great detail in other contexts. One encouraging feature of such pattern theory approach is that it enables the reduction of a PDE to an ODE system and then to an iterative map for the stationary pattern. The sheer appearance of a well-known, chaotic map

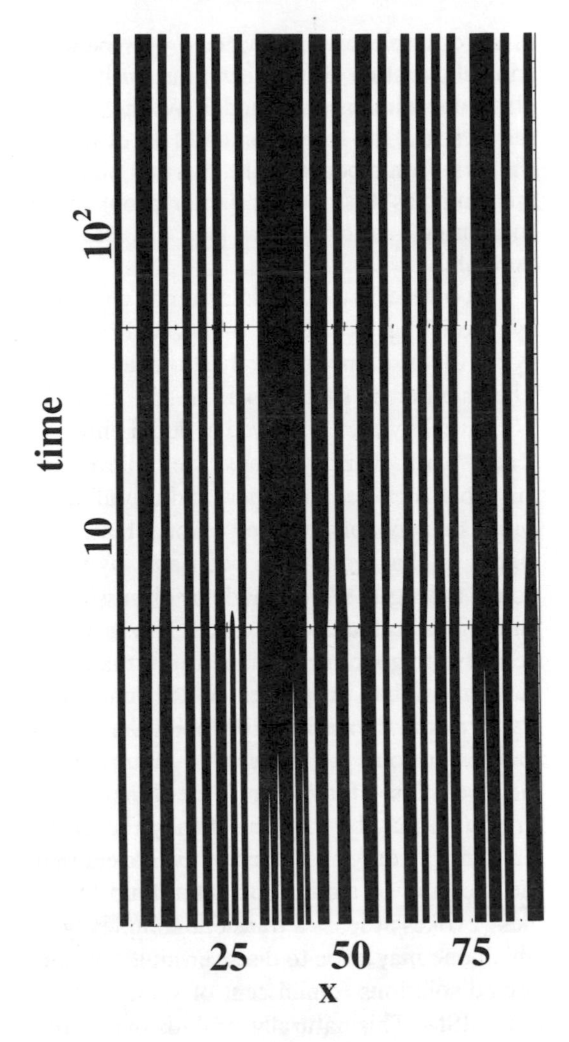

Fig. 3. Same as Fig. 2 but with the addition of spatially periodic forcing to (B2) $a \sin(q x_i)$ with $a = 0.1$ and $q = 3.8$. A spatially chaotic, steady configuration emerges from the randomly distributed initial array of clouds.

does, however, justify the claim that this is a promising approach to modeling complexity. It remains to see how much more realistic an approximation to the interstellar medium such a model can be made.

Conclusion

We have reviewed here a new approach to the problem of a thermally unstable astrophysical fluid in the context of nonlinear dynamics and chaos. The choice of the interstellar medium as an example was motivated by its being a good illustration of the technique in its simplest, one-dimensional model. We have chosen to stress the formation and fate of *spatial patterns,* fronts between the thermal phase in this study, as it is done in similar problems in condensed matter physics [see *(35, 36)*, for example].

In our review of the interstellar complexity modeling, we have mentioned several works that use a nonlinear dynamics approach (the number of such papers is still limited). We have, however, reviewed in detail only one recent work. This is because we believe that this work includes some new and useful ideas and also, probably, due to personal bias. The merits of that work have been already mentioned at length. We conclude by discussing its limitations and future prospects. It is certainly desirable to go to higher dimensions than one, and the global properties of thermally unstable media derived by ERS point the way to such extensions. It is natural to inquire what the evolution of two- or three-dimensional structures may be, but analytically the problem is not an easy one. It is, however, clear that the merging of clouds and intercloud "bubbles" evokes at least a transient complexity in which one may hope to discover multiply connected solutions reminiscent of some aspects of the ISM. This naturally reminds one of the McKee and Ostriker model *(37)*. Sustaining such complexity is a matter of simple forcing, as it is in one dimension. It remains to be seen how far one can get in a second round of such modeling.

As to the many physical effects that should be included in the further enrichment of these models, modeling with moving fronts that stay one dimensional is already in progress. Abstractly speaking, the main result of such enrichments will be to introduce more invariances into the problem; for example, a simple fluid dynamical description will bring in Galilean invariance. This will require the inclusion of more parameters in the characterization of the fronts, thus complicating their temporal dynamics *(38)*. What remains as the difficulty then is the situation of higher dimensionality. At some stage numerical simulation will certainly be useful, and there are others at work in this direction. The work of Chiang and Prendergast *(20)* and subsequent calculations are rich enough in physical effects for many tastes and yet are amenable to numerical techniques, even in three dimensions. There also exist simulations relating to thermal instability in the solar corona [Karpen *et al. (21)* and their previous works] that will provide guidance in such problems. The hope is that such calculations, together with the more traditional theoretical approaches to the ISM physics [see section 4 in *(11)*], can be combined with the new approaches, like those of Scalo and Struck-Marcell and the one reviewed here. This would increase the probability that significant progress in understanding the complex interstellar medium may be achieved.

Acknowledgments

I would like to thank Ed Spiegel and the Astronomy Department at Columbia University for their hospitality. Saul Krasner and the AAAS are also acknowledged for their assistance with the meeting and the publication of this collection.

References

1. Spiegel, E. A., in *Theoretical Approaches to Turbulence*, D. L. Dwoyer, M. Y. Hussaini, R. G. Voigt, Eds. (Springer-Verlag, New York, 1985).
2. Scalo, J. M., in *Interstellar Processes*, D. J. Hollenbach and H. A. Thronson, Eds. (D. Reidel, Dordrecht, Holland, 1987). pp. 349–392
3. Shore, S. N., *Astrophys. J.* **249**, 93 (1981).
4. Shore, S. N., *Astrophys. J.* **265**, 202 (1983).
5. Ikeuchi, S., and H. Tomita, *P. A. S. Japan* **35**, 56 (1983).
6. Regev, O., presented at *NATO ASI Chaos in Astrophysics*, Daytona Beach, FL, 1984.
7. Scalo, J. M., and C. Struck-Marcell, *Astrophys. J.* **301**, 77 (1986).
8. Struck-Marcell, C., and J. M. Scalo, *Astrophys. J. Supl. 64, 69 (1987).*
9. Bishop, A. R., in *Spatio-Temporal Coherence and Chaos in Physical Systems,* G. Gruner and B. Nicolaenko, Eds., *Physica 23D* (North-Hol-

land, Amsterdam, 1986).
10. Bazell, D., and F. X. Desert, *Astrophys. J.* **333,** 353 (1988).
11. D. J. Hollenbach and H. A. Thronson, Eds., *Interstellar Processes* (D. Reidel, Dordrecht, Holland, 1987).
12. Parker, E. N., *Astrophys. J.* **117,** 431 (1953).
13. Zanstra, H., *Vistas Astron.* **1,** 256 (1955).
14. Field, G. B., *Astrophys. J.* **142,** 531 (1965).
15. Zel'dovich, Ya. B., and S. B. Pidel'ner, *Sov. Phys. JETP* **29,** 170 (1969).
16. Defouw, R. J., *Astrophys. J.* **161,** 55 (1970).
17. Balbus, S. A., *Astrophys. J.* **303,** L79 (1986).
18. Shull, in *Interstellar Processes,* D.J. Hollenbach and H. A Thronson, Eds. (D. Reidel, Dordrecht, Holland, 1987), pp. 225–244.
19. Elmegreen, B. G., in *Interstellar Processes,* D. J. Hollenbach and H. A. Thronson, Eds. (D. Reidel, Dordrecht, Holland, 1987), pp. 259–282.
20. Chiang, W-H, and K. H. Prendergast, *Astrophys. J.* **297,** 507 (1985).
21. Karpen, J. T., S. K. Antiochos, J. M. Picone, R. B. Dahlburg, *Astrophys. J.* **338,** 493 (1989).
22. Chieze, J. P., *Astron. Astrophys.* **171,** 225 (1987).
23. Elmegreen, B. G., *Astrophys. J.* **338,** 178 (1989).
24. Elphick, C., O. Regev, E. A. Spiegel, *Mon. Not. R. Astr. Soc.,* in press.
25. Dalgarno, A., R. A. McCray, *Annu. Rev. Astron. Astrophys.* **10,** 375 (1972).
26. Ben-Jacob, E., H. Brand, G. Dee, L. Dramer, J. S. Langer, *Physica* **14D,** 348 (1985).
27. Guckenheimer, J., and P. Holmes, *Nonlinear Oscillations, Dynamical Systems and Bifurcations of Vector Fields* (Springer-Verlag, New York, 1983).
28. Bender, C. M., and S. A. Orszag, *Advanced Mathematical Methods for Scientists and Engineers* (McGraw-Hill, New York, 1978).
29. Lamb, G. L., Jr., *Elements of Soliton Theory* (John Wiley and Sons, New York, 1980).
30. Coullet, P., and C. Elphick, *Physica Scripta* **40,** 398 (1989).
31. Tabor, M., *Chaos and Integrability in Nonlinear Dynamics* (John Wiley and Sons, New York, 1989).
32. Lichtenberg, A. J., and M. A. Lieberman, *Regular and Stochastic Motion* (Springer-Verlag, New York, 1983).
33. Coullet, P., C. Elphick, D. Repaux, *Phys. Rev. Lett.* **58,** 431 (1987).
34. Coullet, P., C. Elphick, D. Repaux, in *The Physics of Structure Formation*, W. Guttinger and G. Dangelmayr, Eds. (Springer-Verlag, New York, 1988).
35. Dee, G., and J. S. Langer, *Phys. Rev. Lett.* **50,** 383 (1983).
36. van Saarlos, W., *Phys. Rev. A.* **37,** 211 (1988).
37. McKee, C. F., and J. P. Ostriker, *Astrophys. J.* **218,** 198 (1977).
38. Elphick, C., O. Regev, E. A. Spiegel, in *"Non Linear Evolution of Spatio-Temporal Structures in Dissipative Systems,"* F. H. Busse and L. Kramer, Eds. (Plenum, NY, in press).
39. Rothman, D. H., and J. M. Keller, *J. Stat. Phys.* **52,** 1119 (1988).

Appendix: Interstellar Fronts and Their Motion

The following expression is used by ERS as an Ansatz:

$$Z(x,t) = k_i(x - x_i) + \sum_{j>i}\left[k_j(x - x_j) - k_j(-\infty)\right] + \sum_{j<i}\left[k_j(x - x_j) - k_j(\infty)\right] + \varepsilon R(x, \varepsilon t), \qquad \text{(A1)}$$

where

$$k_{i \pm 2n}(\zeta) = K(\zeta)$$

and

$$k_{i \pm (2n+1)}(\zeta) = \overline{K}(\zeta)$$

for $n = 0, 1, \ldots,$ and R is a bounded function of x and εt and is a small correction term to this superposition of fronts.

A detailed calculation then gives for the front positions

$$\gamma \frac{dx_i}{dt} = \exp[-\sqrt{2}\Delta(x_{i+1} - x_i)] - \exp[-\sqrt{2}\Delta(x_i - x_{i-1})], \qquad \text{(A2)}$$

where

$$\gamma = \frac{\Gamma}{8\Delta^4}$$

and

$$\Gamma = \sqrt{2}Z_0\Delta + \frac{z_1 z_3}{\sqrt{2}} \log\frac{z_3}{z_1}.$$

This equation does not hold for the two fronts at the extreme ends of the chain, for which one or the other of the two forces must be omitted.

Although the building blocks of this description of a thermally unstable medium are stationary fronts, when several of them exist simultanously it is clear that they move. This motion is the result of the forces acting when several fronts are in interaction. Since the problem is one dimensional, the forces go like separation to the zeroth power, but they do have the range $(\sqrt{2}\Delta)^{-1}$. Moreover, there is a resistance to the frontal motion, measured by the drag coefficient γ. There is, however, no inertia associated to this motion because Eq. 1 does not have Galilean invariance; that is a complication that will appear when fluid motions are included in the theory.

The interactions between the ith kink and its neighboring antikinks are attractive. A kink moves toward the closer antikink and we are led to an annihilation. An isolated cloud does not survive and that may be why some simulations on thermally unstable media do not produce sustained clouds.

To see this in its barest essentials, let $D_i = x_i - x_{i-1}$ be the size of a cloud. From Eq. A1,

$$\gamma \frac{dD_i}{dt} = -2\exp(-\sqrt{2}\Delta D_i) + \exp(-\sqrt{2}\Delta D_{i+1}) + \exp(-\sqrt{2}\Delta D_{i-1}. \quad (A3)$$

Here, $D_{i\pm 1}$ measures the size of the two low-density regions on both sides of the cloud.

To study an isolated cloud of size D, let $D_i = D$ and $D_{i+1} \to \infty$. Thus

$$\gamma \frac{dD}{dt} = -2e^{-\sqrt{2}\Delta D} \quad (A4)$$

This equation implies that an isolated cloud of initial size D_0 disappears in a time

$$t_* = \frac{\gamma}{2\sqrt{2}\Delta}(e^{\sqrt{2}\Delta D_0} - 1), \quad (A5)$$

which, for $D_0 \sim d$ (a characteristic separation), is of order ε^{-1}. The number of fronts decreases in time, therefore, and *phase separation* occurs as in similar problems *(39)*.

If a forcing terms is added in Eq. 1, it becomes a parametric forcing in Eq. 3. Thus Eq. 3 is replaced by

$$\partial_t Z = F(Z,p) + Z\,\partial_x^2 Z + A \sin qx, \quad (A6)$$

where A is an amplitude, generally dependent on Z and its derivatives, and the origin of x is arbitrary. For simplicity, ERS adopt A = BZ, with B constant. Assuming that $B = O(\varepsilon)$, and carrying out the same procedure as in the derivation of Eq. A1, after suitably shifting the x_i by $i\pi$ and rescaling time, Eq. A1 is replaced by

$$\gamma \frac{dx_i}{dt} = \exp\left[-\sqrt{2}\Delta(x_{i+1} - x_i)\right] - \exp\left[-\sqrt{2}\Delta(x_i - x_{i-1})\right] + a\sin(qx_i). \quad (A7)$$

Here a is a constant proportional to B and is $O(\varepsilon)$.

Equations A7 should have steady solutions in the form of a periodic array of kinks and antikinks. However, the period is not arbitrary as for Eq. A2, since the fronts have to be located at the zeros of sin qx. Therefore, the possible periods are $d_0 = 2\pi m/q$, where m is any integer. Such steady, periodic solutions can be seen to be unstable if they do not minimize the periodic function

$$V(x) = -(a/q)[1 - \cos(qx)].$$

In general, any steady solution of Eqs. A7 satisfies

$$0 = \exp\left[-\sqrt{2}\Delta(x_{i+1} - x_i)\right] - \exp\left[-\sqrt{2}\Delta(x_i - x_{i-1})\right] + a\sin(qx_i). \quad (A8)$$

This can be rewritten as a two-dimensional map. Let

$$\Theta_i \equiv qx_i \text{ and } I_{i+1} \equiv \exp\left[-\sqrt{2}\Delta(x_{i+1} - x_i)\right]. \quad (A9)$$

Then Eq. A8 is

$$I_{i+1} = I_i - a\sin\Theta_i,$$

$$\Theta_{i+1} = \Theta_i - \frac{q}{\sqrt{2}\Delta}\log I_{i+1}. \quad (A10)$$

19

A Stochastic Nonlinear Model for Coordinated Bird Flocks

Frank Heppner and Ulf Grenander

Abstract

Certain small birds such as pigeons, starlings, and shorebirds fly in coordinated flocks that display strong synchronization in turning, initiation of flight, and landing. Experimental efforts to find leaders in such flocks have to date failed. We propose that synchronization of movement may be a byproduct of "rules" for movement followed by each bird in the flock. Accordingly, we have developed a computer-simulated bird flock employing stochastic differential equations which demonstrates realistic "flocking" behavior. The functions employed in our model include attraction to a roost, a nonlinear attraction to flockmates, preservation of flight velocity, and an *n*-dimensional Poisson stochastic process. By varying the values of the parameters of the model, we have seen organized flight develop from chaotic milling and an organized flock break up into chaotic flight.

The coordinated flight formations typical of small birds such as pigeons, shorebirds, and blackbirds have been simulated with a model that assumes no leadership within the flock, with rules for movement as a function of the movements of neighbors, and containing a nonlinear component that produces occasional chaotic behavior. Changing the numerical values of the parameters in the flock model produces behavior typical of different bird species.

The apparently coordinated movements of large flocks of certain small bird species such as starlings and small shorebirds flying in cluster flocks *(1)* have fascinated observers since the time of Pliny *(2)*. How do the flocks coordinate their movements and decide when to turn and wheel? Although a "group mind" model has been proposed *(3)* in which the individual nervous systems of the flock members were somehow connected, many investigators have assumed a leadership model, in which a leader directs the movements of the flock *(4)*. Efforts to identify such a leader have thus far been unsuccessful, even using methods that permitted an analysis of the position of individually identified birds in a free-flying pigeon flock *(5)*. An alternate view to a leadership model has been proposed *(6)* in which a "self-generated synchrony" provides the mechanism for coordinated movements. In this proposed mechanism, coordinated turning is the result of individual birds "voting" with their bodies on the flight path the flock should take. When a "critical mass" is reached, the flock as a whole then turns in the direction expressed by the initiators of the turn. Some support for this hypothesis is provided by the observation *(7)* that waves of turning in dunlin (*Calidris alpina*) flocks can be initiated by individuals turning toward the interior of the flock.

We suggest here that the apparently coordinated movements of cluster flocks represent an organization whose collective movement is shaped by relatively simple rules, which determine the movements of its members. We present a computer simulation of a bird flock incorporating nonlinear elements, which produces "behavior" strikingly similar to natural flocks. We further suggest that such a model may be applicable to fish schools and other organized moving groups of animals.

The core of the simulation was initially inspired by a game concept developed by John Conway *(8)* in the 1960s. Conway's game was called "Life," the play of which was that squares on a grid were filled or not filled in subsequent moves of play, depending on the presence or absence of filled neighboring squares in the present move. "Colonies" of cells would be established, grow, send off new colonies, decline, and die, depending on the original state of the game. A significant feature of the game was that there was no self-generated behavior of the cells; what they did in future moves was *entirely* a function of each cell's response to neighbors. Striking patterns would develop, sometimes hundreds of generations after the initial state, which gave the appearance of orderliness or direction, despite a lack of direction in the "rules" of the game.

It occurred to us that a coordinated flock might be built around the same concept, where a behavior of the flock as a whole, e.g., a turn, might arise as a byproduct of rules influencing the movements of individuals, the majority of which might be governed by reactions to neighbors. Because real flocks will maintain an orderly pattern of turning and wheeling for a time, then spontaneously break up, and conversely form a synchronized group after flying without coordination for a time, we felt that the introduction of a nonlinear component as well as a random term in the simulation might account for this behavior.

Our initial model contains the following features: (i) it exhibits the collective behavior of flocking seen in some species, without assuming that the "flock" has a leader; (ii) it also exhibits the somewhat erratic, unsystematic behavior seen in flocks, due to wind gusts, external disturbances, and random causes; (iii) the flock aims for some roosting area; (iv) the flock turns and wheels for an indeterminate time over the roost; (v) individuals can change their position within the flock; (vi) from time to time, the flock can split into two or more "subflocks"; (vii) the flight paths of individual birds in the flock can be approximately parallel and to some degree random; and (viii) the model postulates no other systematic external influence acting on a totality of the "flock" to steer it in a flock-like fashion.

The model is given through a stochastic differential driven by a Poisson process with associated random variables equation, using the notation of stochastic differential equations

$$\begin{cases} dv_i(t) = (F_{home} + F_{vel} + F_{interact})\,dt + dP(t) \\ v_i(t) = du_i(t),\ i = 1,2,..n \end{cases} \quad (1)$$

where $u_i(t) = (x_i(t), y_i(t)$ is the location of bird number i at time t. The equation involves the functions for

homing: F_{home} expresses the tendency towards the roosting area

velocity regulation: F_{vel} controls the velocity of an individual

interaction: $F_{interact}$ measures the influence of one individual on another

randomness: $P(t)$ is a Poisson stochastic process.

The first three terms are deterministic in contrast to the fourth one, which is intended to simulate the effects of wind gusts and random local disturbances. This term is obtained from an n-dimensional Poisson process with associated independent and identically distributed (*i.i.d.*) random variables.

More precisely, our model assumes the following. For the roosting tendency we postulated for the ith component

$$F^{(i)}_{home} = -u_i(t)\, f_{home}\,(|u_i(t)|);\ i = 1,2,\ldots n; \quad (2)$$

where u_i is the vector from the homing zone to bird i and where f_{home} is a scalar valued function (Fig. 1), where r_0 is the minimum radius of a circular homing area. No influence is felt from the homing area if the distance r is greater than r_1.

The velocity control term (Fig. 2) is postulated to be

$$F^{(i)}_{vel} = -v_i(t)\, f_{vel}\,(|v_i(t)|), \quad (3)$$

where v_0 is the preferred velocity.

The third term, $F_{interact}$ (Fig. 3), was the most difficult to choose. If $d_{ij}(t)$ means the

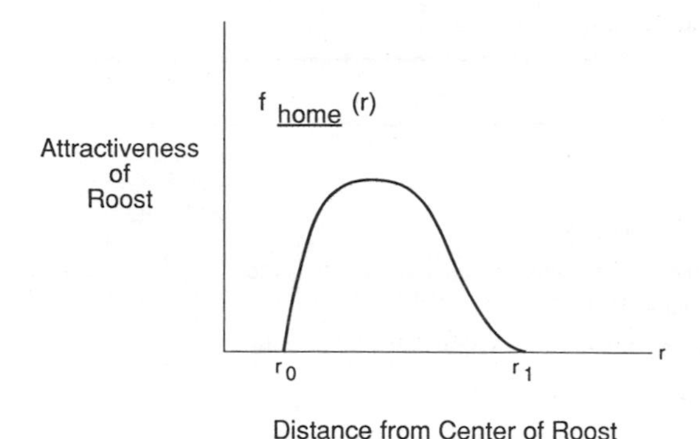

Fig. 1. Attractiveness of the roost.

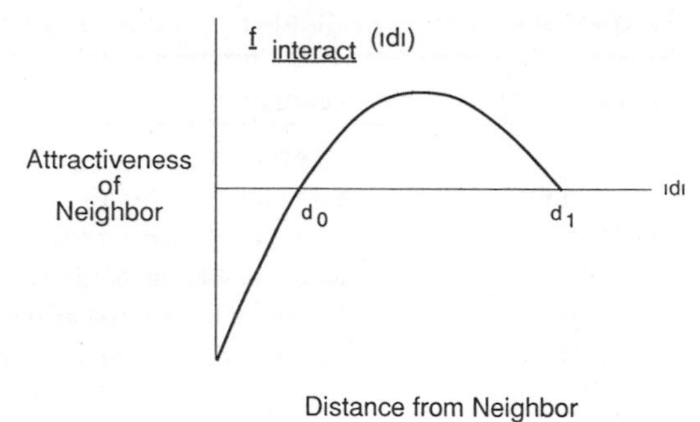

Fig. 3. Interaction function. Birds closer to each other than d_0 are repelled, and avoid collision. Between d_0 and d_i they are attracted, and beyond d_i they are not influenced by the proximity of a neighbor.

vector difference from individual i to j: $d_{ij}(t) = u_j(t) - u_i(t)$ we postulate the ith component to be

$$F^{(i)}_{interact}(t) = \sum_{j=1}^{n} f_{interact}(|d_{ij}(t)|)\, d_{ij}(t). \tag{4}$$

In Eq. 4 the summation is over all j. The interaction dies out if the inter-individual distance $d_{ij} > d_1$.

The random impact term is modeled by an n-dimensional, time-homogeneous Poisson process with stochastically independent components. The ith component of $dP(t)$ is zero unless t happens to be an event of the ith component of the Poisson process. In the latter case,

$$dP(t) = \text{random 3-vector with uniformly distributed components} \tag{5}$$

with a scalar parameter controlling the magnitude of the random vector.

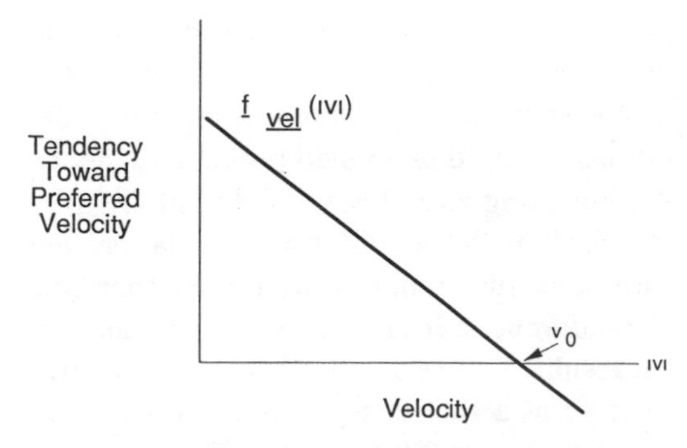

Fig. 2. Velocity control. Birds will accelerate or decelerate to the preferred velocity if one of the other parameters speeds them up or slows them down.

In terms of pattern theory, a striking feature of this model, which differs from other pattern models, is that the connector graph varies in time. It changes as the inter-individual distances change. One other case of this type could be mentioned, namely, for herd leader succession *(9)*.

We simulated the dynamics represented by Eq. 1 and displayed the solution graphically for small values of $n(<20)$ and for different values of the functions $f_{home}, f_{vel}, f_{interact}$, and $P(t)$. This was done by developing software written in the programming language APL, then translating into C, and run on a Silicon Graphics IRIS computer.

In practice, 15 parameters are available (Table 1) to vary the four terms. By changing the values of these parameters, different flock behaviors can be produced.

The appearance of the flock on the computer screen is striking (Fig. 4). By varying the values of the parameters of the terms, various group behaviors can be produced. "Birds" are initially lined up linearly, as on a telephone wire, with equidistant spacing between birds. With the set of values we call "c20.2," the birds initially fly toward the roost, mill around over it for about a minute in real time, then begin to form a cohesive flock which flies clockwise around the roost, "recruiting" stragglers, until all birds have joined the flock. In this flock, after an initial period of shifting positions, the position of each individual eventually stabilizes.

Table 1. Parameters available to modify "flock" behavior.

Parameter	Function
NB	Number of birds
FDISPLAYP	Scalar for force display
$HOMER_0$	Radius at which roost attraction forces begin
HOMERM	Radius of location of maximum attractive homing roost attracting force
HOMERE	Parameter for controlling the shape of the force
HOMERL	Analogous to a Gamma density exponent and Gamma density Lambda
FHOMERP	Scalar for roost attractiveness
$VELOCV_0$	Preferred bird velocity (V_0)
FVELOCP	Scalar for velocity
$INTDISTI_0$	Preferred interbird distance (d_0)
$INTDISTI_1$	Maximum distance at which a neighbor is attractive (d_1)
INTMAXR	Maximum force of repulsion for a neighbor closer than d_0
FINTP	Scalar for attractiveness of neighbor
HLAMBDA	Poisson frequency, Hlambda, for random term
$FRNDP_0$	Scalar for Poisson controlled impact force

In the value set called "Mill 1," the attractiveness of neighbors is increased. Immediately as the program is started, the birds fly toward each other, and form a grouping that resembles a swarm of gnats – individuals maintaining flight paths that contain them within the "ball" represented by the swarm itself, but not displaying any parallelism in their flight paths. This "swarm" then drifts over the roost and does not display the directional flight seen in c20.2.

In "2flock," the attractiveness of neighbors is reduced back to the level of c20.2, but the attractiveness of the roost is reduced. In this situation, the half-dozen birds most distant from the roost form an organized flock with roughly parallel flight paths which flies past the roost at a distance and continues, maintaining a straight direction of flight. The other birds fly toward the roost and form a cohesive group as in c20.2. The result is two coordinated flocks, one circling the roost, the other permanently leaving the roost area.

These patterns were discovered serendipitously, by simply changing values and seeing what happened. As more is learned about the properties of this system, it should eventually be possible to control the flight paths of selected individual birds qualitatively. This would enable the simulation of a predator attack *(10)* on a cluster-flying species and a test of the hypothesis that a turn can be initiated by an individual bird *(6,7)*, followed by a wave of turning in the flock.

Depending upon the choice of parameters, flocks tend to form even from fairly arbitrary initial locations around the roost or to break up from a well-formed flock. In the latter case, the behavior becomes more chaotic for a time, while in the former order emerges out of chaos. Our knowledge of the qualitative properties of this system are still limited. Our current computer experimentation suggests that the system exhibits chaotic regimes. Further study is required to resolve this question.

The model is simple and natural – an obvious extension of the classical *n*-body problem to a stochastic model with unconventional (non-Newtonian) forces. Such models involving stochastic differential equations are extensively used in applied probability theory. A pioneering and closely related model *(11)* also uses stochastic differential equations but with other functional forms and another type of randomness. It also presents some analytical results. Our reason for using this one was to serve as a tool in systematic mathematical experiments on the computer. The results of such experiments qualitatively resemble the flocking behavior of real birds.

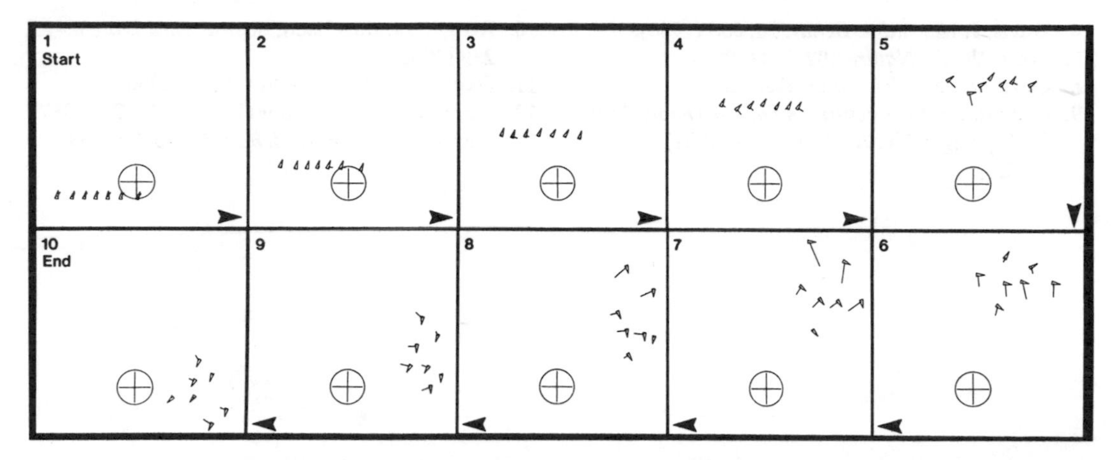

Fig. 4. A series of time-equidistant "frames" from a continuous flock movement. The circled cross is the "roost," triangles are "birds," with the vertex indicating the direction of "flight," and the line segment attached to each bird indicating the vector sum of the four forces—homing, velocity, interaction, and random impact—acting on the bird at the time depicted in the frame.

To our knowledge, the first operating computer model of a bird flock was reported in 1987 *(12)*. Although the mathematical expressions used to prepare this model were not presented, it appears generally to differ from ours in that our birds do not attempt to match velocity with neighbors but attempt to return to preset individual velocities, and our birds are attracted to a central roost. Further, it was not specifically reported in the earlier model that a stochastic differential element was included in its rules for movement, although such an element was necessary to produce flocking in ours. The fact that both models produced flocking under certain circumstances, using different parameters, provides fuel for further investigation.

We plan to extend the code to three dimensions, to give the user several options to choose the way he wishes to see the birds, and to remove the attractions of a roost. We hope to make the resulting code available to other researchers interested in this problem.

This model is mechanistic. Flocking here is the result of attractiveness of neighbors in flight, attractiveness of a roost, or other focal point, and the particular functional forms selected for the force terms in Eq. 1. Indeed, it took considerable trial-and-error experimentation with these functions before flock-like behavior was produced. Early attempts, for example, using Newtonian gravitation forces, failed to produce flock-like behavior. Also, the choice of $P(t)$ as a Poisson-based force succeeded where a Gaussian one did not. Unanswered is the question of why a neighbor should be attractive, and this is where possible selective forces such as predation deterrence *(13)* might enter.

Acknowledgment

We wish to thank Dan Potter for assistance in programming and for improving the algorithm by extensive experimentation.

References

1. Heppner, F. H., *Bird Banding* **45**, 160 (1974).
2. Rackham, H., *Pliny's Natural History*, Book X:XXXV (Harvard University Press, Cambridge, MA, 1933).
3. Selous, E., *Thought-Transference (or what?) in Birds* (Constable, London, 1931).
4. Heppner, F., and J. Haffner, in *Biological and Clinical Effects of Low Intensity Magnetic Fields*, J. G. Llaurado, A. Sances, and J. H. Battocletti, Eds. (Thomas, Springfield, MA, 1972), pp. 147–162.
5. Pomeroy, H., *Three-Dimensional Analysis of Turning Within Airborne Bird Flocks*, Ph.D. dissertation, University of Rhode Island, 1983.

6. Davis, J. M., *Anim. Behav.* **28**, 668 (1980).
7. Potts, W. K., *Nature* **309**, 344 (1984).
8. Gardiner, M., *Sci. Amer.* **223**, 120 (1970).
9. Grenander, U., *Lectures in Pattern Theory*, Vol. III (Springer-Verlag, New York, 1981).
10. Ross, T., and J. Olsen, *Austral. Bird Watcher* **12**, 239 (1988).
11. Okubo, A., *Adv. Biophys.* **22**, 1 (1986).
12. Reynolds, C., *Computer Graphics* **21**, 25 (1987).
13. Parrish, J. K., *Anim. Behav.* **38**, 1048 (1989).

List of Authors

Alfonso M. Albano, *Department of Physics, Bryn Mawr College, Bryn Mawr, PA*

Theodore R. Bashore, *Department of Psychiatry, The Medical College of Pennsylvania, Philadelphia, PA*

Michele Boldrin, *Department of Economics, University of California, Los Angeles, CA*

J. Robert Buchler, *Department of Physics, University of Florida, Gainesville, FL*

Robert L. Devaney, *Department of Mathematics, Boston University, Boston, MA*

Ronald F. Fox, *School of Physics, Georgia Institute of Technology, Atlanta, GA*

Walter J. Freeman, *Department of Physiology and Anatomy, University of California, Berkeley, CA*

Sandra L. Fulmer, *Department of Linguistics, University of Arizona, Tucson, AZ*

Ary L. Goldberger, *Cardiovascular Division, Beth Israel Hospital, Boston, MA*

Ulf Grenander, *Division of Applied Mathematics, Brown University, Providence, RI*

Frank Heppner, *Department of Zoology, The University of Rhode Island, Kingston, RI*

Roderick V. Jensen, *Mason Laboratory, Yale University, New Haven, CT*

Laurence Keefe, *Center for Turbulence Research, NASA–Ames Research Center, Moffett Field, CA; currently at Nielsen Engineering and Research, Mountain View, CA*

John Kim, *NASA–Ames Research Center, Moffett Field, CA*

Peter M. Koch, *Department of Physics, State University of New York at Stony Brook, Stony Brook, NY; Max-Planck-Institut für Quantenoptik, Garching, Federal Republic of Germany*

Arnold J. Mandell, *Department of Psychiatry, University of California, San Diego, La Jolla, CA; currently at the Department of Mathematics and Center for Complex Systems and the Brain Sciences, Florida Atlantic University, Boca Raton, FL*

Jacques Martinerie, *Laboratoire d'Electrophysiologie et de Neurophysiologie Appliquée C.N.R.S., Hôpital de la Salpetrière, Paris, France*

Gottfried Mayer-Kress, *Santa Fe Institute, Santa Fe, NM; Center for Nonlinear Studies, Los Alamos National Laboratory, Los Alamos, NM; Mathematics Department, University of California, Santa Cruz, CA*

Alistair I. Mees, *Department of Mathematics, University of Western Australia, Nedlands, Western Australia*

Parviz Moin, *Department of Mechanical Engineering, Stanford University, Stanford, CA*

Lars F. Olsen, *Department of Biochemistry, Odense University, Odense, Denmark*

Paul E. Rapp, *Department of Physiology, The Medical College of Pennsylvania, Philadelphia, PA*

Oded Regev, *Department of Physics, Technion-Israel Institute of Technology, Haifa, Israel; Department of Astronomy, Columbia University, New York, NY*

David R. Rigney, *Department of Physiology/Biophysics, Harvard Medical School, Boston, MA*

Alvin M. Saperstein, *Department of Physics, Wayne State University, Detroit, MI*

Chera L. Sayers, *Department of Economics, University of Houston, Houston, TX*

William M. Schaffer, *Departments of Ecology and Evolutionary Biology, The University of Arizona, Tucson, AZ*

Michael F. Shlesinger, *Office of Naval Research, Arlington, VA*

Leonard A. Smith, *École Normale Supérieure, Paris, France*

Edward A. Spiegel, *Astronomy Department, Pupin Physics Laboratory, Columbia University, New York, NY*

Reuben Thieberger, *Physics Department, Ben Gurion University, Beer Sheva, Israel*

Greg L. Truty, *Management Information Science Department, University of Arizona, Tucson, AZ*

Charles W. Van Atta, *Department of Mechanical Engineering, University of California, San Diego, La Jolla, CA*

Irwin D. Zimmerman, *Department of Physiology, The Medical College of Pennsylvania, Philadelphia, PA*

Appendix

Several eminent scholars who participated in the chaos symposia at the 1989 AAAS Annual Meeting were unable to contribute chapters to the present volume. This bibliography consists of publications considered by them to be most representative of the material discussed at that meeting.

Blum, L., M. Shub, S. Smale, "On A Theory of Computation and Complexity Over the Real Numbers: NP-Completeness, Recursive Functions and Universal Machines," *Bull. (New Series) Am. Math. Soc.* **21**, 1 (1989).

Ford, J., "Chaos: Solving the Unsolvable, Predicting the Unpredictable!" in *Chaotic Dynamics and Fractals*, M. F. Barnsley and S. G. Demko, Eds. (Academic Press, New York, 1986), pp. 1–52.

Ford, J., "Directions in Classical Chaos," in *Directions in Chaos*, Hao Bai-Lin, Ed. (World Scientific Publishing Company, Singapore, 1988), vol. 1, pp. 1–16.

Ford, J., "Quantum Chaos: Is There Any?" in *Directions in Chaos*, Hao Bai-Lin, Ed. (World Scientific Publishing Company, Singapore, 1988), vol. 2, pp. 128–147.

Ford, J., "What is Chaos That We Should Be Mindful of It?" in *The New Physics*, Paul Davies, Ed. (Cambridge University Press, Cambridge, 1989), p. 348.

Holmes, P., "Nonlinear Oscillations and the Horseshoe Map," *Proc. Symp. Appl. Math.* **39**, 25 (1989).

Holmes, P., "Poincaré, Celestial Mechanics, Dynamical Systems Theory and 'Chaos'," *Phys. Rep.*, in press.

Krishnaprasad, P. S., and J. E. Marsden, "Hamiltonian Structures and Stability for Rigid Bodies with Flexible Attachments," *Archive for Rational Mechanics and Analysis* **98**, 71 (1987).

Meneveau, C., and K. R. Sreenivasan, "Interface Dimension in Intermittent Turbulence," *Phys. Rev. A* **41**, 2246 (1990).

Oh, Y. G., N. Sreenath, P. S. Krishnaprasad, J. E. Marsden, "The Dynamics of Coupled Planar Rigid Bodies. II: Bifurcations, Periodic Solutions, and Chaos," *Journal of Dynamics and Differential Equations* **1**, 269 (1989).

Posbergh, T. A., P. S. Krishnaprasad, J. E. Marsden, "Stability Analysis of a Rigid Body with a Flexible Attachment Using the Energy-Casimir Method," *Contrib. Math.* **68**, 253 (1987).

Smale, S., "Some Remarks on the Foundations of Numerical Analysis," *SIAM Rev.* **32**, no. 2, 211 (1990).

Sommeria, J., S. D. Meyers, H. L. Swinney, "Laboratory Model of a Planetary Eastward Jet," *Nature* **337**, 58 (1989).

Sommeria, J., S. D. Meyers, H. L. Swinney, "Experiments on Vortices and Rossby Waves in Eastward and Westward Jets," in *Nonlinear Topics in Ocean Physics*, A. Osborne, Ed. (North-Holland, Amsterdam, 1990).

Sreenath, N., Y. G. Oh, P. S. Krishnaprasad, J. E. Marsden, "The Dynamics of Coupled Planar Rigid Bodies. Part 1: Reduction, Equilibria and Stability," *Dynamics and Stability of Systems* **3**, 25 (1988).

Sreenivasan, K. R., R. Ramshankar, C. Meneveau, "Mixing, Entrainment and Fractal Dimensions of Surfaces in Turbulent Flows," *Proc. R. Soc. Lond. A* **421**, 79 (1989).

Sussman, G. J., and J. Wisdom, "Numerical Evidence That the Motion of Pluto Is Chaotic," *Science* **241**, 433 (1988).

Welch, G. R., M. M. Kash, C. Iu, L. Hsu, D. Kleppner, "Experimental Study of Energy-Level Statistics in a Regime of Regular Classical Motion," *Phys. Rev. Lett.* **62**, 893 (1989).

Welch, G. R., M. M. Kash, C. Iu, L. Hsu, D. Kleppner, "Positive Energy Structure of the Diamagnetic Rydberg Spectrum," Preprint: submitted to *Phys. Rev. Lett.*

Wisdom, J., and S. Tremaine, "Local Simulations of Planetary Rings," *Astron. J.* **95**, 925 (1988).

Wisdom, J., "Chaotic Behavior in the Solar System," *Nucl. Phys. B (Proc. Suppl.)* **2**, 391 (1987).

Wisdom, J., "Rotational Dynamics of Irregularly Shaped Natural Satellites," *Astron. J.* **94**, 1350 (1987).

Wisdom, J., "Urey Prize Lecture: Chaotic Dynamics in the Solar System," *Icarus* **72**, 241 (1987).

Index